59/60 Structure and Bonding

Actinides – Chemistry and Physical Properties

Editor: L. Manes

With Contributions by
U. Benedict M. S. S. Brooks J. M. Fournier J. Ghijsen
L. Manes W. Müller J. R. Naegele J.-C. Spirlet

With 115 Figures and 53 Tables

Springer-Verlag
Berlin Heidelberg New York Tokyo

ISBN 3-540-13752-1 Springer-Verlag Berlin Heidelberg New York Tokyo
ISBN 0-387-13752-1 Springer Verlag New York Heidelberg Berlin Tokyo

Library of Congress Catalog Card Number 67-11280

Printed in Germany

Typesetting and printing: Schwetzinger Verlagsdruckerei. Bookbinding: J. Schäffer, Grünstadt.
2152/3140-543210

Foreword

Actinides, the chemical elements with atomic numbers ranging from 89 to 103, form the heaviest complete series in the Periodic Table. They are radioelements, either naturally occurring or synthesized by nuclear reactions. Their predominant practical application depends on the nuclear properties of their isotopes: decay, spontaneous or induced fission. Their chemical and physical properties reflect a very complex electronic structure, and their study and understanding are a challenge to experimentalists and theoreticians.

The research programme of the European Institute for Transuranium Elements was, from its very beginning, devoted to both basic research on advanced plutonium containing fuel and to fundamental research on actinide elements. Non-fuel actinide research in Europe started more than 20 years ago with the reprocessing of irradiated actinide samples. Since the first isolation and purification of transplutonium elements, actinide research developed steadily in close contact and cooperation with specialised laboratories in Western Europe and in the United States.

The separation and purification of multigram amounts of ^{243}Am and ^{244}Cm in the late sixties at Karlsruhe allowed studies at a scale interesting solid state physicists. On the basis of the first results obtained, a programme "Actinide Research" at the European Institute for Transuranium Elements was established with the aim of investigating the contribution of 5 f electrons to the chemical bond. This programme comprised the preparation of actinide containing samples (metals, intermetallics, binary compounds), their analysis and characterisation and, finally, the study of bond-related thermodynamical, structural and electronic properties. The experimental effort of the programme was complemented by theory. As the Institute for Transuranium Elements has become a "focal point" of actinide research in Europe, it seemed appropriate to assemble actual experimental and theoretical results to describe the state of the art in form of a monograph on Solid State Properties of Actinides.

W. Müller

General Introduction

The properties of actinide solids have been particularly at the center of scientific attention as soon as their investigation raised the first suggestions that, at least in the first part of the series, the unsatured 5f shell had an important role in establishing the chemical bond of the solid, especially the metallic bond.

This book attempts to give an overview of the state of knowledge currently available on the solid state physics of actinides.

Emphasis will be placed on the basic concepts underlying the treatment of actinide systems. Therefore, the aim of the book is not to be comprehensive in all aspects, but rather to present a coherent picture which may be useful to the beginner in actinides and also to the specialist who might feel that a unified picture can help him in interpreting his results and promote further progress. Appropriate examples are chosen for this purpose.

Theoretical concepts and the description of experimental methods utilized in actinide research are usually scattered in many textbooks and in the general literature. A further aim of the book is therefore to assemble them in a well connected way.

The book is organized as follows:

- Chapter A is intended to be an extended overview, where: (1) the problems of actinide solids are presented (Part I); (2) the conceptual context is indicated in which the problems have been partially solved (Parts II and III); (3) relevant experimental evidence for understanding the solid state behaviour of actinide materials (Parts IV and V) is reviewed. Since the book consists of separate monographic contributions, the authors of Chapter A have attempted to give it a self-consistent character.
- Chapter B focuses attention on the preparation and characterization of samples suitable for actinide solid state research. In this context, it is perhaps worthwhile to remark how actinide solid state physics has been, in these years, a subject in which interdisciplinary cooperation has been strongly needed and achieved, perhaps more than in other fields.
- Chapters C, D and E, discuss thermodynamic and structural properties, magnetism, and photoelectron spectroscopy of actinide systems in their relation to bonding.
- Chapter F presents the most advanced theory of 5f bonding. An effort was made by the author (as in the rest of the book) to present the essential theoretical formalism in a simplified and hopefully understandable manner. At least, the reader should be able to find in this chapter a help to introduce himself to the most refined theoretical treatments which are nowadays of great relevance in actinide solid state physics.

Table of Contents

Chapter A

Actinide Solids
5 f Dependence of Physical Properties

J. M. Fournier[1] and L. Manes[2]

1 University of Grenoble and Centre d'Etudes Nucléaires de Grenoble, Grenoble, France
2 Commission of the European Communities, Joint Research Centre, Karlsruhe Establishment, European Institute of Transuranium Elements, 7500 Karlsruhe, F.R.G.

This chapter is an introduction to the main concepts underlying the present unterstanding of the physical properties of actinide solids.

In the first half of the actinide series the elements behave as 5 f transition metals; 5 f electrons are then described as itinerant in metallic solids. Between plutonium and americium a cross-over from itineracy to localization, i.e. a true Mott transition is observed. Heavier actinides have a lanthanide-like behaviour, the 5 f states exhibiting similar properties to the 4 f states in lanthanides.

The characteristics of 5 f wavefunctions in the actinide atom as well as in actinide solids are reviewed. In the latter, overlapping of 5 f wavefunctions between neighbouring atoms as well as their hybridization with other orbitals (actinide 6 d, or orbitals of non-actinide elements in compounds) are pointed out as being the two phenomena determining physical properties and bonding characteristics, in particular magnetism. Theories describing them, in particular Stoner's and Mott-Hubbard's models, are discussed in the light of their application to actinide physics.

A short summary is presented of the main experimental evidence which justifies the theoretical decription of actinide solids.

Structure and Bonding 59/60

I. Introduction: Problems of the Actinide Series

1. Radon Core and Then What?

Actinide atoms, in their ground configuration, comprise the closed shell electronic structure of the noble gas radon

$$1s^2\,2s^2\,2p^6\,3s^2\,3p^6\,3d^{10}\,4s^2\,4p^6\,4d^{10}\,5s^2\,5p^6\,4f^{14}\,5d^{10}\,6s^2\,6p^6$$

and from three (Ac) to seventeen (Lr) external electrons.

As known from the quantum treatment of the hydrogenoid atom (see, e.g.[1]), these outer electrons may be accommodated in the 5f, 6d and 7s shells.

In the forties Np (atomic number Z = 93) and Pu (Z = 94) were added[2] to the already known Ac (Z = 89), Th (Z = 90), Pa (Z = 91) and U (Z = 92). The central question, for the chemistry of these elements, was:

i. Which shells are filled by these external electrons, and in which order?
ii. Is the actinide series a 6d transition or a 5f lanthanide series?

The answer was by no means straightforward since the first elements (Th, Pa, U) to be studied displayed properties reminiscent of both a transition series and a lanthanide series. The tendency was to expect that the electrons would follow the same order of filling as for the elements from La to Lu, in which the 4f shell is filled, giving rise to the lanthanide series, before the filling of the 5d shell. Consequently, Seaborg[2] proposed the name "actinides" for this 5f series, Ac being the homologous of Ln.

In fact, as early as 1941, Mayer[3] showed by means of a Thomas-Fermi type calculation that the atomic wave functions of the 5f electrons drop suddenly in energy and spatial extension in the vicinity of Z = 92 (i.e. Uranium). This is consistent with the existence of a second lanthanide series, following Ac (Atomic number Z = 89), and going from Th (Z = 90) to Lr (Z = 103). It is customary to place actinides in Periodic Charts of Elements as a homologous series of the lanthanides, meaning that progressive filling of the 5f shell occurs throughout the series.

The results of atomic spectroscopy as well as atomic quantum calculations have made it possible to determine the ground state of the free actinide atoms. These results (see Table 1[4]) (that will be reviewed in the next section of this Chapter) confirm the progressive filling of the 5f shell. From the point of view of the electronic structure of the free atom, therefore, question ii. is solved in the sense of actinides being a series in which the unsaturated 5f shell is progressively filled (only one or two electrons being accomodated in the 6d shell).

Only for element 104 could the 6d shell start to be filled. Chemistry is known for this element (Rutherfordium or Kurchatovium), although only few atoms have been synthesized.

2. A Short Discussion on the Chemistry of Actinides

The lanthanides have, as known, very similar chemical properties across the series. Writing them in a single separate line in the periodic chart intends to convey this information to the reader.

Table 1. Ground state and first excited state configurations, and their energy difference, for lanthanides and actinides

	Ground state	First excited state	Distance (ev)		Ground state	First excited state	Distance (ev)
La	ds^2	d^2s	0.33	Ac	ds^2	d^2s	1.14
Ce	fds^2	fd^2s	0.30	Th	d^2s^2	d^3s	0.68
Pr	f^3s^2	f^2ds^2	0.50	Pa	f^2ds^2	fd^2s^2	0.25
Nd	f^4s^2	f^3ds^2	0.83	U	f^3ds^2	f^3d^2s	0.77
Pm	f^5s^2	f^4ds^2	0.99	Np	f^4ds^2	f^5s^2	0.12
Sm	f^6s^2	f^6ds	1.34	Pu	f^6s^2	f^5ds^2	0.78
Eu	f^7s^2	f^7ds	1.60	Am	f^7s^2	f^7ds	1.80
Gd	f^7ds^2	f^7d^2s	0.79	Cm	f^7ds^2	f^8s^2	0.15
Tb	f^9s^2	f^8ds^2	0.04	Bk	f^9s^2	f^8ds^2	0.92
Dy	$f^{10}s^2$	f^9ds^2	0.94	Cf	$f^{10}s^2$	f^9ds^2	2.11
Ho	$f^{11}s^2$	$f^{10}ds^2$	0.95	Es	$f^{11}s^2$	$f^{10}ds^2$	2.36
Er	$f^{12}s^2$	$f^{11}ds^2$	0.89	Fm	$f^{12}s^2$	$f^{12}sp$	2.42
Tm	$f^{13}s^2$	$f^{12}ds^2$	1.62	Md	$f^{13}s^2$	$f^{13}sp$	2.48
Yb	$f^{14}s^2$	$f^{13}sp$	2.15	No	$f^{14}s^2$	$f^{14}sp$	2.60
Lu	$f^{14}ds^2$	$f^{14}s^2p$	0.51	Lr	$f^{14}s^2p$	$f^{14}ds^2$	0.99

Is the same true for the actinide elements?

In Fig. 1, we have plotted the oxidation numbers of the actinides and of the lanthanides. We see that for the lanthanides the valence 3 is the most stable valence throughout the series. There are exceptions: Ce displays for instance tetravalency in many compounds; Eu and Yb display divalency. These exceptions are understood: e.g., Eu and Yb are at the half-filling and at the filling of the 4f shell, which are stable electronic configurations. There is a tendency for both to share just the two outer 5s electrons in bonding, displaying therefore, divalency, and preserve these stable configurations.

On the contrary, there is a spread of oxidation numbers for the light actinides (at least up to Cm), which, for Pu and Np, range from 3 to 7! After Cm, however, the trivalent oxidation state is always met, and this second half of the actinide series approaches more the behaviour of the lanthanides.

Thus, the implications about the chemical behaviour derived from a Periodic Chart in which the actinides are all placed in one line may be somewhat misleading. Rather, it appears that we should distinguish two parts in the series: one up to Cm ("light actinides"), another one from Cm on ("heavy actinides").

In Fig. 1, the valences of the transition 3d-, 4d-, 5d-series are also plotted. As for what regards the spread of valences, an interesting observation is that, at least for the light actinides (if not for the whole series), there is more similarity between the actinides and the d-transition elements than the actinides and lanthanides.

The oxidation number of an element is used by chemists as a formal quantity, and is a count of bonding electrons. It is so useful that it has generated, as we shall see later, the concept of "metallic valence" when dealing with elemental metals and the metallic bond.

In this simple context, the comparisons established in Fig. 1 are particularly significant. Let us take, e.g., the comparison between Sm ($4f^6 5d^0 6s^2$) and Pu ($5f^6 6d^0 7s^2$).

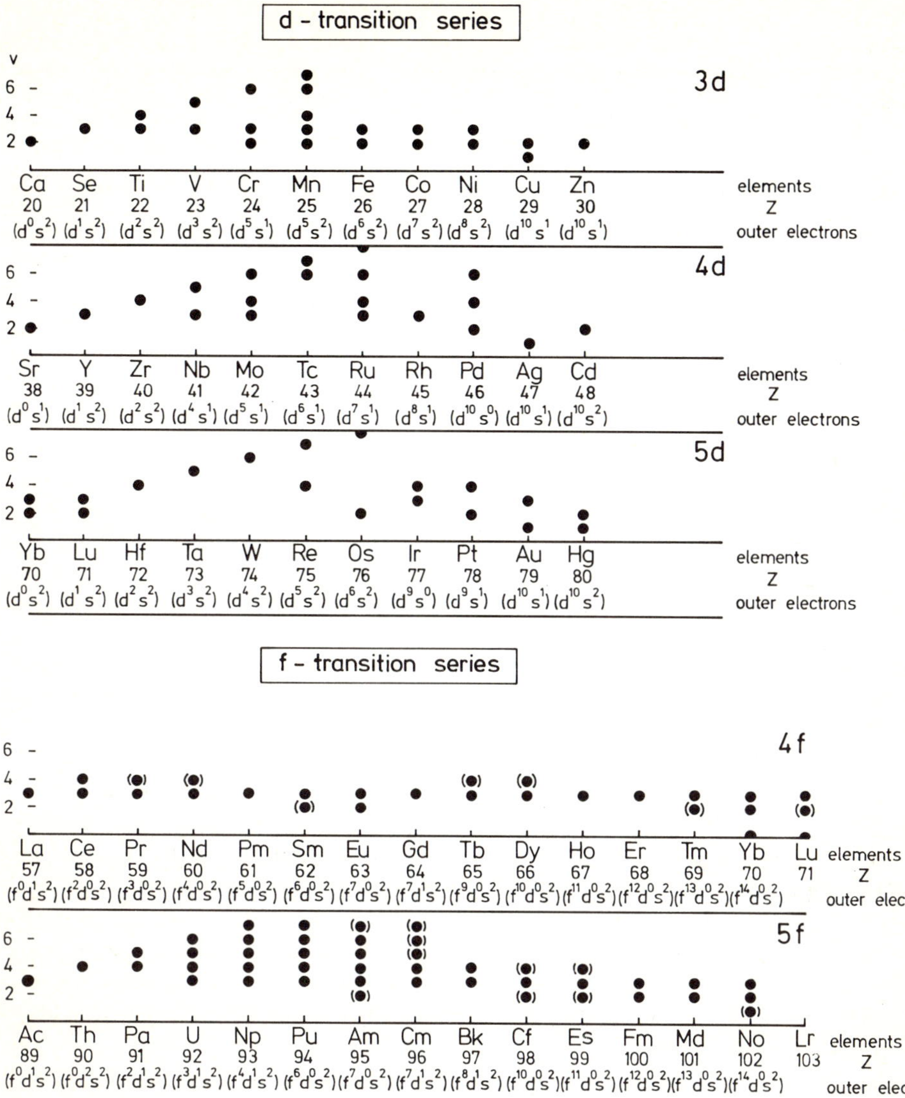

Fig. 1. Oxidation numbers for: – d-transition series', – f-transition series' (lanthanides and actinides); (non-common or uncertain oxidation numbers have been put between *brackets*)

We start by considering that the trivalency of La ($4f^0 6d^1 7s^2$) in its chemistry can be understood with the bonding being performed by the one d and two s outer electrons. In Sm, the trivalency may be understood in the same way with the help of a subsidiary step: the promotion of one f electron to the d shell (from $4f^6 5d^0 6s^2$ to $4f^5 5d^1 6s^2$) a process requiring little energy, and which produces the same $5d^1 6s^2$ outer configurations as La. In fact, for all lanthanides the trivalency may have the same explanation. But complementary to this explanation, another statement can be made: for Sm (and for all other lanthanides as well) there is a 4f core which is essentially non-bonding: these (five) electrons "cling to the atom". One says that, in lanthanides, the 4f electrons retain atomic properties (a conclusion largely supported by physical properties, e.g. magnetism).

If we turn now to Pu, and seek to explain the large spread of valency by the same naive picture, we are forced to accept one or both of two hypotheses:

i. that very little energy may be required to form, from the ground state, a large number of possible configurations by moving 5f electrons to the d shell; in other words many configurations ($5f^x 6d^y 7s^2$) ($x + y = 6$) have similar energies (bonds being formed from 6d and 7s electrons only);

ii. the 5f electrons have themselves a tendency to establish bonds.

Of course, the two hypotheses are not really independent and only the simplicity of the language employed allows us to state them in such a clear-cut form. Nevertheless, they embody a problem for chemists which is analogous to the one encountered in the chemistry of d-transition series, and which has kept researchers busy for a very long time.

3. The Metallic Valence: From Chemistry to a Solid State Case

The concept of valence developed in the preceding section is the basis of the first correlations aiming at a global theory of the actinide metallic bond. These correlations were established between the atomic volumes of actinide elemental metals, and the electronic configuration of the actinide atoms[4–8]. Their aim was to provide a general theory of actinides (i.e. to give an answer to the questions i. and ii. of Sect. A.I.1.) within the framework of a simple model of the metallic bond.

It is known that the cohesion of a metal is ensured by the electrons partially filling a conduction (or valence) band. The wave functions of these "conduction electrons" are Bloch functions, i.e. amplitude modulated plane waves. Even though these wave functions are linear combinations of the electronic wave functions in the isolated atoms, reminiscence of the atomic orbitals is lost (or is eventually contained in the amplitude factor). The conduction electrons are, of course, originally, the outer or valence electrons of the atoms; but in a metal, to describe them as s, p, d or f, i.e. with the quantum number proper to the atomic case, has little meaning. They may be considered to many purposes to be "free electrons".

The simplest model of a metal is therefore the one in which the metal is depicted as an array of ions "glued together" by conduction (quasi-free) electrons. If this is the case, one may define a *"metallic valence"* as being, essentially, the charge left in the ion cores when outer electrons have been stripped off. Conversely, the "metallic valence" can be defined as the contribution of outer electrons each atom gives to the "sea" of bonding conduction electrons.

The loss of outer electrons from the atoms to a conduction band, in such a simple picture, bears a similarity to the redox process which is supposed to occur in molecular bonds, and which is at the basis of the "oxidation state" or "formal valence" concept. If this is the case, the metallic valence should coincide with one of the valences that one encounters in compounds, possibly with the most stable one.

In this model, the binding electrons are often assimilated to a negatively charged, *isotropic* continuum ("jellium"[5]) surrounding the positively charged ions (the ionic charge being the metallic valence v). Due to the isotropy of the bonding, one expects "compact" crystal structures to exist in elemental metals (which is the case for most metals, crystallizing in the twelve-coordinated f.c.c. and h.c.p., or in the eight-coordinated b.c.c. structure – see Chap. C). The shape of the Wigner-Seitz cell[1)] may be safely assimilated to a sphere. The atomic volume V_{at}, coincides with that of the Wigner-Seitz sphere V_{WS} ($V_{at} = V_{WS} = \frac{3}{4}\pi R_{WS}^3$).

In this "jellium" model, the equation

$$R_{WS}^3 = \frac{9}{4}\pi\lambda_F^3 v \tag{1}$$

can be obtained[5)], where λ_F is the so-called Fermi-length of a free electron, i.e. the wavelength of a free electron wave, and v is the metallic valence. The quantity λ_F depends on the detailed description of the total (kinetic and potential) energy of the free-electron. Important contributions are the Coulomb and exchange interactions with the outer electrons of the cores and with other conduction electrons, which add to the electrostatic attraction to the nuclei. However, from (1), one notices that equivalent metals, having approximately the same λ_F, should have the same R_{WS}. If the radii across a transition series are inspected vs Z, information may be gathered on v.

a. The Case of Lanthanides

For lanthanides, the metal radii decrease monotonically with increasing atomic number (Fig. 2), with the three anomalies of Ce, Eu and Yb.

Let us assume that the metallic valence is three for most lanthanides, and two for Eu and Yb. This is consistent with the chemistry of these elements, since (Fig. 1) three is their most common oxidation number. Trivalency has been explained in the preceeding chapter, by assuming the easy promotion of one 4f electron to a 5d level. The conduction quasi-free electrons are therefore of 5d and 6s origin. We may say that the conduction band has a prevalent (s, d) character.

Two other facts comfort the theory. Firstly, most lanthanide elemental metals crystallize in a close-packed structure as expected for the simple model of the metal employed. Secondly, the metals are paramagnetic, and order magnetically (in rather complicated structures). The paramagnetism can be explained by taking into account the magnetic moment of the $4f^n$ configuration which is retained by the ion cores: in fact, paramagnetic moments experimentally determined for metals are almost equal to those of the trivalent, $4f^n$, free ions (in compounds, or solutions). One says, therefore, that the $4f^n$ is *strongly localized* in the lanthanides, meaning that the 4f shell wave functions retain their atomic character.

The trend of radii vs. Z shown in Fig. 2 finds then an explanation consistent with the picture of ion cores, each containing a $4f^n$ non-binding shell. In fact, the $4f^n$ shells screen only imperfectly the outer electrons. Adding one nuclear charge when going from Z to Z + 1, means increasing the central field of the nucleus, without increasing as efficiently

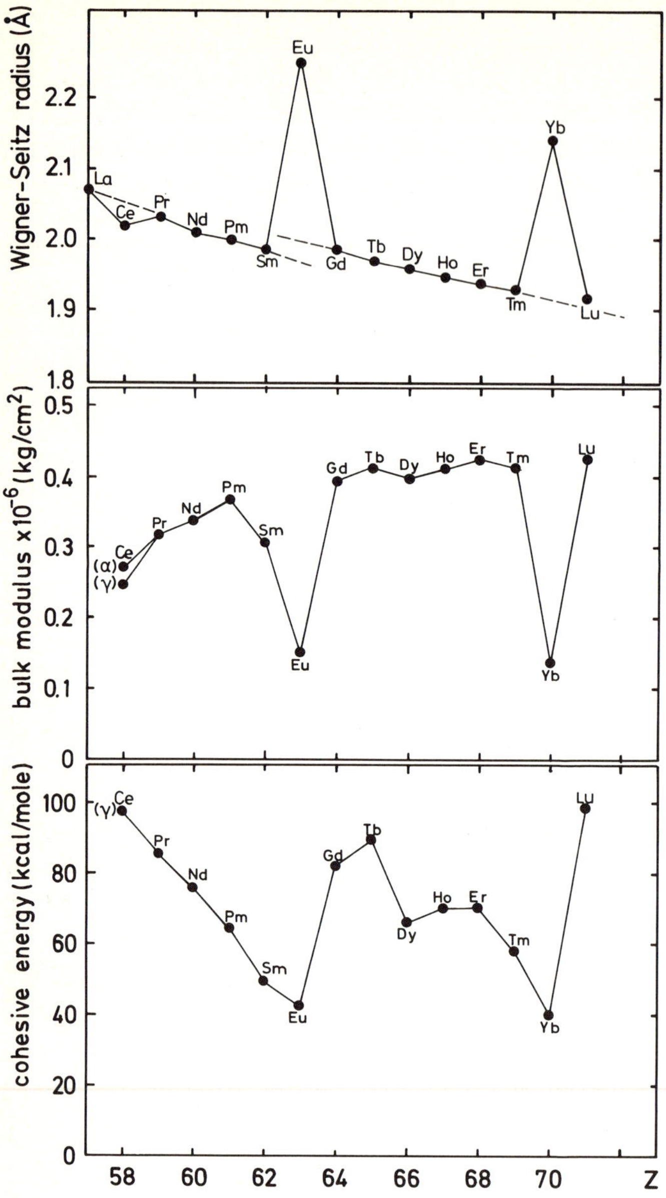

Fig. 2. Wigner-Seitz radii, bulk moduli, and cohesive energies of lanthanides plotted vs atomic number Z (The radius of a Wigner-Seitz sphere is related to the atomic volume by: $V_{at} = 4/3\ \pi\ R_{WS}^3$)

the screening of this field. Hence, the outer electrons being more attracted, the net result is a contraction of the ion cores, explaining the trend of Fig. 2 (*lanthanide contraction*). This effect, deriving from the wave functions of f-electrons, must also occur in actinides, for their 5f electrons. We should expect an *actinide contraction* analogous to the lanthanide contraction.

b. The Case of Actinides

For actinides up to Am, the plot of atomic radii vs. Z is given in Fig. 3. In the same figure, the same plot is given for d-transition elements. The first obvious observation is that the situation does not ressemble at all the lanthanide one. The second observation is that, for the light actinides (up to Pu) the trend which is followed ressembles the (almost

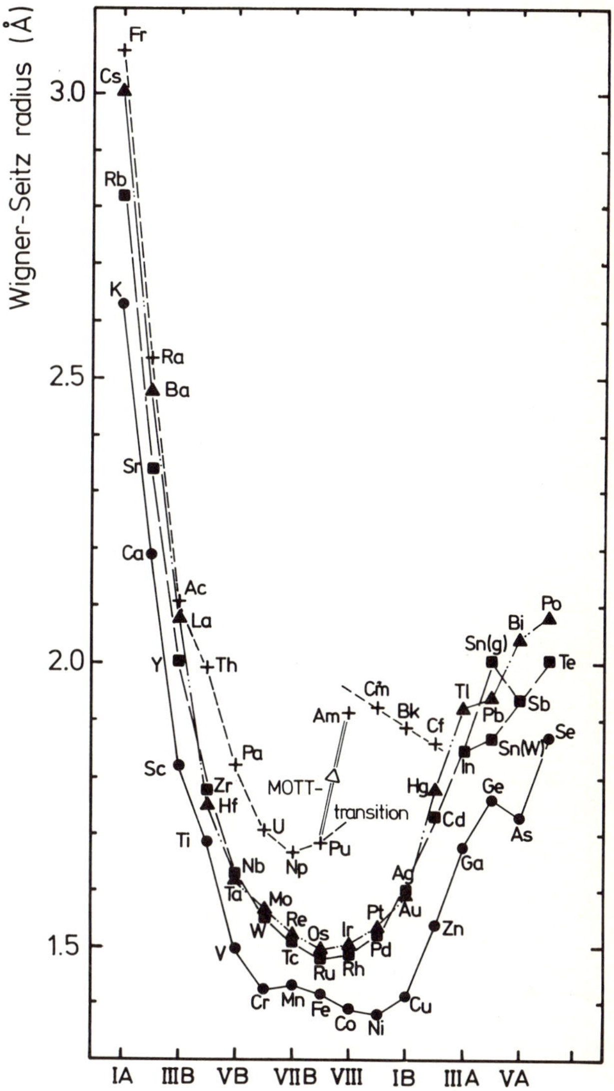

Fig. 3. Wigner-Seitz radii of d-transition metals *and* actinides vs atomic number Z. To the plot, elements displaying empty and full d- and f-shell have been added. In abscissae, the groups of the Periodic Chart of Elements have been indicated (see, e.g. Handbook of Chemistry and Physics). The figure shows the sudden jump in radius between Pu and Am discussed in this chapter, and, more deeply, in Chap. C

parabolic) trend of d-metals. Between Pu and Am, however, one finds a sudden jump. For the heavier actinides, there is a trend which seems to indicate an *actinide contraction*.

Zachariasen[6–10] proposed a theoretical explanation of the observed trend (Fig. 3), focussing on its similarity to the d-transition metals'. This explanation is particularly suggestive for its simplicity, and contains the hypothesis of non-bonding 5f-electrons in the metals of the 5f-series (which Zachariasen calls: the "thoride" series) as it is the case for lanthanides. It has led Zachariasen to the assignment of metallic valences throughout the series, and also to the prediction of them for the (not yet known at the time of his first presentation of the theory) heavier actinide metals. The theory presents the actinide metallic bond as typical of metals belonging to a 6d-transition series, in which, however, for the heavier actinides (see Table 2 from[10]) the 5f shell fills up gradually, causing an actinide contraction.

The theory is described in a very concise way by Kmetko and Hill[11], who have also criticized it in the same paper. Figure 4 is essentially the same figure presented in[11], re-drawn with the latest Zachariasen's assignments in[10].

The measured metallic radii of the metals are drawn, in Fig. 4, on a network of dashed lines specifying the actinide contractions for f configurations progressively increasing by one electron at constant metallic valence (i.e. without change of the number of s, p, d electrons, which are thought to provide the only bonding). (The

Table 2. Zachariasen's[10] assignment of valence v and number of non-bonding 5f electrons f to thorides (actinides)

v + f	Metal		R in Å	v	f
2	Ra		2.293	2	0
3	Ac		1.977[a]	3	0
4	Th	α	1.798	4	0
4		β	ca. 1.80	4	0
5	Pa	α	1.642	5	0
5		β	1.775	4	1
6	U	α	1.542	6	0
6		β	1.548	6	0
6		γ	1.548	6	0
7	Np	α	1.503	7	0
7		β	1.511	7	0
7		γ	ca. 1.53	7	0
8	Pu	α	1.523	6.2	1.8
8		β	1.571	5.4	2.6
8		γ	1.588	5.2	2.8
8		δ & δ′	1.640	4.7	3.3
8		ε	1.592	5.1	2.9
9	Am	α	1.730	3.9	5.1
9		β	1.730	3.9	5.1
10	Cm	α	1.743	3.8	6.2
10		β	1.782	3.5	6.5
11	Bk	α	1.704	4.0	7.0
11		β	1.767	3.5	7.5

[a] Interpolated value

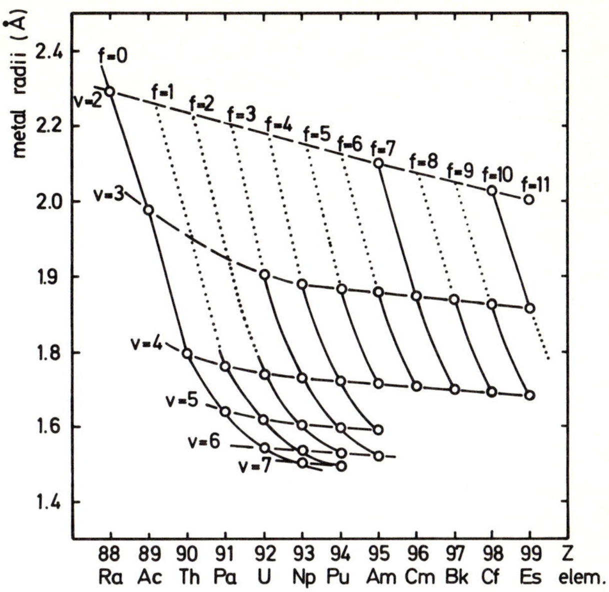

Fig. 4. Kmetko and Hill[11] grid for Zachariasen's theory, redrawn according to Zachariasen's latest assignments[10]

relative amount of contraction is taken from the known bond-lengths in actinide dioxides, ionic solids for which the ionic volumes are well defined[10]).

The full lines of Fig. 4 represent, on the contrary, the variation of atomic radii for (d-like type) metals, in which, for a certain f^n configuration, a valence electron is added. The f^0 curve is a "universal"[11] transition metal curve, on which the experimental radii of Ra (v = 2, $6d^0 7s^2$), Ac (v = 3, $6d^1 7s^2$ interpolated value), Th (v = 4, $6d^2 7s^2$), α-Pa (v = 5, $6d^3 7s^2$), α-U (v = 6, $6d^4 7s^2$), α-Np (v = 7, $6d^5 7s^2$) are seen to fit[8]). The attributions of valencies in the brackets are consistent with the known chemistry of these elements (pentavalent Pa, hexavalent U, heptavalent Np in compounds and solutions).

The electronic configurations given in brackets are consistent with the hypothesis of a metal series essentially of 6d-character.

In Table 2 (taken from[10]), the metallic valences and the number of non-bonding f electrons are then predicted for all known phases of actinide metals.

One major difficulty of this attribution lies in the assignment of a tetravalent state for Am, Cm, and Bk metals. This is not consistent with other physical properties of these metals, as we will see later.

An attempt to meet this difficulty gave rise to other treatments of Zachariasen's type, in which the non-bonding character of 5f electrons was retained[12, 13]). In Ref. 12, e.g., a valence 3 was assigned to Cm. This leads, however to an unrealistic valence close to 3 for Th.

The metallic radii listed in Table 2 are defined as one half the bond-length in the first coordination sphere of the metal structure. They are obtained from the atomic volumes V_{at}, calculated at room temperature for all phases of a metal (see Chap. C), by means of the equation[10])

$$R_{12} = V_{At}^{1/3}/2^{5/6} \tag{2}$$

which adjusts the radius of such asymmetric structures as the α phase of Pu (monoclinic) to the radius R_{12} of the twelve-coordinated close-packed structures. In fact, as we have seen, the simple model of a metal for which the concept of a metallic valence is valid, requires a compact structure (fcc or hcp). For lanthanides, this is normally the case, confirming the validity of the picture of these elemental metals drawn from their atomic volumes. However, the presence in the light actinides (U, Np, Pu, see Chap. C) of very asymmetric structures, points clearly to a lack of validity of the simple model employed and of the consequences drawn from it. Structures such as the monoclinic α-Pu indicate indeed a very high directionality of the bonds, such as could appear if the 5f shell, with its complicated symmetry properties, took part in the bond. Also, it was soon very clear that the magnetism of light actinide metals is far from being as simple as that of lanthanide metals.

These considerations have brought actinide researchers to develop models for actinide metals which are based not only on structural but also on thermodynamic properties (Brewer-type models). Here, the 5f participation in the conduction band is usually taken into consideration. A review of these models appears in Chap. C.

The impact on actinide solid state physics of Zachariasen's ideas has been great. It has triggered a very fruitful effort in structure determination. In the meantime, it has underlined the importance, in the theory of actinides, of knowing exactly atomic volumes as an indicator of the complexities of the metallic bond. The same picture of the variation of atomic radii appears for compounds having partial metallic character. In Fig. 5, for instance, the variation vs. Z of the lattice parameter a_0 of actinide pnictides is compared with the atomic radii of metals. Actinide pnictides have an NaCl type structure. Their

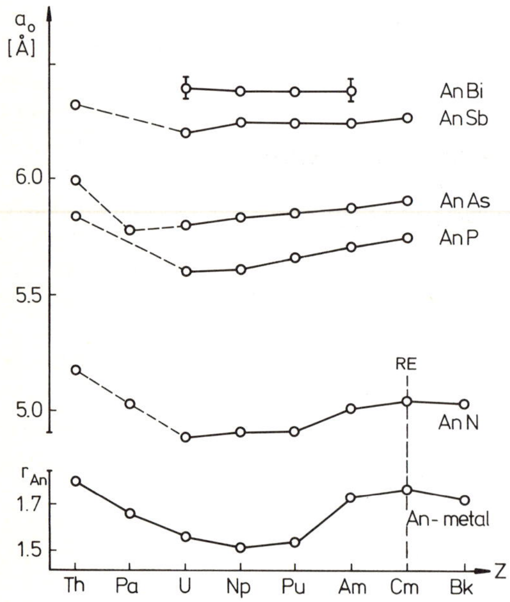

Fig. 5. The lattice constants a_0 of NaCl-structured AnX actinide compounds is plotted vs. atomic number Z and compared with the atomic radii of actinide metals. Notice the 5f-bonding minimum and the set-up of lanthanide behaviour (RE *vertical dashed line*)

bond is essentially metallic with some admixture of ionic and covalent bond, as will be shown in Chap. C.

In Fig. 5 a minimum as in the metals is observed in the curves of the compounds with more metalloidic elements. For AnAs and AnSb this minimum tends to disappear. After the minimum (see AnN), there is a decreasing trend, which can be explained in terms of actinide contraction. Between PuN and AmN, a "jump" is seen, which is similar to the one met in metals (see Fig. 2).

The general shape of these curves at least for pnictides up to AnAs indicates clearly that the metallic bond is dominated by the electronic structure of the actinide atom. In the further discussion of this chapter (and in Chaps. C–F) it will be made clear that the shape is explained, contrarily to Zachariasen's hypothesis, by means of the f participation in the bond.

4. Conclusions

Considerations on the crystal structures and other physical properties of the light actinides have triggered a large effort in quantum calculations for the wave functions of the outer electrons of actinides, including 5f[11,14,15] in atoms as well as in solids.

The problem by which physicists were confronted, was to enclose in a single model the great admixture of different physical and chemical properties which were constantly discovered on actinides, following the constant progress in the preparation of actinide samples and the accompanying progress in experimental methods of investigations. The increasing understanding of the two classes of elements to which actinides bore some ressemblance: d-transition and lanthanides, gave also a constant conceptual input to actinide research. The problem is easily expressed as a third question, following questions i. and ii. of 1.:

iii. are the light actinides a 5f-transition series?

All considerations in the following development of this chapter answer this question affirmatively. This affirmative answer means that the light actinides (up to Pu) constitute a new kind of transition series, as yet unknown, where the 5f wave functions play the role that d-wave functions have in the d-transition metal series. After Pu (with Am (and Cm) occupying an intermediate and particularly interesting position) heavy actinides may on the contrary be described as a new lanthanide-type series.

The rest of the chapter is structured in such a way as to give the reader the possibility to get convinced of this description. The description originates from the peculiar properties of the 5f electron states. Therefore, in Part II, these states will be examined in free actinide atoms. In Part III, the meaning of the atomic information for solids will be discussed: the puzzling problem of atomic-like localized states, which, however, spread enough to form narrow bands and allow a delocalized description. In this part, the conceptual tools to understand the puzzle are provided.

In the Part IV, the experimental results leading to the above description are reviewed.

II. The Free Actinide Atom

1. Introduction

Most results on the free actinide atom came from atomic spectroscopy and from atomic quantum calculations of wave functions and eigenvalues of their outer electrons. This section cannot be an exhaustive review devoted to the theory and interpretation of the very complex spectra of the actinide atoms and ions. We shall recall briefly the theoretical approach used in atomic calculations and then give some of the numerous useful informations that derived from atomic studies for solid state physicists and chemists.

2. The Central Field Approximation (Non-Relativistic)

Energy level calculations of atomic systems, as well as the wave functions are obtained from the Schrödinger equation:

$$H\psi = E\psi \tag{3}$$

The non-relativistic Hamiltonian for an idealized system made up of a point nucleus of infinite mass and charge Ze, surrounded by N electrons of mass m and charge $-e$ is:

$$H = \sum_{i=1}^{N} \frac{p_i^2}{2m} - \sum_{i=1}^{N} \frac{(Ze)^2}{r_i} + \sum_{i>j=1}^{N} \frac{e^2}{r_{ij}} \tag{4}$$

where

$\frac{p_i^2}{2m}$ is the kinetic energy of the electron i (p_i its momentum)

$\frac{(Ze)^2}{r_i}$ is the potential energy of the electron i in the field of the nucleus

$\frac{e^2}{r_{ij}}$ is the Coulomb interaction between the electrons i and j.

A better description of the energy level structure of the actinide atoms is obtained by adding to (3) terms of relativistic origin, which, in fact, represent the magnetic interaction of orbital and spin momenta of the electrons. (They are of particular importance for actinides since they depend on the fourth power of the atomic number Z)[16].

By far the dominant one is the spin-orbit interaction described by a term of the form[16]:

$$\sum_{i=1}^{N} \xi(r_i) l_i \cdot s_i \tag{5}$$

where l_i and s_i are the orbital and spin quantum numbers respectively of the ith electron.

The approximate Hamiltonian is thus given by:

$$H = \sum_{i=1}^{N} \frac{p_i^2}{2m} - \sum_{i=1}^{N} \frac{Ze^2}{r_i} + \sum_{i>j=1}^{N} \frac{e^2}{r_{ij}} + \sum_{i=1}^{N} \xi(r_i) l_i \cdot s_i \tag{6}$$

A more accurate description is obtained by including other additional terms in the Hamiltonian. The first group of these additional terms represents the "mutual magnetic interactions" which are provided by the Breit equation[17)]. The second group of additional terms are known as "effective interactions" and represent, to second order perturbation treatment, interaction with distant configurations[18)]. These "weak" interactions will not be considered here.

Exact solutions of the Schrödinger equation are, of course, impossible for atoms containing 90 electrons and more. The most common approximation used for solving Schrödinger's equation for heavy atoms is a Hartree-Fock or central field approximation. In this approximation, the individual electrostatic repulsion between the electron i and the N−1 others is replaced by a mean central field giving rise to a spherically symmetric potential

$$-\frac{U(r_i)}{e}$$

The basic Hamiltonian for the central field approximation is thus:

$$H_0 = \sum_{i=1}^{N} \frac{p_i^2}{2m} + U(r_i) \tag{7}$$

and the difference $V = H - H_0$, where H is the Hamiltonian given by Eq. (4) may be treated as a perturbation. The solution of Schrödinger equation for the basic Hamiltonian

$$\sum_{i=1}^{N} \left\{ \left(-\frac{h^2}{2m} \right) V_i^2 + U(r_i) \right\} \psi = E^0 \psi \tag{8}$$

yields the zero-order wave functions $\psi^0(K)$ and eigenvalues $E^0(\alpha)$, where α represents a set of quantum numbers $\alpha = (n, l, m_s, m_l)$ where n, l, m_s and m_l have their usual signification. Since the $E^0(\alpha)$'s are degenerate with respect to m_s and m_l the eigenvalues E^0 are characterized by the symbols $(n_1l_1)(n_2l_2)\ldots(n_Nl_N)$. This sequence of numbers specifies the electronic configuration of the system and the energy levels E^0 are called configurations.

a. *Electrostatic and Spin-Orbit Interaction*

Consider the perturbation potential

$$V = H - H_0 = \sum_{i=1}^{N} \left\{ -\frac{Ze^2}{r_i} - U(r_i) \right\} + \sum_{i>j=1}^{N} \frac{e^2}{r_{ij}} + \sum_{i=1}^{N} \xi(r_i) l_i \cdot s_i \tag{9}$$

where the term in brackets, purely radial, causes only a shift of each configuration.

The electrostatic interaction $H_1 = \sum_{i>j=1}^{N} \frac{e^2}{r_{ij}}$ and the spin-orbit interaction $H_2 = \sum_{i=1}^{N} \xi(r_i) l_i s_i$ are different for different states of the same configuration and partly remove the degeneracy of the configurations.

The electrostatic interaction H_1 within a configuration contains a direct and an exchange part.

The spin-orbit interaction H_2 for each electron i can be written in the form:

$$\xi_{n,l}\, \mathbf{l} \cdot \mathbf{s} \tag{10}$$

It is considered as an interaction between the spin and the orbital angular momenta of the electron i, and is represented by the coupling parameter $\xi_{n,l}$.

b. Models of Couplings (For Two-Electron Configurations)

Magnetic and spectroscopic properties of free atoms depend on the interplay of the interactions H_1 and H_2, since they determine the magnetic moment and the energy spectrum of the atom. Models of this interplay (coupling models) are assumed for lanthanide and d-transition elements. We shall examine in a simple way possible couplings, and point out the difficult case of actinide atoms.

In first-order perturbation theory (and restricted for simplicity, to two-electron configurations) the Eq. (9) is simplified if it is possible to neglect one of the two terms H_1 or H_2 as a perturbing term (this is of course possible if the two differ greatly in magnitude). Three cases may therefore be distinguished (couplings):

- $H_1 \gg H_2$: this is the L-S (or Russell-Saunders) coupling. In this case, only H_1 is taken into account in first order perturbation theory. It commutes with L and S where $L = l_1 + l_2$ and $S = s_1 + s_2$, so that each configuration splits into terms being $(2L + 1)(2S + 1)$-fold degenerate.
- $H_2 \gg H_1$: this is the j-j coupling. In this case only H_2 is taken into account in first order perturbation theory. It commutes with the total angular momentum $j_i = l_i + s_i$. The configurations split into levels being $(2j_1 + 1)(2j_2 + 1)$ fold degenerate.
- $H_1 \approx H_2$: this is the so-called intermediate coupling. When the electrostatic and spin orbit interactions are of the same order of magnitude – and this is the case of the actinides – both should be included in first-order perturbation theory.

In this case, the energy matrices are diagonal only in J and M_J and are independent of M_J. Consequently, it is impossible to define a coupling in the configurations under discussion. The lack of a definite coupling is referred to as *intermediate coupling*.

If the atom or ion is situated in an environment of different atoms or ions, as, for instance, ions in a solid, the surrounding ligands exert on it a further interaction, which is called the crystal field interaction H_{CF} and enters Eq. (9). One has to compare H_{CF} with H_1 and H_2, in order to decide whether it is or not a small perturbative term. In actinide solids, it is usually found that H_{CF} is of the same order of magnitude as H_1 and H_2, so that intermediate coupling schemes are necessary which include H_{CF} as well. (For a more exhaustive treatment of couplings in actinides, see Chap. D.)

3. *The Relativistic Central Field Approximation*

It is known[19] that the non-relativistic Schrödinger equation, employed in the preceding subsections, should be substituted by the Dirac relativistic equation if the states of (high speed) electrons need to be described correctly. The Dirac equation is preferable to the Schrödinger equation to calculate wave functions and eigenstates for electrons in all atoms[20], but it is particularly needed when dealing with outer electrons in heavy atoms, such as the 5f, 6p, 6d, 7s electrons in actinides, which are at great distance from the central potential of the nucleus. (We have already stated that the spin-orbit interaction term H_2 which is of relativistic origin, is very large in actinides.)

One result of the use of the Dirac equation in a central field potential (i.e. purely Coulombic, giving rise to hydrogenic wave functions) is that the charge density of relativistic electron states is more contracted towards the nucleus than for the non-relativistic potential. This effect however decreases strongly with increasing angular momentum (e.g., it is greater for s and p states than for d and f states) (for a detailed discussion of relativistic effects see Chap. F). The relativistic binding energies are larger than the non-relativistic, particularly for $J = 1/2$ levels.

Qualitatively, "internal" orbitals are contracted towards the nucleus, and the Radon core shrinks and displays a higher electron density. "External" electrons are much more efficiently "screened" from the nucleus, and the repulsion described by the (Hartree-Fock) central potential $U(r_i)$ in Eq. (7) of the preceding subsection is greatly enhanced. In fact, wave functions and eigenstates of "external" electrons calculated with the relativistic and the non-relativistic wave equations differ greatly.

4. *The Electronic Structure of the Actinide Atoms*

Having examined the principal concepts of the physics of actinide atoms, we shall now relate results obtained from atomic quantum calculations, with special emphasis on 5f states.

a. Atomic Wave Functions

The characteristic structure of the actinide spectra is mainly determined by the special properties of the 5f wave functions; these, in turn, arise from the special nature of the 5f effective potential-energy function U_{eff} (5f). U_{eff} describes the effect on the 5f electrons of the attraction by the nucleus as well as of all the interactions with other electrons. It contains also a centrifugal term, $l(l + 1)/2r$, of particular importance for the properties of 5f wave functions, the effect of which will be illustrated later in more detail.

As discussed above, these properties were first investigated by Mayer[3]. By using selfconsistent Dirac-Fock methods, Desclaux[20] has calculated U_{eff} (5f) as a function of the atomic number Z. In Fig. 6a–c are plotted the effective potential energy U_{eff} as well as the 5f radial charge densities for actinium (configuration $6d^1$, $7s^2$), thorium (configuration $6d^2$, $7s^2$) and protactinium (configuration $5f^2$, $6d^1$, $7s^2$). These figures need some comments; the effective potential energy is the same for 4f and 5f electrons so that the deep minimum around 0.18 a.u. corresponds to the filled 4f sub-shell. Thus one must

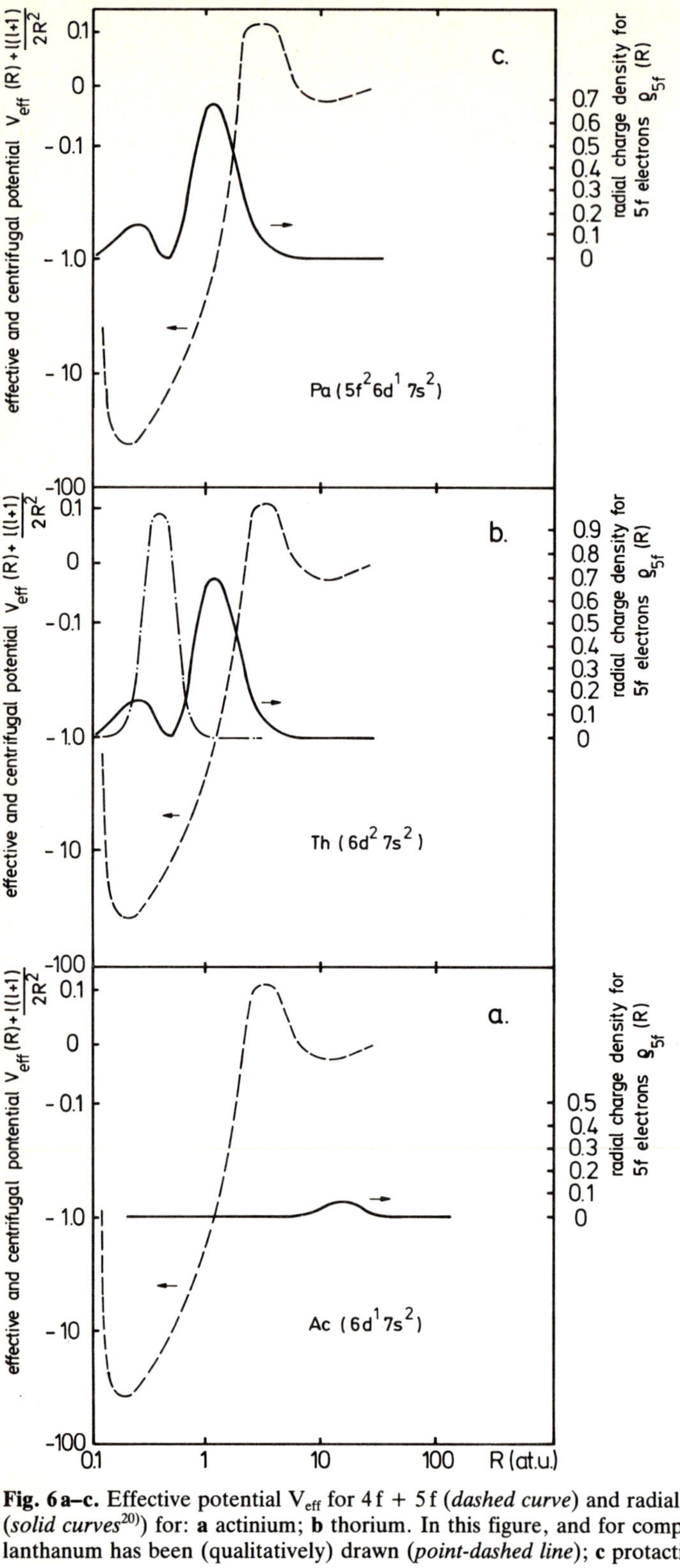

Fig. 6a–c. Effective potential V_{eff} for 4f + 5f (*dashed curve*) and radial charge density $\varrho_{5f} = |\Psi_{5f}|^2$ (*solid curves*[20)]) for: **a** actinium; **b** thorium. In this figure, and for comparison, the $\varrho_{4f} = |\Psi_{4f}|^2$ for lanthanum has been (qualitatively) drawn (*point-dashed line*); **c** protactinium

not only study U_{eff}, but modifications in U_{eff} to see how a second minimum develops sufficiently to contain 5f wave functions. The oscillatory 5f wave functions being orthogonal to the 4f ones, one always finds some 5f density around 0.18 a.u. (negative radial 5f wave function due to orthogonalization). Additionally, one clearly sees a sudden "localization" of the 5f radial wave function starting with thorium. Here "localization" (a term that will be very much employed later) means a contraction of the 5f-charge density in the neighbourhood of the nucleus. Comparing these results with those obtained for lanthanides, one finds that, as far as 5f "localization" is concerned, thorium and not actinium is the analog of lanthanum, that is to say that when comparing actinides and lanthanides, the f localization is delayed by one atomic number.

This is proved also by Figs. 7 and 8. In Figs. 7 and 9, respectively, are shown the expectation value of r, $\langle r \rangle$, and the energy eigenvalue $E_{n,l}$ for external orbitals along the actinide series. Both the spatial extension and the binding energy of the 5f wave functions drop suddenly when going from Ac to Th. At thorium, indeed, the 5f electrons become internal electrons, and the 5f shell is one unsaturated internal shell, in the same way as d-shells are in transition metals and the 4f shell is for rare earths.

Spatial extension, as expressed by the expectation value $\langle r \rangle$, is roughly comparable for 4f and 5f wave functions (Figs. 7 and 8). However, the many-electron wave functions resulting from the solution of the relativistic Dirac equation may also be used to calculate a number of physically interesting quantities, i.e. expectation values of observable

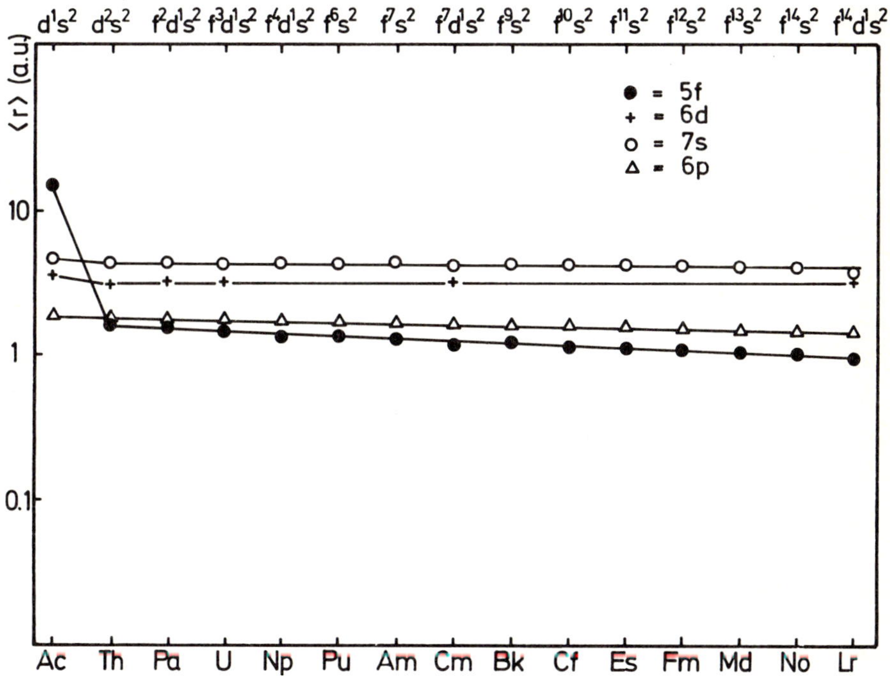

Fig. 7. The expectation value $\langle r \rangle$ for 5f, 6p, 6d and 7s orbitals in actinide atoms (within the ground state configurations indicated on the top of the figure) (The atomic calculations have been performed by Desclaux, J. P.)

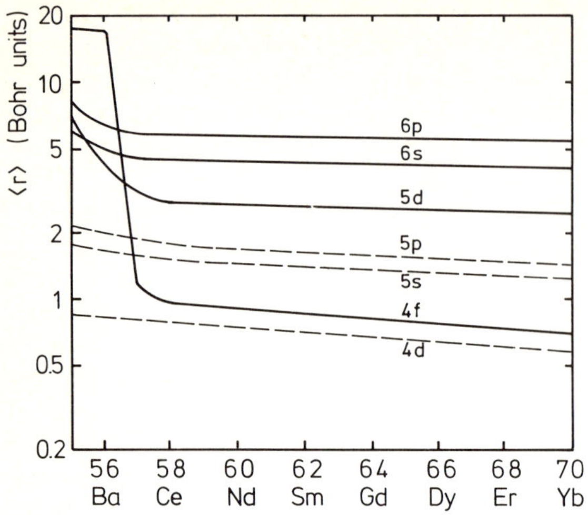

Fig. 8. The expectation value $\langle r \rangle$ for 4d, 4f, 5s, 5p, 6s, 6p orbitals in lanthanide atoms (from[21])

operators like $\langle r^{-3} \rangle$ (needed to estimate the hyperfine interaction), $\langle r^2 \rangle$, $\langle r^4 \rangle$, $\langle r^6 \rangle$ (useful in determining crystal field splitting parameters), and the radial integrals j_i (needed to determine the magnetic form factors).

A very relevant representation of atomic 5f wave functions is given by Freeman and Koelling[15].

Although the $\langle r \rangle$ expectation value is almost the same for 5f and 4f, a plot of the charge density shows that, at a given distance from the nucleus (e.g. 4 a.u.), whereas the 4f charge density dies off very rapidly, the 5f charge density has still a meaningful value. This can be described as the 5f charge density having a longer "tail" than the 4f (similar behaviour is shown by 3d-wave functions). Of course, 4 a.u. is a distance comparable to inter-atomic distances in solids. Figure 10, taken from[15], compares the expectation value

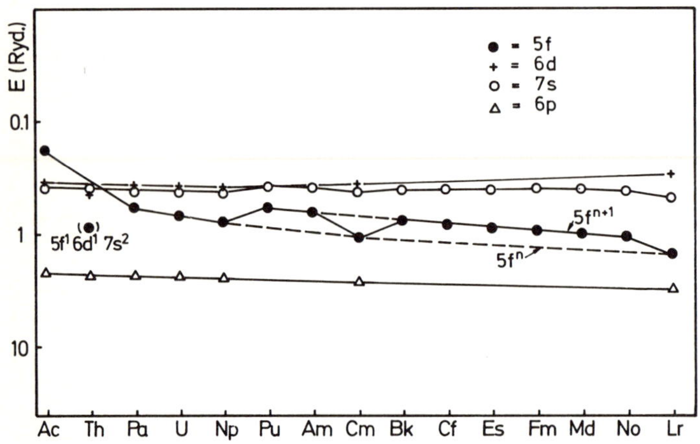

Fig. 9. Eigenvalues for 5f, 6p, 6d and 7s orbitals in actinide atoms (the configurations are the same as in Fig. 7. However a configuration $5f^1\ 6d^1\ 7s^2$ for thorium is also considered) (The atomic calculations have been performed by Desclaux, J. P.)

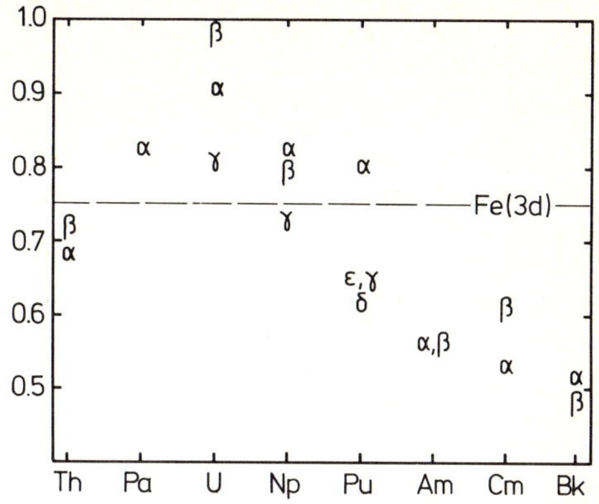

Fig. 10. The $\langle r^4 \rangle^{1/4}/R$ (R: interatomic distance) ratio for different crystallographic modifications of actinide metals is plotted vs atomic number Z. The ratio is a rough measure of f-f overlapping. In the figure, the value for a 3d-metal (Fe) is given for comparison (from[15])

$\langle r^4 \rangle$ (rather $\langle r^4 \rangle^{1/4}$), which is more sensitive to the "tail", with the half internuclear separation R in different phases of actinide metals. The ratio $\langle r^4 \rangle^{1/4}/R$, plotted in this figure, gives an idea of the amount of 5f character between two actinide atoms: roughly, an idea of 5f overlapping in elemental metals.

b. Atomic Eigenvalues and Electronic Configurations of the Atom

Whereas spatial extension as expressed by the expectation value $\langle r \rangle$ is roughly comparable for 4f and 5f wave functions, this is not the case (Figs. 9 and 11) for the energy eigenvalues. It is clear from Fig. 9 that 5f electrons are *not* strongly bound to the nucleus and that 5f, 6d and 7s electrons have similar binding energies. It is only at the end of the actinides series (No, Lr) that the binding energy of the 5f electrons becomes comparable to the binding energy of the 4f electrons at the beginning of the lanthanide series (Ce). Since all outer orbitals (5f, 6d, 7s) have comparable energy eigenvalues, there is a competition between these orbitals.

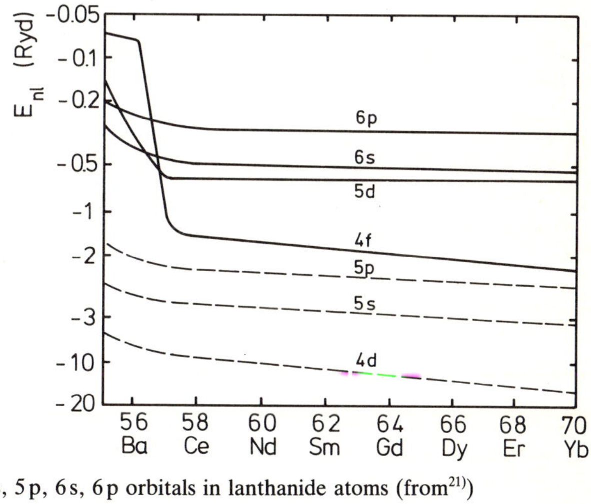

Fig. 11. Eigenvalues for 4d, 4f, 5s, 5p, 6s, 6p orbitals in lanthanide atoms (from[21])

This competition, manifested in the structure of the actinide atomic spectra, also appears as a competition between configurations: there is an accumulation of many configurations, both even and odd, at about the same height, leading to a high density of energy levels (in fact, we had arrived at the same conclusion by inspecting Fig. 1 in Sect. A.I.2).

The energy level values of the lowest spectroscopic term of the electronic configuration of lanthanide as well as actinide atoms, were tabulated by Brewer[22]. Such tables are very useful for phenomenological correlations concerning actinide metals (see Chap. C). From these tables one can obtain Table 1 giving the ground state and the first excited level of the actinide atoms as well as of the lanthanide atoms for comparison[4].

When the f sub-shell is less than half-filled the stabilization of the divalent state (f^{n+1}) relative to the trivalent state is more rapid for the lanthanides than for the actinides. But the behaviour is reversed in the second half of the series. In part, this must be due to the low binding energy of the 6d electrons compared to the 5d.

III. General Concepts in Actinide Solid State Physics

The aim of this part is twofold:
– to give the reader the necessary conceptual tools for the understanding of the rest of the chapter (and of the book);
– to help the reader getting familiar with problems concerning narrow bands (such as the localization vs. itineracy problem), which are central to actinide solid state physics (perhaps even more than for d-transition metals physics).

Since band theory has proved an invaluable tool for the description of the light actinide solids, Sects. 1 and 2 will be devoted to the band description and to its limits, Sect. 3 analysing in more details the band one-electron Hamiltonian, and the ways it can be written in order to take into account electron-electron interaction.

In Sect. 4 and 5, the consequences of spin polarization in unsaturated 5f shells are analysed, and the Stoner and Mott-Hubbard theories briefly reviewed. In this way, concepts which are central for actinides, and which the reader may find only when perusing many textbooks (references of which are duly given) are adequately concentrated.

1. What is Learnt From the Atomic Wave Functions

a. The Actinide Metals

In Part II. Figs. 8 and 11 showed that in lanthanides the 4f electrons can be described as localized so to be treated, in all respects, as atomic electrons. Thus, for instance, in the lanthanide metals, properties related to the conduction band (which has essentially a (5d, 6s) character) may be – in first approximation – separated from the magnetic properties related to the highly localized 4f electrons.

Different is the case of 5f electrons in actinide solids. What we have learnt above about 5f's in the atom, suggests that care must be taken of two important factors:

a) as seen from atomic considerations (Fig. 9), 5f do not differ greatly in energy from 6d and 7s, thus giving rise to many possible electron configuration in the atom, hence to many possibilities of *stable* linear combinations of atomic wavefunctions.
b) although spatial extension of the atomic 5f wave function is roughly comparable to that of 4f wave functions (Fig. 7), this is no longer true at distances equaling some atomic distances, in a metal crystal.

Point b) is illustrated by Fig. 10, that we have already discussed. The probability of finding a 5f electron of one atom in an actinide crystal at positions occupied by outer electrons of other crystal atoms is, therefore, rather high. The tails may overlap with other tails of 5f wave functions, coming from adjacent actinide cores: a phenomenon called *f-f overlapping*. On the other hand this effect seems to decrease from Am on. In Fig. 10, $\langle r^4 \rangle^{1/4}/R$ for α-Pu is comparable to that of Fe, indicating an overlapping of the f wave function in this elemental metal similar to the d-overlapping of a typical 3d-transition metal; for Am, Cm, Bk this parameter is much lower: an indication of the progressive retraction of 5f wave functions in the second part of the actinide series, as we had discussed in Part II.

For the light actinides from a) and b) qualitative consequences may be drawn. It follows, from *f-f overlapping,* that the atomic 5f wave functions will broaden into 5f bands when building the metal. Moreover, since the 5f, 6d and 7s energy eigenvalues are very close (see Sect. A.II) the 5f band will *hybridize strongly* with the 5d and 7s bands.

Contrary to the lanthanide metals, at least in the first half of the series, the conduction band of the actinide metals (bonding band of the metal) will be very complex. It will consist of 6d, 7s and 5f admixtures. The physical properties, even the magnetic ones will be determined by this complex conduction band.

b. The Actinide Compounds

We have particularly emphasized the metallic bond, because the peculiar properties of the 5f wavefunctions are better seen in its context. Also, in compounds, however, *f-f overlapping* as well as *hybridization* play a conspicuous role. Only, the presence of a non-actinide element introduces other factors that have to be taken into account:

– the difference in electronegativity between the actinide and the non actinide atom will favor charge transfer between the two, so that the cohesive energy will contain a part of electrostatic interaction between the ions so formed (Madelung energy, typical of ionic bonds);

– the interactinide separation may be greater than in metals, so that f-f overlapping decreases and so does the metallic bonding;

– hybridization may take place not only with the 6d and 7s states of the actinide atom, but with outer states of the non-actinide metal, causing, eventually, covalent bonding.

All these factors have to be taken into account at the same time, if the electronic structure and the type of bonding of a compound have to be understood.

2. *Localization vs. Itineracy for 5f Electrons*

a. *Physical Properties in the Atomic or in the Band Limit*

The f-f overlapping in light actinides may cause broadening of the 5f wave functions into 5f bands. On the other hand, from Am on, this overlapping having decreased, this effect occurs much less. It follows that physical properties which depend from 5f orbitals may be better understood, in one case, in the band limit, in the other case, in the atomic limit.

We will take as an example one property: magnetism, which is almost completely 5f-dependent.

It is known that an atom containing an unsaturated inner shell displays a net magnetic moment. This is the reason of magnetism being found in transition metal atoms (unfilled d^n shell), in lanthanides (unfilled $4f^n$ shell) and in actinides (unfilled $5f^n$ shell). The magnetic moment is caused by the fact that the electrons fill the shell in such a way as to have the maximum alignment of their spins: a set of rules, Hund's rules, give a very precise prescription[1)].

Hund's rules are a manifestation of the Pauli principle. Electrons of parallel spin in an atom tend to avoid each other, i.e. not to occupy the same positions in space. If one calculates the electrostatic repulsive interaction between two charge densities arising from these parallel spin electrons, this repulsive interaction is lower than for those of electrons of opposite spin: an energy gain is therefore achieved (exchange energy). This gain in energy lowers the total energy of the atom, and makes the maximization of the spins a stabilizing condition.

If an f-band is formed, however, the effect of the atomic Hund's rules (or of the exchange interaction) may weaken. The same physical phenomenon that determines Hund's rules due to the Pauli principle, is still present. However, the exchange energy may be negligible with respect to the high kinetic energy that the electrons assume. If the f-f-overlapping is so large that f-bands become broad bands, only the typical weak Pauli paramagnetism of free electrons, due to an induced magnetic moment, is found.

In many actinide solids, as we shall see, the experimentally determined magnetic properties are explained well by assuming the permanent magnetic moment due to Hund's rules. The f-electrons are considered atomic, and their interaction with the environment is through crystal field forces or weak exchange forces with conduction electrons. Here, the magnetic properties are explained in the atomic limit.

In other solids, Pauli paramagnetism is found to be explained in the band limit. In others, still, an intermediate behaviour is encountered.

b. *The Band Description of Electrons in Narrow Bands*

The reader shall be familiar with the concepts of band-theory (which are illustrated in excellent textbooks[5, 23, 24)]) such as the band dispersion E(k) and the Bloch-state solutions of the equation of motion

$$(T + V(\underline{R}))\psi_k(\underline{R}) = E(\underline{k})\psi_k(\underline{R}) \qquad (11)^1$$

where $V(\underline{R})$ is a periodic potential, formed by the superimposition of the central potentials of the ionic cores (Fig. 12).

If $V(\underline{R})$ is small with respect to T in (11) the periodic potential $V(\underline{R})$ of the ionic cores is that of a lattice of scattering centres. The electronic excitation, described by ψ_k, is a scattered electron wave of high kinetic energy, traveling through the crystal. Many

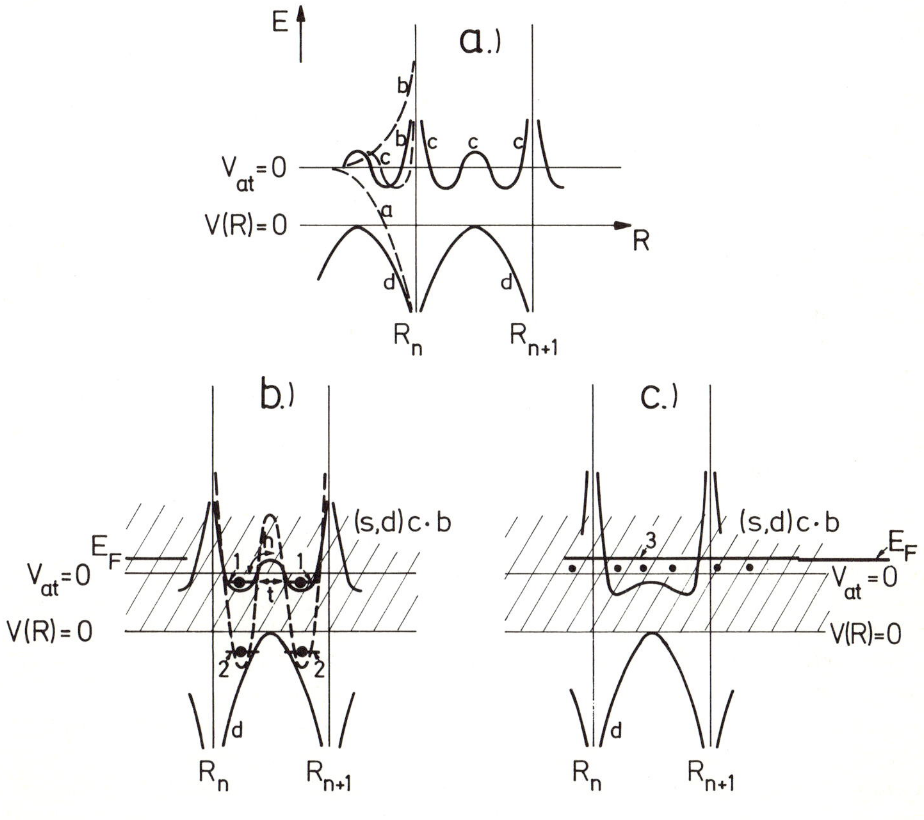

Fig. 12 a–c. Schematic representation of the effective potential V_{eff} and of different possibilities of localized and itinerant states for electrons of high l quantum number. **a**) The *solid line* d represents the periodic potential set-up by the cores R_n and R_{n+1}, which is a superimposition of central potential a (*dashed line*). The *dashed line b* represents the centrifugal potential of kinetic origin $l(l+1)/2R^2$ in an atom, and c (*dashed line*) the effective potential V_{eff} for an atom (compare Fig. 6) and (*full line*) for a solid. **b**) Relative to two shapes of the effective potential V_{eff}, two examples of *localized* state are given: 1. resonant state; 2. fully localized state. Notice that 1. is very near to E_F. h and t represent hopping and tunneling processes. **c**) A narrow band is formed (resonance band), pinning E_F: 3. narrow band

1 For the non-relativistic case (Schrödinger equation), $T = -\nabla^2$. For relativistic case (Dirac equation), $T = c \cdot \alpha \cdot p + \beta mc$[17], where m is the rest mass of the electron, c is the velocity of light. We have preferred to write the T operator in a general form, covering both cases, given the importance of the relativistic approach in band calculations for actinide solids – see Chap. F

properties of the band (e.g. a parabolic E(k) dispersion with deviations only at the Fermi-surface) and of the electrons can be described by the free-electron or quasi-free electron approximations, $V(\underline{R})$ being only a small perturbing term in (11); also, interaction between electrons can be treated in the Fermi-Thomas approximation[5, 23, 24]. All this is, however, mostly valid for rather broad bands, when large overlap of the atomic wave functions of the electrons occurs throughout the crystal (as for the (spd) conduction band in actinide solids).

Narrow bands arise when the overlap of the atomic wave functions is small (as for 5f's). In this case, the dispersion $E(\underline{k})$ is strongly reduced and the bandwidth W becomes very small (zero, in the case of no overlap). The electron charge density, caused by these wave functions, is high in the core region of Fig. 12, and the quasi-particles spend most of their life there, nearly "bound" to the atom. In case the charge density is all confined within the core region (as for 4f in lanthanides), then the bond description loses its meaning and the atomic description holds.

Here we shall briefly discuss: i. the ψ_k of (11); ii. the parameters allowing to choose between a band or an atomic description.

The $\psi_k(\underline{R})$ describing the motion should retain many characteristics of the atomic wave functions, if electrons in narrow bands are treated. If we imagine the solid as a lattice of atomic (or Wigner-Seitz) spheres, each surrounding one core, then a trial solution to (11) may be written in the form:

$$\psi_{k,\alpha}(\underline{R}) = \sum_{\underline{k}} e^{i\underline{k}\cdot l}\phi_\alpha(\underline{R} - l) \tag{12}$$

(l being a lattice vector) where the ϕ_α's constitute some set of wave functions which are solutions of the Schrödinger equation within the atomic sphere (where the central potential of the core is predominant), and the label α denotes some set of atomic quantum numbers. A solution such as (12) is taken in the so called "tight binding" approximation[5, 23, 24], where the ϕ_α's are just atomic orbitals ($\alpha \equiv n, l, m$) and (12) is called a Wannier function. The ϕ_α's have to be joint in a smooth way at the surface of the spheres, so that (12) be continuous and differentiable everywhere in the solid. Because of the difficulties in achieving these conditions, the atomic orbitals are substituted with specially constructed orbitals, called *M*uffin *T*in *O*rbitals (MTO), in very efficient recent methods for band calculations (e.g. LMTO – ASA: *L*inear combination of *M*uffin *T*in *O*rbitals in the *A*tomic *S*phere *A*pproximation (see Chap. F for a short recall of LMTO – ASA formalism and concepts, and[5, 23, 24] for its exhaustive description).

The important point of (12) is, however, that it retains atomic information in the one electron wave functions of (11). In LMTO – ASA, e.g. the bandwidth W will depend essentially on the charge density $|\phi_\alpha|^2$ evaluated at the surface of the atomic sphere, being large when $|\phi_\alpha|^2$ is high (the electron wave is hardly contained within the sphere), small when $|\phi_\alpha|^2$ is low (the electron wave is mostly contained within the sphere).

We may be interested in how much time a band electron spends in the vicinity of a core during its propagation through the crystal. Intuitively, if the attraction by the core potential is not effective (as for free electrons) this time will be very short; if it is very effective, very long. We may figure that the band electron is "hopping" (or "tunneling") from core to core, and the displacement event will have a certain frequency of occurring. The time we are looking for, is the reciprocal of this frequency, and will depend on the range of energies the band electrons have, which is roughly given by the bandwidth W of the band: i.e. the "hopping time" t is about equal to h/W. We see, therefore, that very small bandwidths lead to the electron staying very long times (in the limit $W \to 0$, which describes an atomic level, an infinite time) in the core potential well: we say that the electron is "localized".

If an electron is "localized", abandoning its core has important consequences. An important assumption of (11) is that the $\psi_{k,\alpha}$ are unchanged if we excite one electron from a state $\psi_{k,\alpha}$ to a state $\psi_{k,\gamma}$. This change in "configuration" (see A.II.) within the band does not perturb the other electron states (Koopman theorem[21]).

This is profoundly different from what happens in an atom, where a change in configuration involves always a certain amount of energy.

Suppose that a particular electron is orbiting around a core at a particular moment, and suppose that its motion is that of the ground state configuration of the atom. In order that the electron be displaced to another atom, it must be "excited" or "activated" out of the core and into another core, which, then, will increase its population of one. There are important terms of Coulomb and exchange interaction energies which change in this process. Let us call these changes U.

The wave functions of the two states (the depleted and the enriched atom) are also profoundly changed, and this is in serious contrast with the above mentioned Koopman theorem, showing that the condition it formulates is valid for (11) but not in the atomic limit. The Coulomb interaction term U is either neglected or included in an average way in (11), as we will see later. In reality, we are considering here what happens to other electrons when one electron move, i.e. we are examining qualitatively the correlation between the different motions of the electrons. Averaging out of this correlation is a useful thing if the correlation is small, but may lead to large errors in the opposite case. In reality, given the atomic-like properties of 5f-wave functions (at least in the second half of the series), the correlations are rather large in actinides.

In fact, whether to interpret properties in the atomic or in the band limit depends upon the competition between these two quantities: the bandwidth W, describing itinerant behaviour, and the Coulombic interaction U, describing atomic behaviour.

c. *Effective Potential V_{eff} for an Electron in an Atom and in a Lattice. Localization vs. Itineracy*

The core potential effects on a single electron's motion may be discussed by introducing the concept of the effective potential V_{eff} that this object feels.

We have already mentioned V_{eff} for a 5f electron in the case of an isolated atom (A.II.). We may start from this case, and follow a very valid description employed by Chan and Lam[25], making use of Fig. 12 (which is just a modified version of a figure to be found in [25]).

When writing the Hamiltonian of the motion of an electron in a central potential, a term $l(l + 1)/R^2$ arises from the Laplacian transformation of the kinetic operator[1]. This term combines with the central potential to produce a centrifugal barrier confining the electron to an annular region (see Fig. 12a, dashed line). The centrifugal barrier plays an important role for high l electrons, as the f-electrons are. The amount of confinement of the charge density within the trough depends on how much the electron may *hop over* or *tunnel through* the barrier provided by V_{eff}: which is a function of the height and width of the barrier, but also of the kinetic energy that the electron has. When comparing 4f in lanthanides with 5f in actinides, in the latter[15)] the centrifugal barrier is narrower, and the kinetic energy of the 5f is higher: this explains the long tails of the 5f atomic wave functions[2].

When the atoms are approached, to form a lattice (as in a metal), the effective potential V_{eff} is now the combination of the centrifugal term and the periodic potential $V(\underline{R})$. Troughs and barriers are created in the inter-core zone (see Fig. 12a, full lines).

Two situations may occur for the high l electron (for simplicity, we will refer hereafter to f electrons).

i. The barrier is effective to retain the electron in an annular region of the core most of the time, i.e. in an atomic-like, bound form. The electron is clearly *localized*. Its ground-energy is given by a level situated within the trough: the exact position of it depends on the details of V_{eff} (see Fig. 12b).

ii. The barrier lies below the V_{at}-zero; there is enough f-f overlapping between adjacent atoms (see Fig. 12c), and a narrow band is formed. The electron is said to be *itinerant.*

If the "localized" electron tunnels out through the barrier (state 1 in Fig. 12b) a certain amount of f-f overlapping is present. States like 1 in Fig. 12b are called sometimes *"resonant"* states or *"virtually bound"* states. In contrast with case 2 in Fig. 12b, which we may call of *"full localization"*, the wave function of a resonant state does not die out rapidly, but keeps a finite amplitude in the crystal, even far away from the core[20)]. For this reason, overlapping may take place with adjacent atoms and a band may be built as in ii. (If the band formed is a very narrow band, sometimes the names of *localized state* or of *resonance band* are employed, too. Attention is drawn, however, that in this case one refers to a many-electron, many-atoms wave function of itinerant character in the sense of band theory whereas in the case of resonant states one refers to a one-electron state, bound to the central potential of the core (see Chap. F)).

d. Localized and Itinerant States in the Fermi-Dirac Statistics

In a scheme of available energy states, a population of electrons distributes according the Fermi-Dirac statistics[1, 20)]. The probability f(E) of having an electron in a state of energy E, is, at temperature T

$$f(E) = \left(\exp\left(\frac{(E - \mu_F)}{k_B T}\right) + 1\right)^{-1} \qquad (13)$$

The function f(E) is a step function for metals at 0 K: electrons fill all states up to a well defined energy value E_F, which is called the Fermi energy of the solid.

The Fermi level is the *chemical potential* μ_F of the electron population[1, 20)]. It may vary when the thermodynamic variables of the system are varied, e.g. temperature and

2 This is not due, however, to the orthogonality constraint between 4f and 5f orbitals, contrary to a general belief (private communication from J. P. Desclaux)

pressure; at 0 K, $\mu_F = E_F$ for metals. For insulators, μ_F lies in the band gap between the valence and the conduction bands.

Its position in energy is determined by an integral equation such as:

$$N = \int_0^\infty f(E)\,N(E)\,dE \qquad (14)$$

where N is the total number of electrons.

In (14), N(E) is the *density of state function,* i.e. the number of eigenstates of the electron (in the given energy scheme) between E and E + dE. Equation (14) is then just a conservation condition for the number of electrons N.

N(E) is in general a complicated funtion of E. For free electrons is proportional to $E^{1/2}$, and this is a good approximation for a broad band.

For a narrow band, N(E) is very high, since very many states are confined in a very narrow bandwidth. In an energy level (zero bandwidth) N(E) diverges (it is a delta-function).

Consider now a bound state, of very large negative energy (as 2 in Fig. 12 b). Then, by (13), this state will be occupied (at T not too high) independently of the position of the Fermi level ($f(E) \approx 1$). Its contribution to (14) is just one, as may be seen if N(E) is taken to be a delta-function. Therefore it is better to neglect it in the count of electrons and change the limits of integration of the integral. A bound state is always filled, irrespective of the Fermi level. Conversely, the state does not help at all to define the Fermi level by means of (14). This is essentially an *atomic localized* level.

Different is the case for a state (a *resonant state*) as 1 in Fig. 12 a, lying very close to the Fermi level (E very close to E_F). For this state, (13) gives a probability of occupation which is strongly temperature dependent. How much it determines the position of the Fermi level depends on the contribution which, together with the population of the broad (s-d) conduction band which we have super-imposed in Fig. 12 b, it gives to (14).

These two cases are both cases of *localization,* in the sense that both are relative to an atomic central potential. However, one practically fills independently of the chemical potential μ_F, the other may be emptied easily by perturbing fields, or by thermal motion.

We may then consider the case of the narrow band as in Fig. 12 c. We have represented it as overlapping in energy with the broad conduction band of s and d character, as is often the case for actinide metals (it will be one of the condition for hybridization between the two bands). Such a band has a very high N(E), as we said before. In (14), the high density of state of the narrow band will be very effective in determining the position of μ_F, which will be very close to it (sometimes, one says that μ_F will be "pinned to", the narrow band).

e. *Physical Properties of Conduction Electrons Dependent on $N(\mu_F)$*

The density of states at the Fermi-level $N(\mu_F)$ is responsible for many electronic properties, e.g. the electronic contribution to the low-temperature specific heat of a solid, and the Pauli paramagnetic moment of conduction electrons. The specific heat contribution may be written as[23]:

$$C_{el} = \gamma T = (2\,k_B N(\mu_F))T \qquad (15)$$

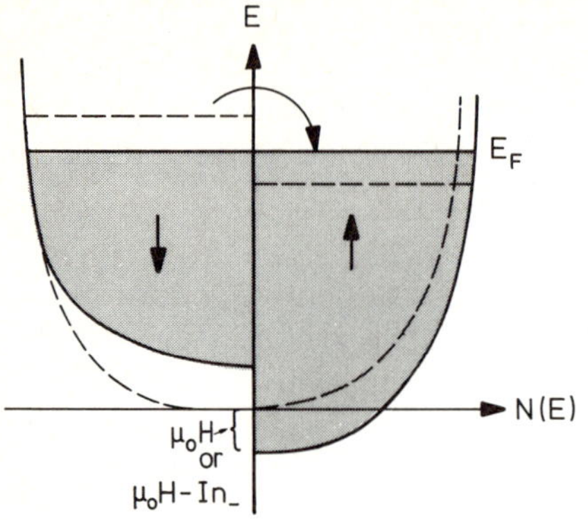

Fig. 13. Modification of N(E) under the effect of a perturbing agent and equalization of the electron population to the same (Fermi) level. The picture illustrates qualitatively ($N(E) \alpha E^{1/2}$ for simplicity) the statistical procedure in the derivation of Pauli paramagnetism and in the Stoner theory

(where k_B is the Boltzmann constant). Specific heat measurements are used to determine γ and hence $N(\mu_F)$ (see Chap. D).

Pauli magnetism is induced in a free electron gas by the application of an external magnetic field H. This magnetic field, acting as a perturbation, creates different energy states for electrons with spin aligned to the field, the concentration of which is n_+, and electrons with spin opposite to the field, the concentration of which is n_-. Also the density of state function will be different for the two populations (see Fig. 13) ($N_+(E)$ and $N_-(E)$).

The two populations however, must have the same chemical potential μ_F, in a Fermi-Dirac statistics (Eq. (13)).

Two integral equations for n_+ and n_-, similar to (14), together with the conditions $n_+ + n_- = N$, are the conservation conditions of the total number of electrons N. They permit to determine n_+, n_-, and μ_F.

A net magnetic moment is obtained which gives rise to Pauli paramagnetism:

$$M = \mu_0 (n_+ - n_-) = \mu_0^2 H N(\mu_F) \tag{16}$$

(μ_0 is the Bohr magneton) and which is proportional to the density of state at the Fermi level $N(\mu_F)$.

We have described in some detail these two properties. Measurements of the magnetic susceptibility and of the electronic specific heat give very clear information, from $N(\mu_F)$, on the presence of an f narrow band. They will be discussed in more detail in Chap. D.

Spectroscopic methods, especially photoelectron spectroscopy, permit to obtain the same information directly. They will be discussed in Chap. E.

3. *The One-Electron Hamiltonian and the Local Density of State Approximation*

a. *Introduction*

All the discussion of the preceding subsection has supposed the Hamiltonian (11) to be a good description of the energy of the electron population in the solid. In reality, the rapidly moving excitation waves of the huge negatively charged cloud constituted by the electrons create numerous interaction fields. The position (and momentum) of one electron is strongly correlated to the positions (and momenta) of the other electrons. The Hamiltonian (11) treats one-electron wave functions as if they were independent: in this way, the problem is brought to a relatively easy solution. In reality, the electrons Hamiltonian is:

$$H\phi(R_1 \ldots R_N) = \left(\sum_{i=1}^{N} T_i + V_{ext}(R) + V_{int}(R_1 \ldots R_N)\right)\phi(R_1 \ldots R_N) = E\phi(R_1 \ldots R_N) \tag{17}$$

where $\phi(R_1 \ldots R_N)$ and $V_{int}(R_1 \ldots R_N)$ are many-body functions, depending on the 10^{23} positions of the electrons, and $V_{ext}(R)$ is the external potential set-up by the cores (see Fig. 12).

The problem is to reduce (17) to (11), with appropriate approximations, though preserving information about the interactions within the cloud. We will examine in this subsection the methods employed to this task.

b. *From the Hamiltonian (17) to the Hamiltonian (11) in the Hartree-Fock Limit*

Suppose that $V(R_1 \ldots R_N)$ has the form:

$$V(R_1 \ldots R_N) = \frac{1}{2}\sum_{i \neq j} \frac{e^2}{R_i - R_j} \tag{18}$$

i.e. it consists of the sum of all the two-body Coulomb interactions between electrons.

Let us choose now a set of one-electron wave functions, e.g. of Bloch type. We suppose that $\phi(R_1 \ldots R_N)$ is the Slater-Fock determinant[1, 23–25] of these wave functions:

$$\phi(R_1 \ldots R_N) = \begin{vmatrix} \psi_{k_1}(R_1)\psi_{k_1}(R_2) & \ldots\ldots\ldots & \psi_{k_1}(R_N) \\ \psi_{k_2}(R_1)\psi_{k_2}(R_2) & \ldots\ldots\ldots\ldots\ldots & \\ \psi_{k_N}(R_1) & \ldots\ldots\ldots\ldots\ldots & \psi_{k_N}(R_N) \end{vmatrix} \tag{19}$$

An expression such as (19) is a linear combination of products of the one-electron wave function, and leads to a separation of (17) in a set of N independent equations for the N electrons. It embodies the antisymmetry required by the Pauli principle.

The variational method is then used to minimize the expectation value of total energy: $E = \langle\phi|H|\phi\rangle$ under small variation of the ψ's in (19), and subject to the normalization condition of ϕ: $\langle\phi|H|\phi\rangle = 1$. (This may be done by employing the method of Lagrangian undetermined multipliers).

The result of the variational method leads to the Hartree-Fock non-linear integro-differential equations for a one electron wave function ψ_k

$$(T + V_{ext}(R) + V_d(R))\psi_k(R) + E_{xc}(R, R') = E(k)\psi_k(R) \tag{20}$$

where the E(k) are Lagrangian multipliers.

V_d is the direct Coulomb interaction potential:

$$V_d\psi_k(R) = 2e^2\int dR \frac{n(R, R')}{(R - R')}\psi_k(R) \tag{21}$$

where $n(R, R) = \sum_k |\psi_k(R)|^2$ is the diagonal term of the so-called density matrix[20]. V_d can be identified with the Coulomb interaction potential which all other electrons apply on $\psi_k(R)$, since $n(R, R) = n(R)$ is simply the global electron density at R due to all other electrons. E_{xc} is due to the Pauli principle, and is the exchange integral. It can be written as:

$$E_{xc} = -2e^2\int dR' \frac{n(R, R')}{|R - R'|}\psi_k(R') \tag{22}$$

where $n(R, R') = \sum_k \psi_k^*(R')\psi_k(R)$ is the off-diagonal term of the same density matrix[23]. We have chosen here to hint at the "density matrix" concept, typical of many-body theories, in order to stress that E_{xc} is still a two-electrons operator. (For the many body derivation of (20), see[28]).

Notice that the integrand of (22) does not depend only, as all the other terms of (22), on the positional coordinate R of the $\psi_k(R)$. One says that (22) is a *non-local* term, and this fact considerably complicates calculations unless approximations are performed to make it *local* (so to reduce (20) to the form of the Hamiltonian (11)).

The Slater-Fock determinant (19) describes the motion of independent Fermions. The interaction operators (21) and (22) obtained therefore represent respectively the repulsive Coulomb field created by the averaged distribution of all electrons and the exchange contribution due to the antisymmetry property of the Fermions' wave functions: they are the analog of H_1, the electrostatic interaction within a configuration of an atom appearing in the Hartree-Fock central field approximation of II. There is no analog in (20) of other important couplings, such as spin-orbit coupling (which might be important for atomic-like wave functions, such as the 5f) or the interaction between the magnetic fields of the spins of the electrons, having the effect of creating some net spin in the band (spin-polarization) (also important for the filled 5f bands).

c. *The Local Density Approximation (LDA) in the Band Formalism*

Hohenberg and Kohn[29] have proved generally that the total ground state energy E of a collection of electrons in the presence of an externally applied potential V_{ext} (e.g. the valence electrons in the presence of the periodic potential due to the cores in a lattice), when no net magnetic moment is present, depends only on the average density of electrons $n(\underline{R})$. By this proof, n(R) becomes the fundamental variable of the system (as it is in the Thomas-Fermi theory[23, 28]). Variational minimization of the most general form of E, with respect to n lends to the Hartree-Fock equations formalism.

For the non-local term E_{xc}, defined in (22), Kohn and Sham[30] suggested a functional similar to that of the Thomas-Fermi theory. In this way, Eq. (20) is completely reduced to functionals of n(R) and then to a *local* form.

The problem is now how to compute n($\underline{R}$). A recipe frequently adopted is to build up n(R) by superimposing atomic wave functions of electrons in neighbouring atoms. (They need therefore to be correctly calculated in atoms, since they are the starting point for band calculations[15, 28]):

$$n(R) = \sum_{i=1}^{N} |\psi_i|^2 \tag{23}$$

and equation (20), completed reduced now to (11), is solved self-consistently, building up new sets of ψ_i's and V(R), until variations in the ψ_i's are considered negligible.

Most modern band-calculations for non-magnetic solids are performed in the LDA approximation, which is extended also to the relativistic formulation (see Chap. F). Care is taken in the choice of the set of ψ_i's. A particular problem exists in connecting the atomic wave functions ψ_i's, calculated in a central potential, in the inter-core region (see Fig. 12) of the solid. It is beyond the scope of this Chapter to go deeper into these details, which will be discussed further in Chap. F.

d. LDA Approximation for Actinide Solids

The success and the shortcomings of band calculations performed in the LDA approximation have been excellently reviewed by Koelling[31]. Essentially, the discussion follows points which have been already qualitatively described, concerning the competition between U and W. It is worthwhile to resume shortly these arguments here.

The LDA formalism considers only the ground state total energy E of the electron gas and its dependence on the local density n(R). In particular, it does not discuss the atomic wave functions ψ_i to be used in (23) in order to obtain a good description of n(R). Also, excited states of the solid are not included, so that, eventually, LDA is good in describing ground state properties of solids (it is true that for most problems of interest in this book, ground state properties such as cohesive energy and structure properties are the most important); however, we shall see in Chap. E that results of photoelectron spectroscopy, a very powerful method, may need a correct understanding of excited final states).

As we have seen in A.II, the ψ_i's that we choose for (23) are *orbital* (e.g. 5f) wave functions of electrons within one particular configuration of the atom (i.e. well defined states $5f^x 6d^y 7s^z$ of the atom shell). We call x, y and z the occupation numbers of the orbitals in that particular configuration ν, and use for them the general notation n_ν. Now, to each configuration ν, belongs, as we have seen, an atomic eigenvalue $E_l^{(\nu)}$ of the electrons in the l-orbital, corresponding to an (atomic) wave function $\psi_l^{(\nu)}$.

If we assume (see Chap. C) a continuous, non-integral occupation of the orbitals in configurations, $E_l^{(\nu)}$ may be expressed as[31]

$$E_l^{(\nu)} = \frac{dE}{dn_{\nu,l}} \tag{24}$$

($E_l^{(\nu)}$ so defined is different from the atomic eigenvalues of the Schrödinger (or Dirac) equation in a central potential, which we discussed in Part II, since the latter are defined for integral occupation of the orbitals.)

It can be shown that the band treatment (band limit) is valid when the change in $E_f^{(v)}$ is negligible when $n_{v,l}$ is varied. (This is a formulation of Koopman's theorem already cited.) If not, the change in U in the core is such that the band limit can be abandoned.

If x is the (fractional) occupation number of the 5 f orbital in the core, we may re-state the condition of applicability of the band treatment in the following way (Koelling[31]):

$$\eta = x\overline{U}/W < 1 \tag{25}$$

(when $\overline{U}$ now is a *partial* quantity: dU/dx).

This is a *crucial point* for actinide solids. Koelling[31] shows that in α-U, η is extremely small, indicating a good delocalization of f-electrons. In γ-U, where the An–An distance is greater, η is larger ($\eta < 0.5$): the band picture is perhaps less suited for 5 f states. For Am, Cm and the heavier actinides, η is strongly dependent on the configuration we choose for the ψ_i's. For ionic solids (e.g. oxides) the problem becomes important: η becomes very high, due to the charge transfer and stabilization of open structures, with consequent very little f-f-overlapping.

This leads to a second remark, which concerns hybridization, of which we have also shortly spoken. In choosing the ψ_i's in (23), we may "mix" to the 5 f wave functions the 6 d, 7 s, wave functions by taking "hybridized" orbitals wave functions as a starting point.

Hybridization will be discussed in more detail later on. The point we want to make here is that, due to hybridization, the bandwidth W of the bands having f-character may be greatly increased. This, of course, affects the condition (25).

A last remark: the LDA approach disregards spin-polarization phenomena. Taking into account this effect (already indicated by Koelling[31] as the future direction of development for band calculations) has turned out to be of paramount importance at the centre of the actinide series. We shall examine it in the next section.

4. *Narrow Band Magnetism and Spin-Polarization*

We have discussed and distinguished two limiting cases for 5 f wave functions:

i. *complete localization:* 5 f are atomic; the unsaturated shell fills according to Hund's rule, and displays a net magnetic moment. 5 f do not participate to metallic electrical conduction. Van Vleck or Curie-Weiss magnetism (see Chap. D) is expected and also ferromagnetic or anti-ferromagnetic phenomena, explained in the frame of Heisenberg like theories.

ii. *complete itineracy:* 5 f are Bloch states in a broad band, retaining very little of their atomic character. Pauli paramagnetism is expected and metallic conduction.

Difficulties arise when the d- and f-electrons are in an intermediate state between i. and ii. This is the case for narrow bands (Fig. 12 c). This intermediate state of affairs has been at the center of theoretical and experimental investigations for d-transition solids, especially metallic solids[32, 33].

One important point is whether narrow bands would display permanent magnetic moments and undergo magnetic collective phenomena. This depends clearly upon their bandwidth and will lead again to the problem localization vs itineracy. In band calculations, new ways have to be looked for, since the set of hypotheses examined previously, which hold for non-magnetic solids, must be corrected for spin-polarization.

These concepts are very important also for actinide solids. In particular, the Stoner model for collective magnetism in narrow bands, the frame in which d-transition metals' and actinide metals' magnetism has been discussed, will be reviewed briefly.

a. The Stoner Model for Band Magnetism

The Stoner model, which we discuss in its simplest form, describes the formation of a net magnetic moment for itinerant electrons in a narrow band by adding to the basic Hamiltonian (11) of band theory a perturbing term accounting for total energy increase in case the itinerant electrons spins are not polarized (as in an atom for states not maximizing the spin).

Therefore, considering Bloch states, a *positive* exchange term coupling opposite spins is added to the Hamiltonian (11);

$$H_{xc} = I \sum_{k,k'} n_{+}(k) n_{-}(k') \qquad (26)$$

where $n_{+}(k)$ and $n_{-}(k')$ are occupation numbers of Bloch states with wave numbers k, k', and spin-up (+) and spin-down (−), and I is an interaction parameter (Stoner parameter).

When a magnetic field H is applied, two energies are available for electrons:

$$E_{k+} = E(k) - \mu_0 H + I n_{-} \qquad (27)$$

$$E_{k-} = E(k) - \mu_0 H + I n_{+} \qquad (27\,a)$$

$$E_{k-} > E_{k+} \qquad (27\,b)$$

An exchange splitting of the band levels is produced for every k, which is proportional to the magnetization: $\Delta = I(n_{+} - n_{-})$.

To derive the magnetic susceptibility the argument then follows lines which are similar to the treatment of Pauli paramagnetism. Essentially (see Fig. 13), the two populations must have the same μ_F, and this causes a disequilibrium leading to a net magnetic moment $(n_{+} - n_{-})$. The paramagnetic susceptibility is given as:

$$\chi = \frac{\mu_0^2 N(\mu_F)}{1 - I N(\mu_F)} \qquad (28)$$

in which the Pauli moment is enhanced by a factor:

$$S = (1 - I N(\mu_F))^{-1} \qquad (29)$$

(Stoner enhancement factor). This solution is evidently unstable when:

$$I N(\mu_F) \geq 1 \qquad (30)$$

(which is the Stoner condition). For $I N(\mu_F) = 1$, there is a pole in the paramagnetic susceptibility, i.e. a magnetic transition. The Stoner condition discriminates therefore whether a narrow band solid, at an appropriate critical temperature T_c, would order magnetically.

For solids in which $IN(\mu_F)$ is very near to 1, often, although no magnetic order occurs, long-range fluctuations of coupled spins may take place, giving particular form to properties such as the (Stoner enhanced) magnetic susceptibility χ, the electrical resistivity, and the specific heat of the solid. Spin fluctuations have been observed in actinides, and will be discussed in more detail in Chap. D.

In the Stoner product (30), we find as a factor $N(\mu_F)$, which can be measured independently. The other factor I is an important parameter which we shall discuss briefly.

b. The Stoner Parameter I and Spin-Polarization

Let us resume here the results of the model. A Hamiltonian such as (11), obtained after all the approximations of the electron interaction described in LDA, provides a ground state energy for the electron system which we term $E°(n)$ (we have omitted for brevity the k-dependence, giving rise to the band spread, and indicated explicitly the fundamental dependence on the charge density n(R)). However, Wohlfarth[32)], already in 1953, had analyzed the quantum mechanical derivation of the total energy of a d narrow band (using the so-called "tight-binding" approximation (see[23)]), proper for atomic-like wave functions such as the d wave functions) and found, in the solution, Coulomb and exchange terms dependent on the spin-polarization, and leading to a lowering of the total energy. These terms are mostly intra-atomic in character and have physically the same origin as the terms in the atomic Hamiltonian leading to Hund's rule[32)]. They correct the Hamiltonian (11) as in equation (26), the Stoner parameter I accounting for them.

If we allow for these terms, the energy of the system is to be written in a more complex form: $E = E(n, \xi)$, where we have chosen as a parameter $\xi = (n_+ - n_-)/n$, i.e. the relative magnetization. We make two observations:

i. since $n_+ + n_- = n$, the two numbers n_+ and n_- substitute the total n as the fundamental variables of the Hamiltonian of the system. It has been shown[34)] that one can proceed as we did in LDA, but developing a formalism of spin-polarized band calculations having local spin densities $n_+(R)$ and $n_-(R)$ as the fundamental variables. This formalism (called "local spin density approximation (LSDA)) which has been extended[35)] to the relativistic case, is one of the major recent achievements of theory, and has been extensively applied for transition metals. We shall see (Chaps. C and F) that it is of fundamental importance in the actinides[36)].

ii. following the notation of Brooks and Johansson[36)], $E(n, \xi)$ may be partitioned in two separate contributions:

$$E = E°(n) + \Delta E(n, \xi) \tag{31}$$

$\Delta E(n, \xi)$ is the effect of spin-polarization. Wohlfarth[32, 33)] had already shown very early that $\Delta E(n, \xi)$ may be expressed as an expansion in ξ:

$$\Delta E = n^2 I \xi^2 (1 + \tfrac{1}{2} A \xi^2 + \tfrac{1}{4} B \xi^4 \ldots) \tag{32}$$

The Stoner model retains only the first term, proportional to ξ^2 (molecular field approximation which leads to terms such as (26) in the Hamiltonian).

As a consequence of $\Delta E(n, \xi)$, one has

A. The splitting of the band energy E° into two sub-bands, for spin-up and spin-down electrons.

B. A permanent magnetic moment and, possibly, collective magnetic phenomena.

The Stoner product (30), calculated across the actinide series for homologous compounds, may interpret (or predict) the magnetic behaviour of these solids, and hence suggest a localized or itinerant picture (see Chap. D), provided that we know I for the different actinides across the series, since $N(\mu_F)$ is measurable (e.g. through specific heat measurements) and roughly reciprocal to the bandwidth W ($N(\mu_F) \propto W^{-1}$). I is not directly measurable and must be calculated. (Of course, the discussion above shows that the two quantities are not really independent, since the interactions determining I also play a role in determining the bandwidth, and hence $N(\mu_F)$.)

The importance of Stoner's theory goes beyond the treatment of magnetism in a transition series. The spin-polarization term ΔE in (31) competes strongly with E_0 and therefore influences strongly the cohesive and structural properties of a metal (or metallic compound), which are ground state properties. We have already seen (A.I) that the variation across the series of atomic radii has been at the center of much theoretical thinking (Figs. 2, 3, 5). Also the variation of the cohesive energy of metallic actinide compounds has been at the center of interesting models and theoretical correlations, that we shall discuss in Chap. C.

Brooks and Johansson[36)], by calculating I and ΔE have been able to show that these curves are very much affected by the spin-polarization energy ΔE, which, in fact, appears as the most important contribution in cohesion for actinide elements in which the beginning of localization of 5 f orbitals occurs (at the half-filling of the $5f^n$ shell) and for which spin-polarization plays a major role (see Chaps. C and F).

5. *The Mott-Hubbard Metal-Insulator Transition*

A concept related to the localization vs. itineracy problem of electron states, and which has been very useful in providing a frame for the understanding of the actinide metallic bond, is the Mott-Hubbard transition. By this name one calls the transition from an itinerant, electrically conducting, *metallic* state to a localized, *insulator's* state in solids, under the effect of external, thermodynamic variables, such as temperature or pressure, the effect of which is to change the interatomic distances in the lattice.

This transition has been emphasized by Mott[37)] for the case of localized impurity states in a semiconductor, forming a metallic band at some concentration of impurities (i.e. at some average distance between the impurities). It is referred to very often as the Mott (or Mott-Hubbard) transition.

The reason why it is a useful concept in actinide solid state physics (as in d-transition metals) is that it describes the effect of overlapping in forming a band: therefore, in d or f unfilled bands, a critical value of the inter-atomic distance should determine whether the electron states will be acting as localized (insulator's) states, or as an unfilled (metallic) band. If we refer to the concept of f-f-overlapping, it is thinkable that across a series such as the actinides the f-f-overlapping might come to a critical value distinguishing between band and localized behaviour.

In order to understand the concept, it is customary to present a sort of "paradox" arising from the solution of the Hamiltonian (11) in the tight-binding approximation (Eq. (12)).

Suppose we approach M atoms of an element having an unfilled outer shell, disposed in a lattice. The point may be made clear if we suppose to approach hydrogen atoms (outer shell $1\,s^1$). Equations (11) and (12) would predict the broadening of the electron state in a half-filled s-band, which should therefore allow metallic behaviour. Apparently, this would happen for any inter-atomic distance a and, therefore also at infinite distance. What would change is, of course, the bandwidth, which is determined by matrix elements dependent on the interatomic distance a but the metallic behaviour, depending essentially on the fact that the electrons have available energy states within the band, should occur also at distances where the atoms may well be supposed to be isolated.

This happens for the essential simplicity of the Hamiltonian (11) and of the solutions (12). We have seen that (11) does not represent well electron-electron interactions. At very large inter-atomic distance, the effect, for instance, of intra-atomic interactions, such as exchange, might be such as to stabilize a solution in which electrons are localized within the atoms: the energy scheme would be that of an insulator, where conduction can be obtained only through excitation of the electrons out of the ground state into some conduction band. At rather small distance, on the other hand, the attractive effect of a core potential can be very strongly reduced by the screening of other ionized electrons, as can be seen in the Thomas-Fermi model[23, 28]. Mott[37] has derived a critical distance separating the two behaviours within the frame of the Fermi-Thomas model.

Suppose we take an electron out of one of the hydrogen atoms, leaving behind a H^+ ion. The electron will be re-attracted to the ion by a potential $-e^2/R$. It will be bound in the ground state for this potential, which has a radius $a_H \approx h^2/m\,e^2$. If, however, there are enough ionized electrons in the lattice, they will screen the electron from the positive ion core, according to a potential $-(e^2/R)\exp(-\lambda R)$, where λ is a screening parameter in the Thomas-Fermi model[23, 28].

The electron evidently cannot be bound again to the positive ion if the screening radius λ^{-1} is smaller than the radius of the bound state:

$$a_H/\lambda^{-1} < 1 \tag{33}$$

This constitutes a critical condition for the Mott-transition. The condition may be also written in terms of the charge density n[38]:

$$n^{1/3}\,a_H = x_{Mott} \tag{34}$$

where x_{Mott} is a typical value very near to 1/4 (for doped semiconductors, x_{Mott} has been found to vary between 0.20 and 0.25).

Hubbard[39, 40] treats the same problem emphasizing instead the effect of intra-atomic exchange interaction. The argument is as follows. In a partially filled band (*itinerant state*) the spin of an atom is a quantity that fluctuates randomly in magnitude and direction. The characteristic time of the fluctuation is related, as we have seen, to the bandwidth W, and is of the order of h/W. The spin, propagating with the moving electron, is hopping from atom to atom. Therefore, in the average, there will be atoms excessively filled and atoms excessively depleted in the solid.

One speaks of "polar states" being formed, referring to the ionized states of the atoms being excessively filled or depleted.

If, however, some correlation effect exists avoiding formation of excessive spin density in an atom (for instance making a state with aligned spins stable because of Hund's principle) this state of alignment in the atom may persist much longer than the band hopping time h/W of the electrons. The fluctuations, typical of the band, will tend to even out, and the electrons will stay in the atomic ground state of the core (*localization*).

In order to take into account these intra-atomic terms, and in a way similar to the Stoner's model, Hubbard[39, 40], see also[41] adds to the Hamiltonian (11) an exchange interaction term:

$$H_{exch} = U_H \sum_i n_{i+} n_{i-} \tag{35}$$

where U_H is the intra-atomic exchange parameter, n_{i+} and n_{i-} are occupation numbers of the atom i with spins up and down, and the effect is extended to the whole lattice by summing on all lattice cores. H_{exch} is evidently a positive contribution to the total energy, since Hund's rules would favor alignment of spins within a core.

Without going into the complicated details of the many-body treatment of the Hamiltonian (35)[40, 41], we present in Fig. 14 the results in terms of eigenvalues of the system.

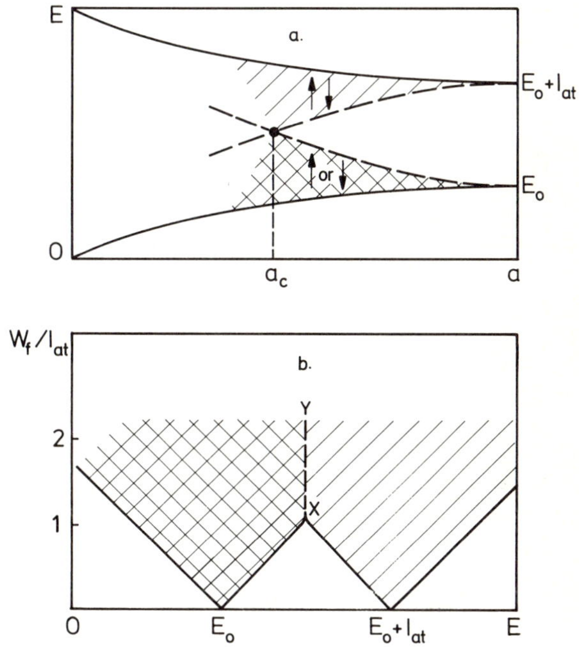

Fig. 14a, b. Hubbard sub-bands for the localized (↑ or ↓) and polar states in a narrow band solid **a)** band spread as a function of interatomic distance: at a_c, the Mott-Hubbard transition occurs. **b)** (from[40]) bandwidth W_f of the sub-bands to interaction parameter I_{at} ratio, as a function of energy: at X, the Mott-Hubbard transition occurs

We take the model of a monovalent element containing just one electron in its outer shell (the "hydrogen" lattice described above).

We consider first (Fig. 14 a) what happens at very large distances. The Hamiltonian (11) (without the correction (35)) would then give rise to a very narrow band (a level). With the correction (35), the band splits into two separate sub-bands (two energy levels) E_0 and $E_0 + U_H$ (see Fig. 14 a). These two sub-bands containing each M (and not 2 M) states, represent, respectively, a state in which each core holds one spin and a state in which half of the cores hold two antiparallel spins, and the others empty (polar states). The two sub-bands are separated by a gap which is exactly U_H. Without excitation to the highest sub-band, in this conditions the lower sub-band is fully occupied. It represents the insulator's state in which all electrons are sitting in the cores, i.e. all electrons are fully localized.

When the cores are approached, the sub-bands split, acquiring a bandwidth, and decreasing the gap between them (Fig. 14 a). At a definite inter-core distance, the sub-bands cross and merge into the non-polarized narrow band. At this critical distance a_c, the narrow band has a *metallic* behaviour. At a_c, the system transits from insulator to metallic (Mott-Hubbard transition). Since some electrons may acquire the energies of the higher sub-band, in the solid there will be excessively filled cores containing two anti-parallel spins and excessively depleted cores without any spins (polar states).

In terms of energies, Hubbard shows[40)] that critical parameter for this behaviour is the ratio U_H/W. This is represented in Fig. 14 b, taken from[40)]. The transition occurs when:

$$U_H \approx W \tag{36}$$

As far as magnetic phenomena are concerned, if the system is all in the lower sub-band (i.e. localized) then the treatment for localized spins holds, and exchange forces between them may provoke eventually ferro- and anti-ferromagnetism. If the system is itinerant in the unsplitted band then treatments of magnetism in narrow bands such as Stoner's are valid. Hubbard[39, 40)] derives also a condition for collective magnetism which is similar to the Stoner's one:

$$I_{eff}\, N(\mu_F) \geq 1 \tag{37}$$

where I_{eff} is an effective exchange parameter containing both U_H and the bandwidth W and thus intra-atomic and interatomic correlations.

6. *Conclusions*

To resume briefly Part III it has been shown:

i. that the atomic wave functions and eigenvalues of 5f electrons suggest the occurrence of two phenomena in the bond of actinide solids: *f-f overlapping* between adjacent actinide atoms, and *hybridization* of 5f with 7s and 6d electrons of the actinide atom as well as with outer electrons of non-actinide atoms in compounds;
ii. the solid state properties are essentially dictated by the 5f behaviour: *localized* (i.e. atomic-like) or *itinerant* (i.e. band-like); definitions of the two behaviours have been provided;

iii. which behaviour is predominant depend on a delicate total energy balance for the solid: intra-atomic (U) vs. inter-atomic (W) energy terms;

iiii. due to the unfilled character of 5f shells, an important intra-atomic term entering the balance derives from the possible energy gain the system obtains by aligning spins according to Pauli principle requirements (spin-polarization); Stoner and Mott-Hubbard criteria for spin-polarization have been discussed.

In Part IV, a rapid survey is given of experimental evidence collected on actinide solids. The application of the concepts i. to iiii. helps in ordering this evidence in a unified description.

IV. The Actinide Metals

Armed with the general concepts developed in the previous section, we proceed now to the experimental evidence of 5f itinerant vs. localized behaviour in actinide metals.

1. *Experimental Evidence of 5f-Band Behaviour in Lighter Actinides*

The experimental evidence for band behaviour of the 5f electrons up to plutonium is obtained through[4]:

- *Atomic volume:* there is a large transition-like decrease up to plutonium; from americium on, on the contrary, a lanthanide-like contraction (cf. Figs. 2 and 3);
- *Crystal structures:* they are complex with many allotropic modifications having, some of them, low symmetries, evidencing therefore, a 6d-5f covalent bonding (see Chap. C);
- *Cohesive properties:* they are anomalous as evidenced in the variation of melting temperatures, vapor pressures, enthalpies of sublimation (see Fig. 11 of Chap. C) and dissolution and bulk moduli vs. atomic number Z (to be discussed in Chap. C);
- *Magnetic properties:* metals up to plutonium are Pauli magnets and Curie-Weiss behaviour starts with curium;
- *Spectroscopic properties:* Photoemission from the valence band of uranium and plutonium indicates a 5f band at the Fermi level[43] (see Chap. E).

Thus, the theoretical approach to light actinide metals must be made through band theory even for the 5f electrons. The calculation of energy bands for actinide metals has to face several complications:

- Necessity of a relativistic treatment.
- Convergence problems due to the presence of high angular momentum states.
- Validity of the band description for decreasing 5f bandwidth.

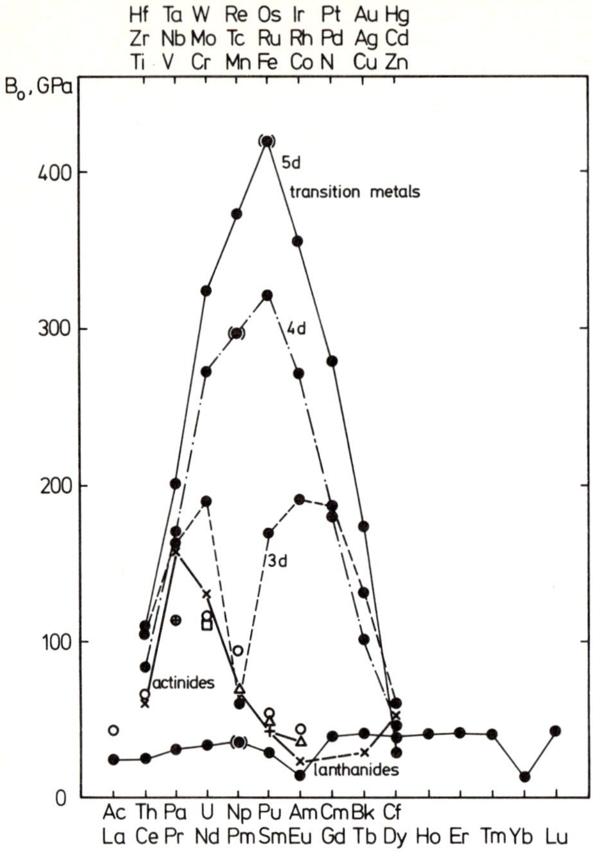

Fig. 15. Experimental and calculated bulk moduli of the actinides[42], compared with those of lanthanides and of transition metals. The calculated values for actinides (0, 0) peak for uranium, whereas the experimental bulk-moduli (x, □, +, △) peak for protactinium (original references to be found in[42])

2. *The Mott-Like Transition Between Plutonium and Americium*

Whereas there are experimental evidences for the band behaviour of 5f electrons up to plutonium (Z = 94) (see Table 3), the same criteria show that suddenly americium (Z = 95) behaves like a normal lanthanide having well localized 5f electrons ([4] and Chap. C):

- Its atomic volume and d.h.c.p. crystal structure are similar to those of light lanthanides; this is also the case for its cohesive properties (ΔH_S = 285 kJ/mole and T_M = 1170 °C).
- Its constant paramagnetism is well explained by a Van Vleck susceptibility, the ground state being the non-magnetic J = 0 due to the $5f^6$ configuration; this interpretation is well confirmed by the very low electronic specific heat. Accordingly, band calcula-

Table 3. Density of states at the Fermi level for actinide metals: from band calculations (model); from the electronic contribution γ to the specific heat; from magnetic susceptibility measurements. The increasing values indicate a decreasing 5f bandwidth pinned at E_F; for americium metal (not shown) there is a sudden decrease in $N(\mu_F)$

An	Th	Pa	U	Np	Pu
$N(\mu_F)$ (model) er/eV at	1.3	1.5	3	3.5	4.2
$N(\mu_F)$ from γ	2	2.5	4.5	6.3	9
$N(\mu_F)$ from χ	1.1	6	8.6	15.6	12

tions[11, 15] show that a band treatment of the 5f electrons is no longer valid from americium on. This has led Johansson to view this sudden change as a Mott-like transition of the 5f electrons along the actinide series[38, 44].

It can be stated that up to plutonium, 5f electrons are in the conduction band and have no magnetic moment; from americium on, 5f electrons are localized and carry a magnetic moment.

Thus, forgetting about the superimposed "normal" 6d-7s conduction band, there is a transition from a non-magnetic 5f metal to a magnetic 5f insulator (americium itself being non-magnetic only because $J = 0$).

To put it on a more quantitative basis, Johansson uses the expression for the critical point of the Mott transition as reformulated by Hubbard[40] in terms of the bandwidth W_f and the polar state formation energy U_H (or effective intra-atomic correlation) (Eq. (36)).

- if $W_f \leqq U_H$ the polar state formation energy is large enough to prevent hopping of the 5f electrons: they are magnetic and insulating.
- if $W_f \geqq U_H$ 5f electrons will be free to move in the metal: they are non-magnetic and metallic.

A simple application of Hubbard's criteria to the actinide metals' series, with a non-hybridized W_f, would tend to indicate Pu as the first localized metal. Two complications, however, occur for the actinide metals:

- 5f are degenerate,
- there is a superimposed 6d-7s conduction band. This presence imposes some considerations.

In the atomic limit, the intra-atomic Coulomb correlation is given by:

$$U_{at} = E(f^{n-1}d^2s^2) + E(f^{n+1}s^2) - 2E(f^n s^2) \tag{38}$$

which is readily obtained from atomic spectra (cf. Sect. (A.II)). Due to the conduction band, the screening will be different in the metal, leading to $U_H \neq U_{at}$.

Whereas Johansson, following Herring's argument[45] assumes that U_H should always be less than U_{at}, Herbst and Watson[46] from relativistic band structure calculation found that while this is true for light actinides, the situation is reversed above Pu, thus favoring the transition between Pu and Am.

Table 4. Comparison of U_{at}, U_H, W_{hyb} in the actinide metals' series from Th to Am. In the last row, only the qualitative departure from 1, rather than a well defined value for W/U_H is indicated, since evaluations different from those selected, are reported in the literature for U_H and W. Notice the peculiar position of Pu metal and the "localized" W/U value for Am metal

An	Th	Pa	U	Np	Pu	Am
U_{at} (ev)	1.5	1.6	2.3	2.6	3.5 ± 1	5 ± 1
U_H (ev)	2.2	1.5	1.8	1.8	4	4
W_{hyb} (ev)	5	4	4	3	2	0.1
W/U_H	>1	>1	>1	>1	≲1	<1

Another effect of the conduction band concerns W_f: the hybridized 5f bandwidth W_{hyb} is larger than the unhybridized one, W_{unhyb}.

Thus these two effects are additive in pushing the Mott-like transition toward higher atomic number and in making such a sharp transition possible.

Results are reported in Table 4. This simple tabulation focusses on the transition that occurs along the actinide series. It makes very clear, that plutonium metal is in a very crucial position (which is confirmed experimentally) having thus the most extraordinary behaviour in the whole of the periodic table.

3. The Polymorphism of Plutonium Metal

In fact, a change in behaviour of the f-band is shown when examining the magnetic and electric properties of Pu-metal. This behaviour indicates that, in its low temperature phase, the f-states contribute strongly to the bonding conduction band of Pu (thus explaining the asymmetric monoclinic structure of this phase). In the high temperature phase, however, the conduction band is mostly of s and d character (thus explaining its more symmetric bcc structure).

The low temperature α-phase of plutonium is clearly on the 5f-itinerant side (constant Pauli paramagnetism, rather large electronic specific heat suggesting a 5f bond). But when we go to the high temperature δ-phase, there is a 18% atomic volume expansion. Such a change is very large and may be compared with the 17% volume change at the α-γ transition in cerium metal under pressure. Johansson describes this transition in cerium as a Mott transition from localized to itinerant behaviour for the 4f electron of Ce[47)]. The number of 5f electrons involved in Pu being about 5, it is difficult to directly compare the (temperature steered) α-δ transition in Pu with the (pressure steered) α-γ transition in Ce. Nevertheless it is clear that such a large expansion necessarily lowers the f-bandwidth and on a Hill plot (see later) δ-Pu is at a critical place. Most interestingly, in agreement with strong 5f band narrowing from α to δ-Pu, it has been found that[48)]:

- the low temperature electronic specific heat of δ-Pu, extrapolated from measurements on δ-Pu phases stabilized at room temperature with different concentrations of Al, is larger than that of α-Pu, ($\gamma = 53 \pm 10$ mJ/mole K^2 instead of $\gamma = 22$–25 mJ/mole K^2), indicating a very narrow band pinned at μ_F
- in stabilized δ-Pu alloys, the magnetic susceptibility saturates at low temperature, following a T^2 law, as predicted by spin fluctuation theories and experimentally observed in UAl_2[49)] (see also Chap. D).

4. Physical Properties of Actinide Metals up to Pu

The physical properties of actinide metals up to Pu – including the magnetic properties – are all governed by the complicated 5f-6d-7s conduction band.

A pioneering work for their qualitative understanding was the covalent treatment of the 5f-6d hybridization due to Friedel[50]. Like p orbitals, f orbital are antisymmetric ($\psi(-R) = -\psi(R)$); thus by linear combination with symmetric d orbitals, they give rise to highly directive hybrid orbitals (like s-p hybrids). Thus in Pa, U, Np and Pu metals each atom has four nearest neighbours more or less coplanar and at 90°. Let's take the example of α-U and put the nearest neighbours along the x and z axes. Necessary hybrid orbitals for such positions are:

$$\varrho_{x\pm} = a_s + a_d(2x^2 - y^2 - z^2) \pm a_{f1}(x^2 - 3z^2)y \pm a_{f2}(x^2 + z^2 - 4y^2)x \qquad (39)$$

$$\varrho_{z\pm} = a_s' + a_d'(2z^2 - x^2 - y^2) \pm a_{f1}'(z^2 - 3x^2)y + a_{f2}'(z^2 + x^2 - 4y^2)z \qquad (39\,a)$$

This 6d-5f hybridization is at the origin of the complex crystal structure of the light actinide metals. In the model density of states, there is a bonding band filled with 4 electrons from Pa to Pu and an antibonding band containing 20 states progressively filled from U to Pu (Fig. 16). Such a simple scheme is able to explain the very low density of states for Pa as well as its rapidly increasing value from U to Pu, in reasonable numerical agreement with values deduced from the electronic specific heat coefficient γ (Table 3).

It allows also to explain, from covalency effects, the anisotropy of transport properties[4]:

- It is this hybridization which constitutes the main differences between early actinide metals and transition metals' transport properties;
- It is this hybridization which causes the large delay for the appearance of magnetism, as was clearly demonstrated by the work of Jullien and Coqblin[51].

Detailed comparison between band structure calculations and experiments has been possible only for Th and to a less extent α-U. It is doubtful that such a level of sophistication will be reached for the low temperature, low symmetry, phases of Np and Pu metals.

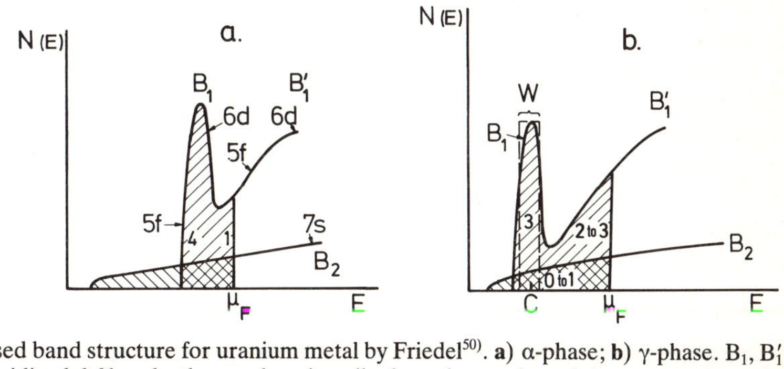

Fig. 16a, b. Proposed band structure for uranium metal by Friedel[50]. **a)** α-phase; **b)** γ-phase. B_1, B_1' and B_2 indicate hybridized d-f bands; the numbers inseribed are the number of electrons per atom in these hybridized bands. In **b)**, the B_1 narrow band is simulated with a square band, with bandwidth W and center C

However, a considerable deepening of the theoretical treatment of the actinide metals was recently presented by Skriver and al.[52]. By means of LMTO (Linear Muffin-Tin Orbitals) band calculations they were able to calculate correctly the cohesive properties of the metals. Within the so-called Atomic Sphere Approximation[26,27,53], a ferromagnetic spin-polarization was allowed, in order to simulate the possibility of localization of the 5f electrons. The calculated equilibrium atomic volumes are compared with those found experimentally in the most dense crystal structures. For Th, Pa and U the agreement is most satisfactory (there may be compensating approximations, however). Only when coming to Np and Pu one can notice significant deviations between the calculated and experimental values. They are most easily interpreted as due to correlation effects which are not included in the calculation: and it is significant that the binding is overestimated. However, there is no spin-polarized solution for Np and Pu at zero pressure: but for somewhat expanded volumes (negative pressure) a net magnetization is calculated. Arriving at Am, the fully spin-polarized solution exists at normal pressure and has the lowest energy. Spin-polarization has a drastic effect on the atomic volume, indicating an almost non bonding character of the 5f electrons. (These results are also discussed in Chap. C and F).

5. *Physical Properties of Actinide Metals From Am on*

We assume, in this case, that the conduction band has become "normal" (that is, it has no longer any 5f character). Thus, physical properties may be usefully compared with those of the lanthanides. In Table 5 we report known basic properties (metallic radii, crystal structures, melting temperatures and enthalpies of sublimation) of the transplutonium metals.

Examination of this table confirms what could already be guessed from atomic calculations: only Am, Cm and to a less extent Bk can be considered similar to normal

Table 5. Ground state properties of "heavy actinide" metals (from Am to Es), emphasizing the lanthanide-like character of this part of the series

	Am	Cm	Bk	Cf	Es
Metal radii L.T. phase	1.72	1.74	1.70	1.69	
Metal radii H.T. phase	1.73	1.78	1.77	2.03	~2.0
Crystal struct. L.T. phase	d.h.c.p.	d.h.c.p.	d.h.c.p.	d.h.c.p.	
Crystal struct. H.T. phase	f.c.c.	f.c.c.	f.c.c.	f.c.c.	f.c.c.
Melting Temperatures (°C)	1170	1350	990	900	860
Enthalpies of sublimation (kj/mole)	285	385	295	195	140

lanthanides. With Bk, a tendency toward divalency begins to appear, as indicated by the anormalous low melting point (see also Fig. 1).

Thus, although it is safe to assume that the 5f states are localized states, excited f states are never far above the Fermi energy. They will, therefore, influence the electronic properties and high temperature phases of Cf and Es (which, with a metal radius R = 2.0 Å seem to have attained divalency).

The situation above described has important consequences in the understanding of the physical properties not only of heavy actinide metals, but also of their compounds. We can expect e.g. that the occurence of valence fluctuations should be quite common in Am, Bk and Cf compounds (to limit ourselves to the sufficiently stable isotopes) as well as in Cf metal itself[4, 54].

Curium metal was found to be antiferromagnetic[55]. Its Curie-Weiß behaviour above the Neel point is in good agreement with the localized picture ($\mu_{eff} = 8\,\mu_B$ as in Gd metal). In the case of Bk, again the Curie-Weiß law gives the expected effective moment value[56].

6. *Superconductivity and Magnetism in Actinide Metals*

Proceeding along the 5f series, superconductivity under atmospheric pressure (Th, Pa), superconductivity under high pressure (U), exchange reinforced Pauli paramagnetism without superconductivity (Np, Pu), superconductivity under atmospheric pressure (Am) and finally magnetic ordering (Cm, Bk) are successively encountered. We remind that in the 4f series, La is superconductor as well as Ce under high pressure.

Some years ago, it had become customary to speak of f-band superconductors. Behind these words, there was the assumption that narrow f-bands might give rise to f level, or f band, polarization[57].

It has now become apparent that f-band superconductors can be understood in a normal way, like d-band ones[4]: we have just to think about density of states, bandwidth lattice unstabilities and electron correlations to interpret it. This was clearly summarized by Smith[58] after the discovery of superconductivity of Pa and Am. Superconductivity of Am is made by the J = 0 ground state of its six 5f electrons; as was stressed by Johansson[59], this could also be the case for stabilized trivalent europium metal. The importance of lattice instabilities (like in A-15 high T_c superconductors) was put forward by Fournier[4] who showed that the very large $\delta T/\delta p$ slope for U was mainly due to a very large change in the electron-phonon coupling associated with the low temperature phase transition.

Thus, the occurence of superconductivity and magnetism along the actinide series is now fairly well understood.

V. The Actinide Compounds

1. *The Hill Plots: Usefulness and Weaknesses*

In A.IV, when introducing the Mott-like transition between Pu and Am metals, we have discussed the localization problem as a function of the number of 5f electrons. In the case of compounds there is the additional possibility to study the localization as a function of inter-atomic distance. This was first done systematically by Hill[14)] who observed that magnetic ordering exists in light actinide compounds (with U, Np, Pu) only above some critical An–An spacing (~ 3,4 Å). Such plots are now commonly referred to as Hill plots. The qualitative simple explanation of how and why it works is the following: the broadening of 5 f bands in metallic actinide compounds is mainly due to direct overlap of 5 f wave functions, which depends directly on the interactinide spacing: if the 5 f band is too large, the density of states will be too low and the Stoner criterion for band ferromagnetism (see A.III, Eq. (30)) will not be fulfilled. I is roughly constant within the actinide NaCl series and approximately equal to 0.012 eV[60)] (see also Chap. D and F). With decreasing bandwidth, the Stoner criterion can be fulfilled even in the case of uranium compounds. We thus explain the Hill plot on the basis of Stoner theory for band magnetism.

A somewhat different interpretation has been given by Johansson[61)] who applied the Mott-Hubbard theory of localized versus itinerant electron behaviour also to compounds. This interpretation differs from the above one mainly in that it assumes complete localization for magnetic compounds, and that at a certain critical inter-atomic distance we have to switch our description from a metallic state to an insulating one for the 5 f electrons (see Eq. (42)). In Eq. (42), a_H is substituted by a convenient measure of the spatial extension of the 5 f orbital, the expectation values $\langle R^2 \rangle^{1/2}$ (analogous to $\langle R^4 \rangle^{1/4}$ of Fig. 10) and x_{Mott} is calculated from the R_{ws} radii of actinide metals (Fig. 3). The result is given in Table 6.

In Figs. 17 to 19 are presented Hill plots for U, Np and Pu metallic compounds. In the case of protactinium recent experimental and theoretical investigations[62)] indicate that no magnetic ordering can be expected in metallic compounds: the Stoner criterion can never be fulfilled even for quite large interactinide spacing. In the same study quantitative theoretical grounds are given to Hill plots: the calculation of the Stoner criterium (Table 7) shows that it is not fulfilled for UC but just with UN in agreement with Fig. 17. In fact, measurements of the magnetic moment of UN under pressure clearly show the critical situation of this compound[63)].

Plutonium is a borderline case in that the 5 f band is quite narrow but a non-magnetic ground state is easily obtained either from a band or a localized point of view. In a band model, the 5 f band is split by the strong spin-orbit coupling into a 5/2 and a 7/2 sub-bands. The 5/2 sub-band (containing 6 states) is almost filled for plutonium and thus the density of states is quite low. From a localized point of view, the strong crystal field may often lead to a Γ_1, non-magnetic, singlet ground state for tetravalent ions (e.g. PuO_2, Pu_2C_3). In the case of chalcogenides there are two alternatives to explain the absence of magnetism: either a Γ_1 ground state assuming Pu^{+4} ions or a mixed-valence situation Pu^{+3}/Pu^{+2} which will be non-magnetic because Pu^{+2} is non-magnetic (J = 0). Clearly Pu chalcogenides deserve more studies.

Table 6. A comparison of the Mott parameter x_{Mott} in the actinide and in some lanthanides (from Ref. 61)

	Th	Pa	U	Np	Pu	Am	Cm	Bk	Ce(α)	Pr	Nd	Pm
x =	0.28	0.28	0.28	0.28	0.26	0.22	0.21	0.20	0.21	0.18	0.17	0.16

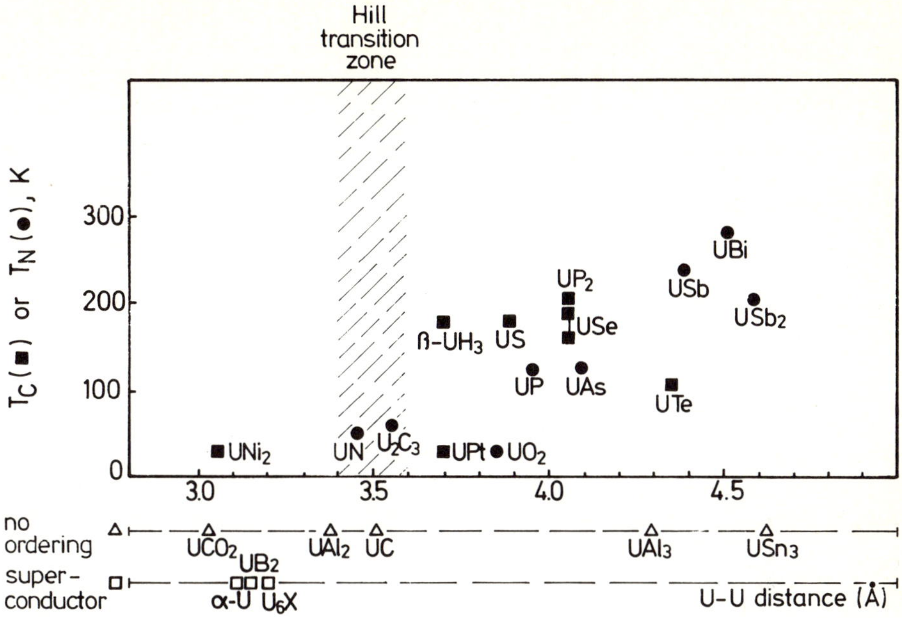

Fig. 17. (Modified) Hill plot for uranium binary compounds

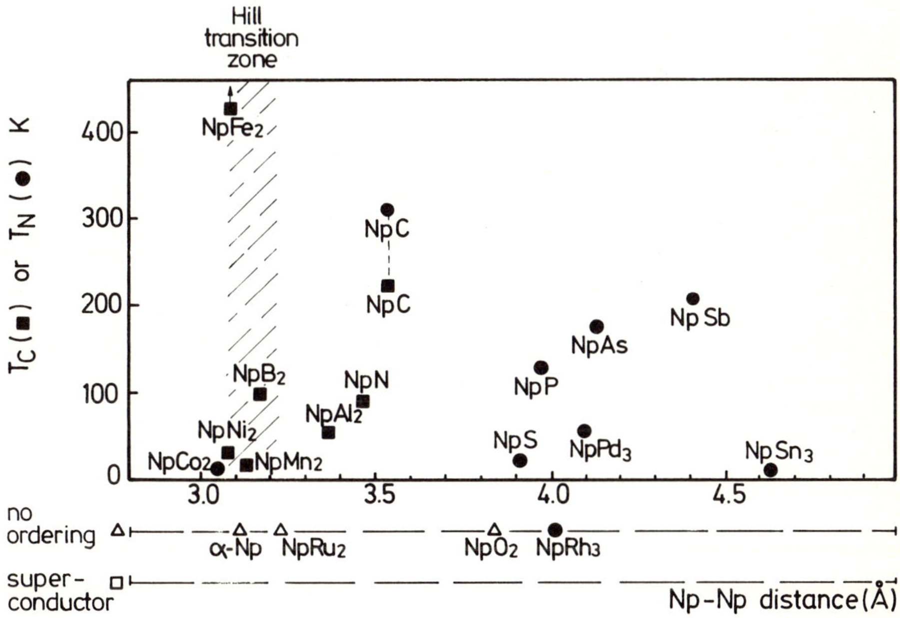

Fig. 18. (Modified) Hill plot for neptunium binary compounds

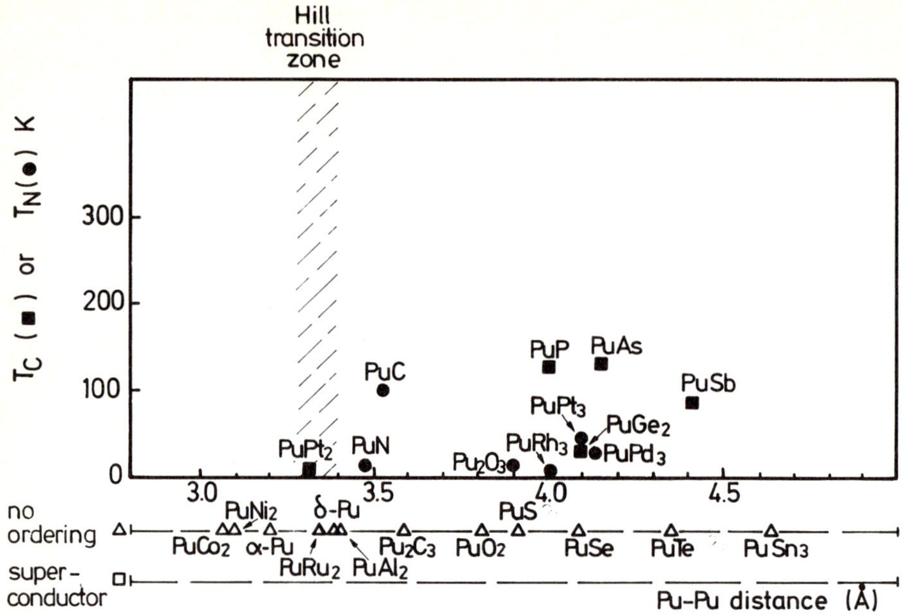

Fig. 19. (Modified) Hill plot for plutonium binary compounds

No Hill plots can be drawn for transplutonium compounds for obvious reasons: very few are known, Am compounds are essentially non-magnetic ($J = 0$) and all known results suggest well localized 5f states.

As noted by Hill himself[14)], the weakness of the Hill plots comes from the crude assumption that direct overlap of 5f wave functions is the only parameter governing the 5f bandwidth.

The very fact that such plots are so successful is in favor of such an assumption. From a band theoretical approach, the second important parameter is hybridization. We thus expect that anomalies in the Hill classification will be indicative of remarkable hybridization situations: Hill plots are useful even for cases where they do not work.

Table 7. The density of states at the Fermi level and the Stoner product in some NaCl-structure binary compounds of actinides

	$N(E_F)$	$IN(E_F)$
ThN	3.6	0.09
ThAs	5.7	0.10
PaN	15.3	0.17
PaAs	33.7	0.54
UC	29.8	0.69
UN	104.6	1.23
UAs	237.2	3.20
USb	241.1	2.85
US	193.9	2.17
UTe	172.3	1.69

A good example is given by intermetallic compounds with the $AuCu_3$ crystal structure: the interactinide spacing are very large (4 Å or over) and they should all be magnetic whereas less than half of the compounds studied thus far order magnetically.

Detailed studies – band structure calculations[52)], de Haas-van Alphen effect[64)] and polarized neutron diffraction[65)] – have evidenced the strong hybridization of 5f bands either with p anions bands (USi_3, UGe_3, USn_3) or 4d bands (URh_3, UIr_3).

On the other side, UNi_2 should not be magnetic since the interactinide spacing is very small (3.0 Å) and Ni must have a filled 3d band as in Laves phases with lanthanide. Thus, the fact that UNi_2 is a ferromagnet while α-U metal is a Pauli paramagnet may have its origin in dehybridization: in UNi_2, 5f wave functions are less hybridized with 6d ones since 6d electrons are transferred to the 3d band and there is no hybridization with the Ni-3d wave functions since the Ni-3d band is then filled[66)].

2. *Band Formation in Actinide Compounds*

In the preceding sections, we have rapidly reviewed the concepts that are involved in the band formation of actinide metals. We would like to point out what more is involved in the band formation of actinide compounds. This is very obvious: the anion valence band. In fact, the hybridization with anion states which we presented as the main correction to the simple Hill scheme is indeed the central question involved in detailed band structure calculations in actinide compounds. We pointed out in the previous paragraph the case of UGe_3; we would like here, as an example, to compare somewhat UO_2 and NaCl compounds of uranium. As confirmed by recent photoemission studies[67–70)], UO_2 has well localized 5f states whereas NaCl compounds have a narrow 5f band pinned at the Fermi level. Nevertheless the U–U spacing is the same in UO_2, UP and US. This difference may be understood in terms of charge transfer versus f-p hybridization.

In UO_2, uranium is tetravalent (O is in fact strongly electronegative; thus the 6d-7s conduction band is empty and oxygen 2p states are far below 5f states (5 eV)). The hybridization is therefore very small and UO_2 is an insulator with 5f states in the gap[71, 72)]. In NaCl compounds, pnictogens and chalcogens are less electronegative and the charge transfer is less pronounced; U is essentially trivalent (with the possible exception of US); the p states of chalcogens are less tighly bound (4 eV below 5f states) and this is even more pronounced for pnictogens (~2.5 eV below 5f states). Thus the hybridization with p states increases for chalcogenides and is even stronger for pnictides (see Chap. C). This is very well shown by band calculation[59)]. There are two consequences:

- 5f states broaden into a narrow band which is close to the Fermi level giving rise to a high density of states at the Fermi level and to a situation quite favorable for mixed valence behaviour; the type of magnetic ordering is much influenced by the f-p hybridization[73–76)].
- From a chemical point of view, 5f electrons will participate in the chemical bond as evidenced by their very high melting points when one compares them with the ones of lanthanide pnictides and chalcogenides.

As can be observed in Fig. 19, the borderline of Pu is much clearer now than at the time of the Hill's original proposal, mainly because new compounds have been studied and numerous compounds for which states may be safely assured to be localized are not magnetic, even for Pu–Pu distances greater than 4.0 Å. Thus such a plot should be used with caution for Pu compounds and should not serve as a good indication for delocalization of f states. Really the behaviour of Pu compounds seems more related to localized magnetism.

VI. Conclusions

In the first part of this chapter, it was anticipated that the light actinides should be considered as constituting a new transition metal series with the 5f shell behaving much in the same way as d shells in the d-transition metals series'. This is verified for the elemental metals, the evidence having been examined with some detail in Part IV. The spatial extension of the 5f wave functions, is responsible mostly for this: f-f overlapping, in elemental light actinide metals, produces narrow f bands, which then hybridize with the s and d broad bands, and pin the Fermi level in their neighbourhood. Thus, the properties of the light actinide metallic solids (not only of the elemental metals) will differ from those of the lanthanides mostly because of a 5f participation in the cohesion. The interactinide distance becomes an important parameter to be held into account whenever the ground state properties are to be analysed or predicted. Correspondingly, in these solids the localized atomic magnetism is not established. Neither is, in elemental metals, band magnetism, since the 5f bands broaden, because of f-d hybridization. The usefulness of Hill's plots (Part V) as a representation of localization vs. itineracy despite their numerous failures (which also have to be kept in mind) derives from having taken the occurrence of localized magnetism and the interactinide distance as the natural frame to represent these two aspects. However, the properties deriving from cohesion (Figs. 2, 3, 5 and 15) and their comparison with those of lanthanides, are as good a frame and historically the first to have attracted attention.

Figure 20 (from[77]), represents the composition of the total ground state energy of bonding electrons (hence, the cohesive energy) in light actinide elemental metals. It is seen in this figure that, in the thermodynamics of the light actinide metallic bond, itinerant 5f electrons, rather than 5f localized electrons are advantageous to cohesion, since the total energy is decreased. Also other features are visible in this figure: that 5f-itineracy produces a smaller equilibrium atomic volume than 5f non bonding localization, since in the latter (typical of the 4f in lanthanides) the conduction bonding band is ensured by the much more delocalized 7s and 6d bands.

The Mott-like transition, a central concept for the description of the actinide metal series, causes the sudden increase of the atomic volumes, encountered when between Pu and Am (Fig. 3). All other properties indicate the onset of a 5f localized behaviour at Am (see Part V): the 5f pressure, which had contained to smaller values the equilibrium interactinide distance, suddenly gives in, with the withdrawal of the 5f's within the atomic core. The occurrence of such a transition within a series characterized by an unsaturated shell, is a unique phenomenon of the actinide series. In lanthanides, it does not occur except perhaps under pressure: in cerium metal the approaching of cerium atoms induces suddenly the itineracy of 4f orbitals and a sudden volume collapse – see Chap. C. Neither it occurs in d-transition metal series, where the atomic volumes have an almost parabolic behaviour when plotted vs. Z (see Fig. 3 and Chap. C). The current

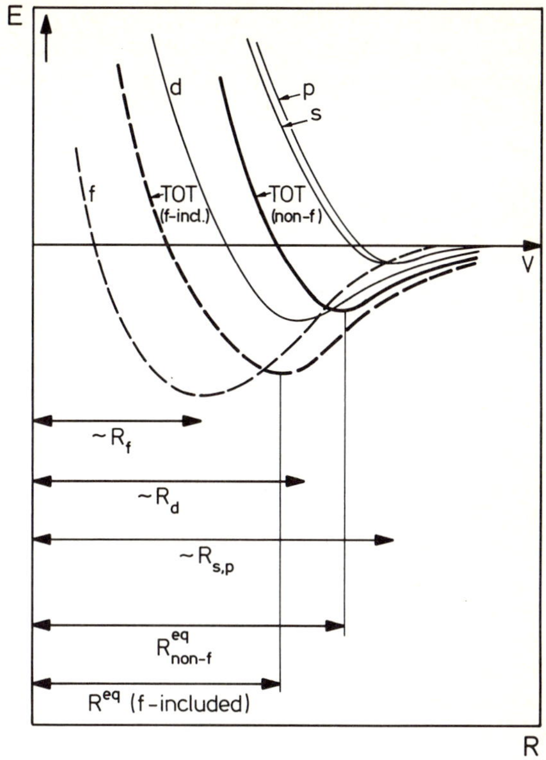

Fig. 20. Schematic representation of the s, p, d and f partial contributions to the total energy of electrons in the conduction band of a light actinide metal. The different R's denote the radial extension of the different contributing orbitals. R^{eq} (f-included) and $R^{eq}_{non\text{-}f}$ refer to the equilibrium volumes when the 5f electrons are itinerant and when they are non-binding (from Ref. 77)

explanation of the Mott-like transition is that the spin-polarization atomic term becomes suddenly important when at americium, so that in the delicate balance between the interatomic U_H and W descriptive parameters (defined elsewhere in this chapter) of models such as the Stoner and Hubbard models (see Part III), U_H predominates. (See, however, a very recent paper by Harrison[78], which, within the frame of the Anderson model, describes the same Mott transition without referring specifically to the spin-polarization energy, but to a different U, or localization energy, separating two resonant 5f states, one localized and one delocalized.)

The general concepts illustrated in this introductory Chapter constitute the guidelines of the book.

In Chap. B, attention is drawn to chemical aspects of preparation of actinide samples, especially single crystals: an important necessity for actinide solid state physics, which has, in the last decade, reached the maturity of other fields in materials science, therefore presenting a need of well characterized and pure samples.

In Chap. C, the thermodynamic and structural outlook of the bond, which had been the matter of discussion in Part A of this chapter, is further developed, and the model formalism, which takes advantage of the well known Friedel's model for d-transition metals but is inspired by the results of refined band calculations, is presented for metals and compounds. Also, a hint is given of the problems which are related to the non-stoichiometry of actinide oxides, such as clustering of defects. Actinide oxides present an almost purely ionic picture: nevertheless, covalency is present in considerable extent, and is important for the defect structure.

Chapters D and E are central to the book: Chapter D is devoted to present (although, in a rather short way) methods and results of magnetic measurements. Also, we thought that an up to date collection of magnetic data would help the reader of this book, and perhaps inspire new lines of research.

In Chap. E, photoelectron spectroscopic methods, in recent times more and more employed to the study of actinide solids, are reviewed. Results on metals and on oxides, which are representative of two types of bonds, the metallic and ionic, opposite with respect to the problem itineracy vs. localization of 5f states, are discussed. In metals photoemission gives a "photographic" picture of the Mott transition between Pu and Am. In oxides, the use of photoelectron spectroscopy (direct and inverse photoemission) permits a measurement of the intra-atomic Coulomb interaction energy U_H.

Finally, in Chap. F, a description of the most refined theoretical methods, as employed in actinides, is given, and its results for actinide solids are discussed. In the first part of the chapter, the author starts discussing actinides by presenting ... the zirconium atom: a striking beginning, purposely introduced in order to show the transition metal properties of 5f wave functions.

VII. References

1. Cohen-Tannoudji, C., Dim, B., Laloë, F.: Mecanique quantique I et II, Collection Enseignement des Science, Paris (1977)
2. Seaborg, G. T.: Chem. Eng. News *23,* 2190 (1945)
3. Mayer, M. G.: Phys. Rev. *60,* 184 (1941)
4. Fournier, J. M.: Thèse de Doctorat d'Etat, Univ. of Grenoble, CNRS Report no AO 10568 (1975)
5. Harrison, W. H.: Electronic Structure and the Properties of Solids, The Physics of the chemical bond, Freeman and Co., San Francisco (1980)
6. Zachariasen, W. H.: Acta Crystallogr. *5,* 19 (1952)
7. Zachariasen, W. H.: ibid. *5,* 660 (1952)
8. Zachariasen, W. H.: in: The Metal Plutonium (eds. Coffinberry, A. S., Miner, W. N.) University of Chicago Press, 1961, p. 99
9. Zachariasen, W. H.: in: The Law of Mass Action Centenary Volume, Norweg. Acad. Science, Oslo 1964, p. 183
10. Zachariasen, W. H.: J. Inorg. Nucl. Chem. *35,* 3487 (1973)
11. Kmetko, E. A., Hill, H. H.: Metallic valencies of the early transactinium metals: theoretical or empirical correlations, in "Plutonium 1970" (ed. Miner, W. M.) Met. Soc. AIME, New York 1970, p. 233
12. Cunningham, B. B., Wallman, J. C.: J. Inorg. Nucl. Chem. *26,* 271 (1964)
13. Weigel, F., Trinkl, A.: Radiochim. Acta *10,* 78 (1968)
14. Hill, H. H.: The early Actinides: the periodic system's f electron transition metal series in "Plutonium 1970" (ed. Miner, W. M.) Met. Soc. AIME, New York 1970, p. 2
15. Freeman, A. J., Koelling, D. D.: Electronic Energy Band Structure of the Actinide Metals, in "The Actinides: Electronic structure and related properties" (eds. Freeman, A. J., Darby, Jr., A. J.), Academic Press, New York 1974, Vol. I, p. 51
16. Condon, E. V., Schockley, G. H.: The Theory of Atomic Spectra, Cambridge University Press 1951
17. Bethe, H. A., Salpeter, E. E.: Quantum Mechanics of one and two-electrons atoms, Springer Verlag 1959

18. Backer, R. F., Goudsmit, S.: Phys. Rev. *46,* 948 (1934)
19. Rose, M. E.: Relativistic electron theory, Wiley, New York and London 1961
20. Desclaux, J. P.: Atomic and Nuclear Data Tables, Vol. *12,* 310 (1973)
21. Goldschmidt, Z. B.: Atomic Properties (free atoms) in "Handbook of the Physics and Chemistry of Rare Earths", Vol. 1 (eds. Gschneider, K. A., Jr., Eyring, Le Roy), North-Holland, Amsterdam 1982, p. 3
22. Brewer, L.: J. of the Opt. Soc. of America *61,* 1101 (1971)
23. Ziman, J. M.: Principles of the Theory of Solids, second edition, Cambridge University Press 1972
24. Kittel, C.: Introduction to Solid State Physics, John Wiley and Sons, New York 1967, 3rd Edition
25. Chan, S. K., Lam, D. J.: Crystal Field Theory in "The Actinides. Electronic structure and related properties" (eds. Freeman, A. J., Darby jr., J. B.), Academic Press, New York 1974, Vol. I, p. 1
26. Andersen, O. K.: Sol. St. Communications *13,* 133 (1973)
27. Andersen, O. K.: Phys. Rev. *B 12,* 3060 (1975)
28. Ziman, J. M.: Elements of Advanced Quantum Theory, Cambridge University Press 1969
29. Hohenberg, P., Kohn, W.: Phys. Rev. *136* (B 864), 540 (1964)
30. Kohn, W., Sham, L. J.: ibid. *140* (A 1133), 535, 540 (1965)
31. Koelling, D. D.: J. de Physique *40* (Proc. of the 3rd Int. Conf. on the electronic structure of actinides, Grenoble 1978) C-117 (1979)
32. Wohlfarth, E. P.: Reviews of Modern Physics *25,* 211 (1953)
33. Edwards, D. M., Wohlfarth, E. P.: Proc. Roy. Soc. *A 303,* 127 (1968)
34. von Barth, V., Hedin, L.: J. Phys. C: Solid State Phys. *5,* 1629 (1972)
35. Rajakopal, A. K.: J. Phys. *C 11,* L 943 (1978)
36. Brooks, M. S. S., Johansson, B.: J. Phys. F: Met. Phys. *13,* L 197 (1983)
37. Mott, N. F.: Phil. Mag. *6,* 287 (1961)
38. Johansson, B.: ibid. *30,* 469 (1974)
39. Hubbard, J.: Proc. Roy. Soc. *A 276,* 438 (1963)
40. Hubbard, J.: ibid. *A 281,* 401 (1964)
41. Rickayzen, G.: in: Green's Functions and Condensed Matter, p. 274, Academic Press, New York 1980
42. Benedict, U. et al.: J. of Mag. and Mag. Mats. *29,* 287 (1982)
43. Naegele, J. et al.: Phys. Rev. Letters *52,* 1834 (1984).
44. Johansson, B.: Phys. Rev. *B 11,* 2740 (1975)
45. Herring, C.: Magnetism (eds. Rado, G. T., Suchl, H.), Academic Press, New York, Vol. IV (1966)
46. Herbst, J. F., Watson, R. E.: Phys. Rev. Lett. *34,* 1395 (1975)
47. Johansson, B.: Phil. Mag. *30,* 469 (1974)
48. Stewart, G. R., Elliott, R. O.: in: Actinides 1981, LBL-12441, 207 (1981)
49. Trainor, R. J., Brodsky, M. B., Culbert, H. V.: Phys. Rev. Let. *34,* 1010 (1975) and Fournier, J. M.: Sol. St. Com. *29,* 111 (1979)
50. Friedel, J.: J. Phys. Chem. Solids *1,* 175 (1956)
51. Jullien, R., Galleani d'Agliano, E., Coqblin, B.: Phys. Rev. *B 6,* 2139 (1972)
52. Skriver, M. L., Andersen, O. K., Johansson, B.: Phys. Rev. Lett. *41,* 42 (1978)
53. Andersen, O. K., Jepsen, O.: Physica *91 B,* 487 (1977)
54. Johansson, B.: in: Valence Instabilities and Related Narrow Band Phenomena, (ed. Parks, R. D.), Plenum Press, New York, p. 435 (1977)
55. Fournier, J. M. et al.: Physica *86 B,* 30 (1977)
56. Fujuta, D. D.: Thesis, Univ. of California Rep. no URCL, 19507 (1969)
57. Kuper, C. G., Jensen, M. A., Hamilton, D. C.: Phys. Rev. *134,* A 15 (1964)
58. Smith, J. L.: 3rd Conf. on superconductivity in d-/f-band metals, LaJolla, Ca. USA 1979
59. Johansson, B., Rosengren, A.: Phys. Rev. *B 11,* 1367 (1975)
60. Brooks, M. S. S.: in: Rare Earths and Actinides, J. Mag. and Mag. Mat. *29,* 257 (1982)
61. Johansson, J. B.: in: Proceedings 2nd Int. Conf. Electr. Str. Actin., Wrochaw, Poland 1976
62. Brooks, M. S. S. et al.: Physica *102 B,* 84 (1980)
63. Fournier, J. M. et al.: Physica *102 B,* 282 (1980)

64. Arko, A. J., Koelling, D. D.: Phys. Rev. *B 17,* 3104 (1978)
65. Lander, G. H. et al.: Bull. Amer. Phys. Soc. *17,* 338 (1972)
66. Fournier, J. M.: 10eme Journées des Actinides, Stockholm 1980
67. Eastman, D. E., Kuznietz, M.: Phys. Rev. Let. *26,* 846 (1971)
68. Keller, J., Erbudak, M.: J. Physique *C 4,* 23 (1979)
69. Baptist, R., Naegele, J., Baer, Y.: ibid. *C 4,* 40 (1979)
70. Evans, S., Faraday, J. C. S. *73,* 1341 (1977); Naegele, J.: J. Physique *C 4,* 169 (1979)
71. Schoenes, J.: Physics Reports *63,* 302 (1980)
72. Brooks, M. S. S., Kelly, P. J.: Solid State Comm. *45,* 689 (1983)
73. Brooks, M. S. S., Glötzel, D.: Physica *102 B,* 51 (1980)
74. Brooks, M. S. S.: J. Phys. F: Met. Phys. *14,* 639 (1984)
75. Brooks, M. S. S.: ibid. *14,* 653 (1984)
76. Rossat-Mignod, J., Burlet, P., Vogt, O.: Physica, *102 B,* 237 (1980)
77. Johansson, B., Skriver, H. L.: J. of Magn. and Magn. Mats. *29,* 217 (1982)
78. Harrison, W. A.: Phys. Rev. *B 29,* 2917 (1984)

Chapter B

The Preparation of High Purity Actinide Metals and Compounds

W. Müller and J. C. Spirlet

Commission of the European Communities, Joint Research Centre, Karlsruhe Establishment, European Institute for Transuranium Elements, Postfach 2266, 7500 Karlsruhe, F. R. G.

Considerable progress has been possible in the determination of chemical and physical properties of actinide solids as a result of improved preparation and purification methods.

The growing demands for well characterized samples have given the incentive to overcome the difficulties presented to the actinide chemist by the high radioactivity hazard and comparative rarity of the actinide elements.

In this chapter, preparation and purification methods are reviewed. In view of the expected role of 5 f electrons in the metallic bond of actinides, methods for the preparation of metals have been particularly studied. There has also been important progress in the preparation of simple binary compounds. Special emphasis has been given to the growth of single crystals, particularly needed for the most refined physical techniques.

The chapter underlines the interdisciplinary nature of actinide solid state research.

1. Introduction

As seen in the first chapter, the study of the solid state properties of actinides and their compounds is advancing rapidly, since theoretical and experimental solid state physicists are increasingly interested in the peculiar behaviour of 5f electrons, which cause solid state properties similar to those of d transition elements in the first half of the series and to those of 4f lanthanides in the second half[1)].

This has been made possible by recent achievements. The most relevant are indicated below.

- The determination of such solid state properties that are influenced by radiation damage or decay heat has been facilitated by the availability of long-lived isotopes of the less common actinides (Table 1).
- Modern techniques of preparation, refining and crystal growth have been adapted to the special precautions for handling radionuclides.

The preparation and investigation of most actinide metals and compounds require the use of high purity inert atmosphere. Such an atmosphere can be maintained in the double glove-box system presented in Fig. 1. The inner glove-box contains high purity argon, the outer glove-box is filled with dry nitrogen. A vacuum lock chamber allows materials to be introduced into the argon atmosphere, the impurity content (O_2, N_2, H_2O) of which is kept in the ppm range.

- Methods of characterization of the solid, e.g. by X-ray diffraction techniques, have been improved and complemented by multielement analysis of impurities.
- Special encapsulating techniques have been conceived for physical measurements of actinide containing samples.

High purity actinide metals are subject to sophisticated investigations of bonding related properties; they are starting materials for the synthesis of compounds.

Among the actinide compounds the interest is concentrating on binary compounds of simple structure (e.g. 1:1 compounds with elements of the groups V and VI of the periodic table) for which the theoretical treatment is rather advanced, and on intermetallic (e.g. Laves-) phases.

Table 1. Decay properties and availability of actinide isotopes for solid state studies[1)]

Nuclide	Half-life (years)	Availability	Problems
^{227}Ac	2.2×10^1	mg	γ(daughters), α
^{231}Pa	3.2×10^4	g	α, γ(^{231}Ac)
^{237}Np	2.1×10^6	g-kg	α
^{239}Pu	2.4×10^4	g-kg	α, heat
^{242}Pu	3.9×10^5	g	n(sf)
^{241}Am	4.3×10^2	g-kg	α, γ(60 keV), heat
^{243}Am	7.4×10^3	g	γ(^{239}Np), α
^{244}Cm	1.8×10^1	g	α, γ, n(α, n), heat
^{248}Cm	3.6×10^5	mg	n(sf)
^{249}Bk	0.9×10^0	mg	β
^{249}Cf	3.5×10^2	mg	α
^{252}Cf	2.6×10^0	mg-g	n(sf), γ, α, heat
^{253}Es	5.5×10^{-2}	$\ll$ mg	α, heat

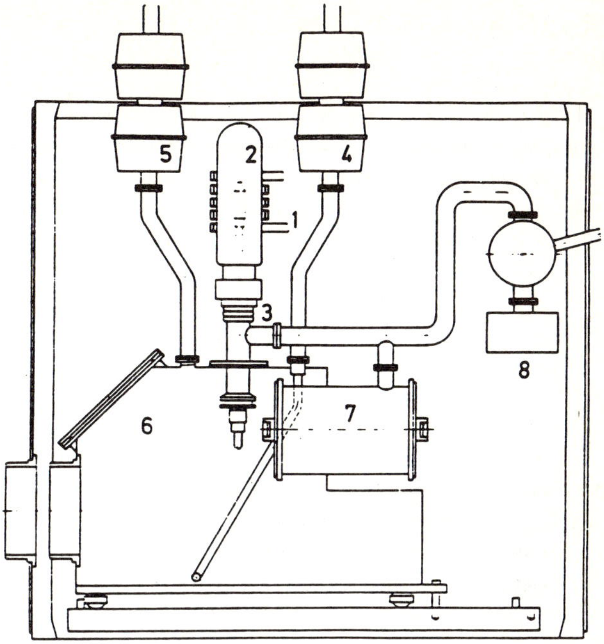

Fig. 1. Double glove-box system for the preparation and refining of actinide metals: *1.* RF heating system; *2.* Water cooled quartz vacuum furnace; *3.* Box filled with circulating nitrogen; *4.* + *5.* Argon in and out filters → filter; *6.* Stainless steel box filled with circulating argon; *7.* Vacuum lock chamber; *8.* Pump

Since uranium continues in being one of the most interesting elements of the actinide series, available in sufficient quantity and purity without necessitating special handling precautions, it is obvious that frequently crystal growth techniques have been developed with uranium compounds.

Previous results of the preparation chemistry of actinide elements have been reviewed in detail by F. Weigel[2]. In the following chapter, the actual state of the possibilities for the preparation and refining of actinide metals will be described; the principles and trends in synthesis and crystal growth of (simple) actinide compounds will be shown.

2. Actinide Metals

a) Preparation

Actinide (An) metals are prepared by reduction or thermal dissociation of their compounds[3]. The metallothermic reduction of halides is still in use, especially for microscale preparations. In this procedure, anhydrous actinide halides, in general fluorides, are exposed to the vapour of alkaline or alkaline earth metals, e.g.

$$AnF_3 + 3\,Li = An + 3\,LiF \tag{1}$$

Excess reductant and alkali halide are evaporated from the reaction mixture leaving, however, non volatile impurities in the metal, including oxygen from incompletely dehydrated starting halides.

The latest developments in the preparation of anhydrous actinide halides and their reduction to metals at the microscale have been reviewed by Haire[4].

The preparation of larger quantities of high purity actinide metals is being based increasingly on separation or purification via evaporation of the actinide metal[5]. In these methods, actinide compounds (oxides or carbides) are reduced by metals forming non-volatile oxides or carbides under conditions where the actinide metals can be volatilized:

$$An_2O_3 + 2\,La = La_2O_3 + 2\,An \quad (2)$$

$$AnO_2 + Th = ThO_2 + An \quad (3)$$

$$AnC + Ta = TaC + An \quad (4)$$

The choice of the starting compound of the actinide (oxide or carbide) and of the reductant is determined by the vapour pressure of the actinide metal. For a given temperature, the vapour pressures of the actinide metals (Fig. 2) span a ratio of more than 10^{10}. As the vapour pressure of La is similar to those of Ac, Cm, Pu, only the more volatile

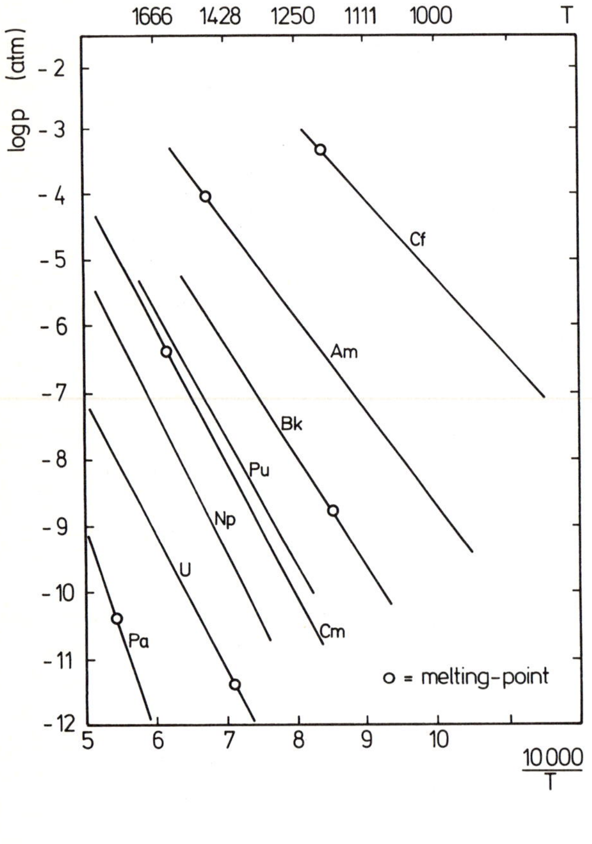

Fig. 2. Vapour pressure p of the actinide metals

actinide metals, Am, Cf, Es[6, 7] are prepared in pure form by this reductant. Th has been used in addition to prepare Am, Ac and Cm[8], whereas tantalothermy of carbides (obtained by the carboreduction of oxides) has been applied for U, Np and Pu[9].

The vapour pressure ratio of actinides to noble metals is also the basis of the actinide metal preparation by thermal dissociation of intermetallic compounds. Such intermetallic compounds of An and noble metals can be prepared by hydrogen reduction of a mixture of an An oxide and a finely divided noble metal (Pt, Ir...)[10]; in the absence of noble metals, hydrogen reduction of An oxides is impossible. Am and Cm metals have been obtained by thermal dissociation of their intermetallic compounds with Pt and Ir[11].

High purity Th and Pa, the least volatile actinide metals, can be prepared by thermal dissociation of their iodides, which form readily by reaction of iodine vapour with carbide:

$$ThC + 2\,I_2 \rightarrow C + ThI_4 \tag{5}$$

$$ThI_4 \xrightarrow{1300\,°C} Th + 2\,I_2 \tag{6}$$

The iodine vapours act as transporting agent in this modified version of the van Arkel procedure[12–14]. By replacing the resistance heated dissociation wire by an induction heated metal sphere (W or Th, respectively) large samples of Th or Pa crystals can be obtained[9].

b) Refining

As many physical properties of the actinide metals depend significantly on the sample purity, refining of the metals is mandatory. The choice of the refining methods is determined by the chemical reactivity of the actinide metal in the presence of the constituents of air, by high temperature reactions with crucible materials, by the specific radioactivity and the availability of the actinide elements.

The different methods of actinide refining[5] are based in part on experience in refining rare earth metals[15]. In these methods, actinide metals and their impurities undergo selective phase transitions like evaporation and condensation, melting and dissolution which result in a separation of the constituents of the sample to be purified.

Vacuum melting is particularly useful for eliminating volatile impurities from actinide metals of relatively low vapour pressures (Th, Pa, U, Np, Pu). These impurities are either present in the starting material or have been introduced during the metal preparation step, from reductant and crucible material in metallothermic processes or from transporting agents in the van Arkel process. Oxygen and nitrogen are, in general, not eliminated. If possible, contact with the crucible should be avoided by use of levitation techniques.

Efficient purification is achieved by selective evaporation and condensation. This technique is applicable to actinides of medium volatility i.e. Am or Cm[6, 16]. The volatile impurities are eliminated by selective condensation of the actinide metal, less volatile impurities are left in the crucible. The efficiency of this refining method is determined by the relative evaporation ratio, which for two elements A and B equals the ratio of their activities at a given temperature.

$$\alpha(T) = (a_A p_A / a_B p_B)(M_B/M_A)^{1/2} \quad (7)$$

where: a = activity coefficient
p = vapour pressure
M = atomic mass

As this ratio (Fig. 3)[16] tends to decrease with increasing temperature, the best separations are achieved at the lowest temperatures, which on the other hand, lead to low evaporation rates. The refining temperature is a compromise between the ratio of the vapour pressures and evaporation kinetics: e.g. 1120 °C Am, and 1600 °C for Cm evaporation; and 900 °C for Am and 1300 °C for Cm condensation[16].

Van Arkel refining by transporting and thermally dissociating actinide iodides has been applied to Th and Pa metals. Bulk metal instead of (brittle) wire can be obtained by dissociating the iodides on a radio-frequency heated sphere of the corresponding actinide metal.

If the actinide metal is available in sufficient quantity to form a rod or an electrode, very efficient methods become applicable: electrorefining, zone melting, solid state electrotransport. Thorium, uranium and plutonium have been refined by electrolysis in molten salts[17, 18]. An electrode of impure metal is dissolved anodically in a molten salt bath, e.g. in LiCl-KCl eutectic; the purified metal is deposited on the cathode as a solid or a liquid.

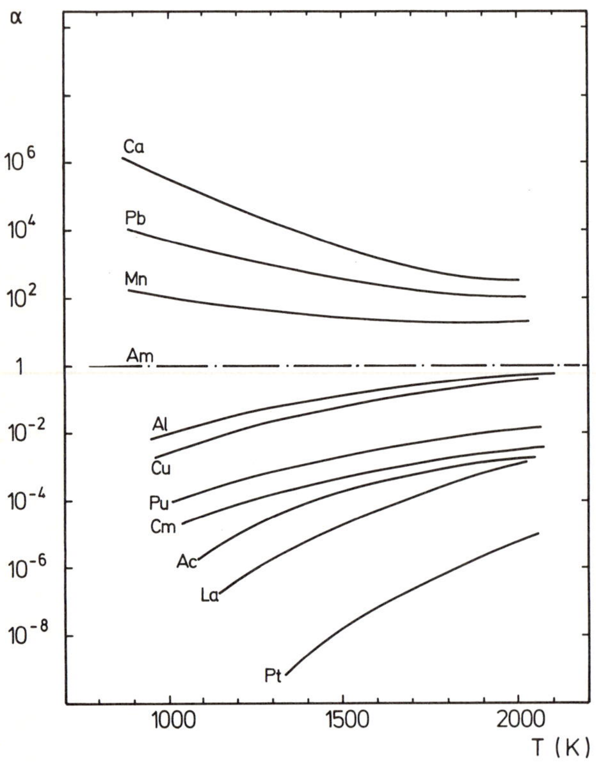

Fig. 3. Relative evaporation coefficient α of relevant elements in a system (impurities, reductant etc.) for Am preparation

In zone melting[15], a narrow molten zone is passing several times through a thin metal rod. Metallic impurities and carbon dissolve in the liquid and migrate with the molten zone to one end of the rod, oxygen and nitrogen move to the opposite side. Refining in an electric field (Solid State Electrotransport Processing, SSEP) becomes possible for actinide metals with low vapour pressures (Th, Pa, U, Pu, Np), if the metal rod can be kept below its melting temperature ($T = 0.8\ T_m$). Oxygen, nitrogen and carbon are carried with the electron flow, metallic impurities move in the opposite direction, resulting in a purification of the center part of the rod. Both, zone melting and SSEP require a good vacuum. Refining of U by SSEP (1600 °C, $2 \cdot 10^3$ A/cm^2, 10^{-8} Torr) resulted in a resistivity ratio R(300 K)/R(4.2 K) = 200; improving the vacuum to 10^{-12} Torr yielded Th with a resistivity ratio of 4200[19].

3. Actinide Compounds

Trends in the preparation of actinide compounds have been summarized by one of the authors[20].

a) Compounds With Group VII Elements

The preparation and properties of numerous actinide halides have been described by D. Brown[21, 22]. Although the oxidation numbers of actinides in halides can vary from II to VI, most solid state studies are limited to di-, tri- and tetrahalides.

They are synthesized from the elements, e.g.

$$Am + 3/2\ I_2 = AmI_3 \tag{8}$$

$$Th + 2\ Br_2 = ThBr_4 \tag{9}$$

or prepared in reactions involving compounds, e.g.

$$Cf_2O_3 + 6\ HBr = 2\ CfBr_3 + 3\ H_2O \tag{10}$$

$$ThC + 2\ I_2 = ThI_4 + C \tag{11}$$

$$Pa + 2\ HgCl_2 = PaCl_4 + 2\ Hg \tag{12}$$

Halides of divalent actinides are obtained by reaction of the metal with a stoichiometric amount of HgX_2 (X = Cl, Br, I):

$$Am + HgX_2 = AmX_2 + Hg \quad {}^{23)} \tag{13}$$

or by reduction of trihalides with hydrogen

$$CfBr_3 + 1/2\ H_2 = CfBr_2 + HBr \quad {}^{24)} \tag{14}$$

Large single crystals (> 1 cm^3) of UCl_4, UBr_4, $ThCl_4$ and $ThBr_4$[25] have been grown by slow cooling of the melt.

b) Compounds With Group VI Elements

Actinide oxides have oxygen to metal ratios of in general 1.5 (cubic, hexagonal, monoclinic An_2O_3) to 2 (cub. $AnO_{2\pm x}$); the composition An_3O_8 is known for uranium and neptunium. Claims for the identification of solid americium monoxide[26] are not confirmed; in analogy to lanthanide element behaviour[27], actinide monoxide synthesis might be achieved in a high pressure reaction.

Well defined oxide phases can be obtained by thermal dissociation of oxalates, by controlled oxidation of compounds or actinide saturated ion exchangers or by reduction of higher oxides with hydrogen. Thermal dissociation of compounds often results in oxides of low density; high (almost theoretical) density oxides can be prepared in "sol-gel" processes.

Single crystals of actinide oxides are grown from the melt[28] or from solutions in molten salts[29,30], electrolytically deposited in molten salts[31], obtained by sublimation[32] in chemical transport reactions[9]. Single crystals of uranium and neptunium dioxide have been obtained in the presence of $TeCl_4$ as transporting agent[33,34]: the starting oxides are transported in a temperature gradient between 1075 °C and 975 °C (Fig. 4); prisms of UO_2 (Fig. 5) and cubes of NpO_2 (Fig. 6) were obtained. The O/U ratio of non-stoichiometric uranium oxides (UO_{2+x}) is maintained during the transport in HCl gas[35].

In the actinide compounds with the other group VI elements (X = S, Se, Te), the X/An ratio varies from 1 to 3. These actinide chalcogenides show predominantly the composition AnX (cub.), An_3X_4 (cub., orthorh.), An_3X_5 (orthorh.), AnX_2 (tetr.) and AnX_3 (orthorh.) which is also the order of decreasing thermal stability.

The chalcogenides are synthesized either by reaction of the chalcogen vapour with the actinide metal or the hydride (Faraday-method), or in reaction with compounds:

$$U + HgX = UX + Hg \quad ^{36)} \qquad (15)$$

$$Bk_2O_3 + 3\,H_2S = Bk_2S_3 + 3\,H_2O \qquad (16)$$

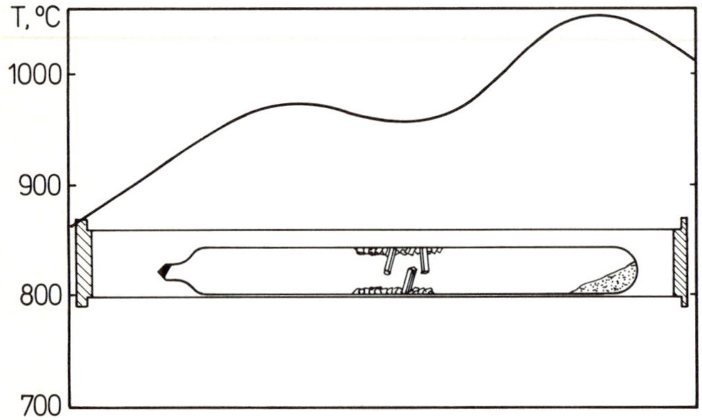

Fig. 4. Preparation of oxide single crystals by chemical vapour transport in a thermal gradient

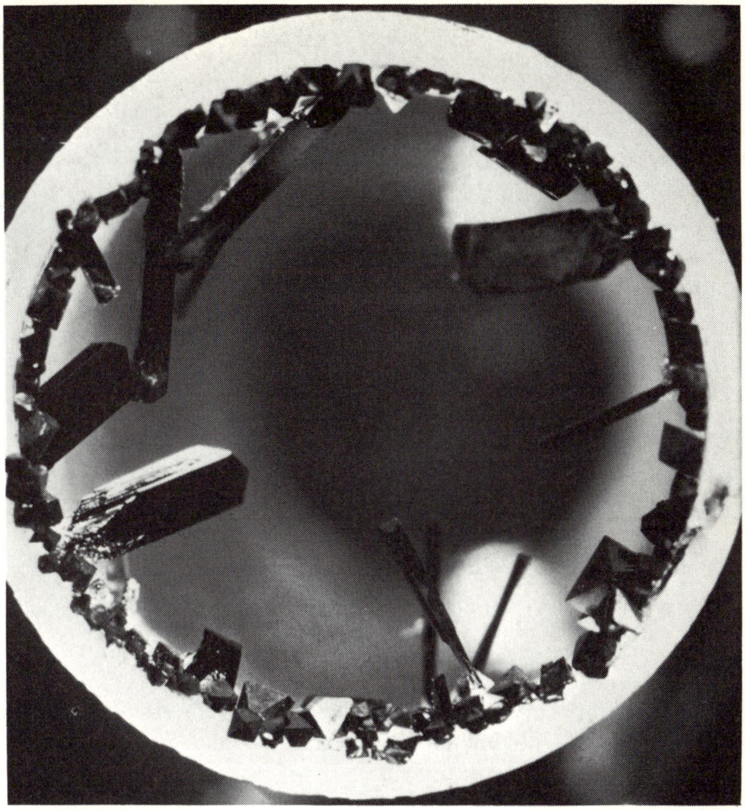

Fig. 5. UO_2 single crystals prepared by chemical vapour transport with $TeCl_4$

Uranium trichalcogenides (Fig. 7) are unstable above 500 °C, uranium dichalcogenides are unstable above 1000 °C. Transuranium chalcogenides are even less stable[37)]: actinide tritellurides $AnTe_3$ (An = Pa, Pu, Am, Cm) dissociate between 400 and 500 °C; ditellurides between 600 and 700 °C.

Single crystals of uranium chalcogenides have been obtained in transport reactions with Br_2, I_2, $TeCl_4$ or AgBr[36)] as transporting agents. Single crystals of thorium and uranium dichalcogenides have been deposited in a thermal gradient of 50 °C/10 cm at 850 °C with I_2 as transporting agent[9)].

c) Compounds With Group V Elements

The compositions of the most numerous actinide compounds with elements of the group V of the periodic table (X = N, P, As, Sb, Bi) belong to the types AnX_2, An_3X_4, AnX. These pnictides can be synthesized in solid-gas-reactions with actinide hydride, or with metal powder obtained by thermal dissociation of hydrides.

Nitrides are formed in reactions with nitrogen or ammonia; e.g. PaN at 1000 °C, the transuranium nitrides NpN, PuN, AmN and CmN at 700 °C. Some transuranium mono-

Fig. 6. NpO_2 single crystals prepared by chemical vapour transport with $TeCl_4$

pnictides (AmP, CmP, AmAs) can directly be synthesized in a Faraday-reactor between 500 and 600 °C. Uranium monopnictides are obtained by stepwise dissociation of dipnictides (UP_2, UAs_2, USb_2):

$$UX_2 \rightarrow U_3X_4 \rightarrow UX \quad (17)$$

Protactinium dipnictides (X = As, Sb) have been synthesized by reaction of As or Sb vapour with metal hydride at 400–700 °C[38]. Pa_3As_4 was obtained by thermal dissociation of $PaAs_2$ at 840 °C, Pa_3Sb_4 from the corresponding dipnictide at 1200 °C. Monopnictides of protactinium were not obtained by thermal dissociation of higher compounds. The diantimonides of the transuranium elements Np, Pu, Am dissociate between 700–800 °C into the monocompounds. Monopnictides of the higher transuranium elements have been obtained at the μg scale with ^{248}Cm[39], ^{239}Bk[40] and ^{249}Cf[41] by thermal dissociation.

From these observations it appears that with increasing atomic number of the actinides

- the temperatures necessary for the direct synthesis of monopnictides decrease
- the thermal dissociation of higher pnictides to monopnictides is becoming easier.

Single crystals of actinide pnictides have been grown from the melt and from the vapour. Single crystals of uranium and thorium nitride have been drawn in an arc from a metal rod in the presence of (20 atm) nitrogen[42]; UBi_2[43], USb_2[44], $PuSb_2$[45] single crystals have been obtained by slow cooling from An-X melts with 5–20 a/o An. Large size UAs- and USb-single crystals are obtained in a recrystallization process by keeping the polycrystallized starting material for several weeks at temperatures just below the melting point[46]. The method was successfully applied to the preparation of single crystals of PuSb[47], PuBi, PuAs, NpAs.

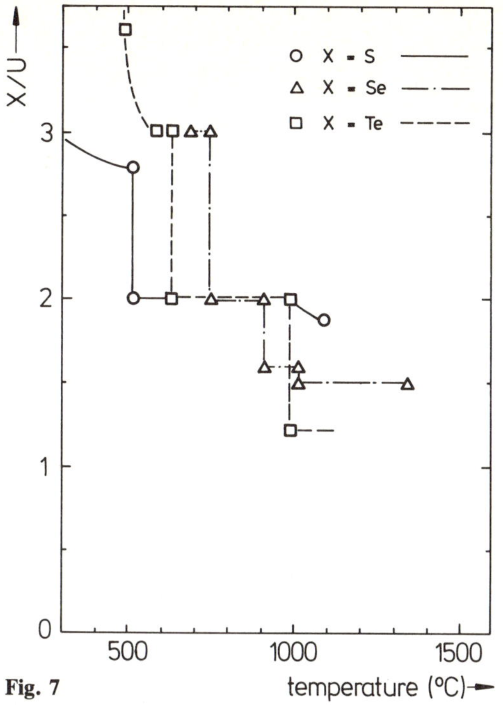

Fig. 7

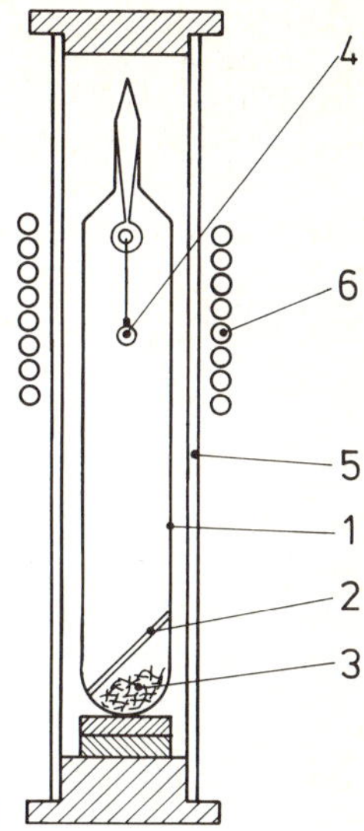

Fig. 8

Fig. 7. Thermal stability of uranium chalcogenides[20)]

Fig. 8. Apparatus for the preparation of single crystals by the Van Arkel procedure: *1.* quartz ampoule; *2.* tube with iodine; *3.* feed material; *4.* RF heated W sphere or plate; *5.* resistance furnace; *6.* RF coil

Chemical transport with iodine as transporting agent can be generally applied for the growth of actinide pnictides in single crystal form:

$$AnX_m + nI_2 \underset{T_2}{\overset{T_1}{\rightleftarrows}} AnI_{2n} + mX \quad (T_1 > T_2) \tag{18}$$

According to this transport reaction, single crystals of UP_2, UAs_2, U_3P_4, U_3As_4[42, 48)] and $NpAs_2$[49)] have been prepared by submitting polycrystalline pnictides to iodine vapour and transporting them in a quartz bulb along a temperature gradient from T_1 to T_2. Attempts to prepare single crystals of thorium pnictides, uranium antimonides or bismutides resulted in the attack of the quartz walls at temperatures of only 800 °C[50)]. In order to prepare $ThSb_2$, the quartz walls had to be coated with carbon. Due to the high temperatures required, monopnictides could not be prepared by this method.

Therefore, and to avoid possible reactions with the quartz wall, it was attempted to combine synthesis and crystal growth of actinide pnictides in a modified van Arkel process[51, 52)]: Actinide metal or carbide – the latter obtained by carboreduction of the oxide – are heated in the presence of the pnictogen and of the transporting agent at the

Table 2. Synthesis temperature of actinide pnictides by the Van Arkel process[20)]

Compound	T_1 (°C)	T_2 (°C)
$ThSb_2$	600	1200
Th_3Sb_4	600	1450
ThSb	600	1700
U_3As_4	470	1300
UAs	470	1650
USb_2	540	1200
U_3Sb_4	540	1450
USb	540	1800
$PaAs_2$	400	1000
Pa_3As_4	400	1500
PaAs	520	2000

temperature T_1 (Fig. 8); the pnictides are deposited at the temperature T_2 on an induction heated, polished tungsten plate. The temperature T_1 determines the vapour pressure of pnictogen and of the iodine, T_2 determines the composition of the pnictide; the transport rate is determined by T_2-T_1. Table 2 contains conditions for the preparation of some actinide pnictides in single crystal form (Fig. 9). By this method ("modified van Arkel-process") PaAs was synthesized for the first time and deposited in form of single crystals[52)].

d) Compounds With Group I–IV Elements

These compounds have a metallic character or are intermetallics. Actinide carbides (mono-, sesqui-, dicarbides) are prepared by carboreduction of the oxides:

$$AnO_2 + 3\ C = AnC + 2\ CO \tag{19}$$

or by reaction between hydrocarbons and actinide metal (powder):

$$An + CH_4 = AnC + 2\ H_2 \tag{20}$$

Direct synthesis is applied to the preparation of other compounds with elements of the first main groups of the periodic table: $AnSn_3$, AnB_2, $AnBe_{13}$.

The reaction of hydrogen with bulk actinide metals yields, in general, hydride (cubic AnH_{2+x}, hex. AnH_{3-x}) powders because of the low density of the reaction product. In the case of the plutonium-hydrogen system, the hexagonal (LaF_3-) phase covers the range of compositions between $PuH_{2,75}$ and $PuH_{2,95}$; there are indications for an orthorhombic (YF_3-) phase between the composition $PuH_{2,95}$ and PuH_3[53)].

Compact (bulk) plutonium hydride ($Pu_{1,95}$ to $PuH_{2,65}$) has been obtained by hydriding under pressure α-Pu in the plastic range (350–400 °C)[54)]. Protactinium trihydride is less stable than uranium trihydride[55)].

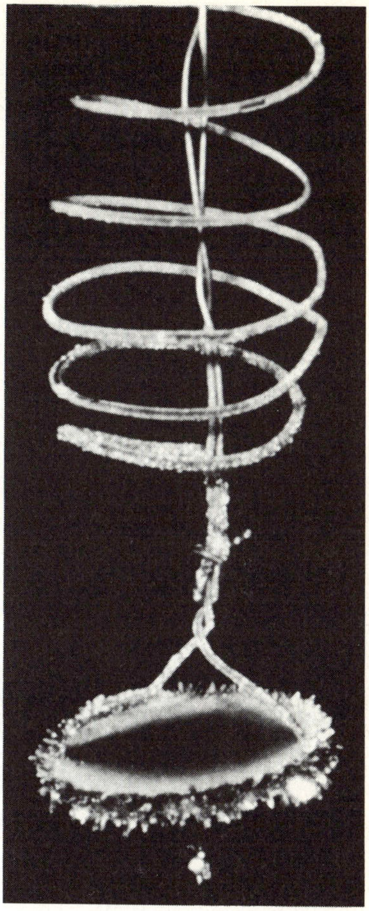

Fig. 9. Crystals of U_3Sb_4 obtained by the Van Arkel procedure

e) Compounds With d-Transition Metals

The compounds with elements of the platinum group can be prepared by direct synthesis, which requires the availability of actinide metals. Such intermetallics can, however, also be obtained by "coupled reduction"[10] of actinide oxides with hydrogen in the presence of finely divided noble metals:

$$AnO_y + yH_2 + x\,Me = AnMe_x + y\,H_2O \tag{21}$$

Single crystals of e.g. UPd_3 have been prepared by floating zone technique[56]. Large and perfect single crystals of e.g. UCo_2 or UNi_5[57] with high purity have been drawn from a levitated melt (Fig. 10).

Fig. 10. Pulling of a single crystal of UNi_5 from the levitated melt by the Czochralski method

4. Characterization

Actinide metal samples are characterized by chemical and structure analysis. Multielement analysis by spark source mass spectrometry (SSMS) or inductively coupled argon plasma (ICAP) emission spectroscopy have lowered the detection limit for metallic impurities by 10^2 within the last two decades. The analysis of O, N, H by vacuum fusion requires large sample, but does not distinguish between bulk and surface of the material. Advanced techniques for surface analysis are being adapted for investigation of radioactive samples (Fig. 11)[1].

Due to the low penetration depth of X-rays in heavy element samples, XRD patterns are not always significant for the structure of the bulk. In analogy to observations made on rare earth metal foils, fcc phases observed occasionally after heating and interpreted as high temperature phases of the actinide metals, might be the product of a reaction between the metal surface and residual gas leading, e.g. to hydride etc.

1 The adaptation for radioactive samples described in Fig. 11 has been performed by J. Naegele[59], basically for the investigation of surface and bulk properties of actinide solids such as discussed in Chap. E. In that chapter, a more detailed description for the characteristics of the system is given

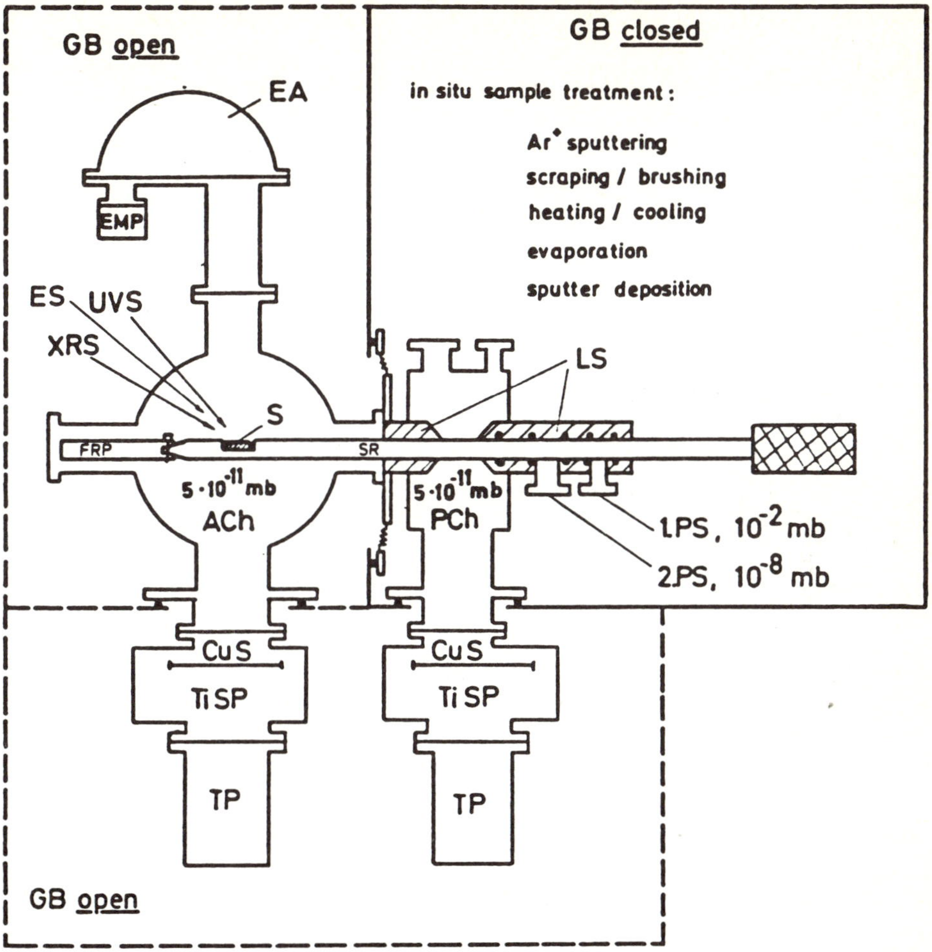

Fig. 11. Scheme of a photoemission spectrometer (ESCA) installed in a glove-box system: *ACh* analysing chamber; *CuS* copper shield; *EA* electron energy analyser; *EMP* electron multiplier; *FRP* final rod piece, can be demounted from SR; *GB* glove box; *LS* lock system, differentially pumped; *PCh* preparation chamber; *S* sample; *SR* sample rod; *TiSP* Titanium sublimation pump; *TP* turbopump; *UVS* UV (rare gas discharge) source; *XRS* X-ray source (Mg or Al anode); 1. PS first differential pumping stage; *2. PS* second differential pumping stage

The presence of decay products of the actinides can occasionally be used as an internal standard when the date of the last separation is known, e.g. in Pa the Ac content calculated as 12 μg/g was observed to be 13 μg/g at the time of the analysis[58)].

The chemical composition of the compounds, until now characterized predominantly by X-ray diffraction methods in analogy to lanthanide or actinide compounds, is increasingly being determined by standard analytical procedures. These methods (e.g. coulometry, spectrophotometry) have to be adjusted to the quantity of material available and effects of radioactive decay have to be taken into account. As with metals, chemical impurities are identified by multielement analysis (SSMS, ICAP). The identification of single crystals is facilitated by the GANDOLFI technique[60] which yields X-ray powder diffraction patterns by rotating the single crystals around two axes.

5. Conclusion

During the last years, the possibilities of preparing well characterized, high purity actinide metals and compounds have improved. Preparation of metals and compounds is increasingly based on methods applying reaction or purification steps via the vapour phase. Single crystals of halides are grown from the vapour or the melt. Single crystals of oxides, chalcogenides and pnictides are obtained in transport reactions with I_2 or $TeCl_4$ as transporting agent. Actinide pnictides are deposited in a modified van Arkel process or by recrystallization ("mineralization") at temperatures close to the melting point. Single crystals of intermetallics can be drawn from a levitated melt.

6. References

1. Müller, W.: in: Lanthanide and Actinide Chemistry and Spectroscopy (ed. Edelstein, N.), ACS Symp. series 131, p. 183 (1980)
2. Weigel, F.: in: Handbuch der Präparativen Anorganischen Chemie (ed. Brauer, G.), Band II, Ferdinand Enke Verlag, p. 1117, Stuttgart 1978
3. Lesser, R., Peterson, J. R.: in: Transurane (ed. Koch, G.), Band 31, Teil B 1, Gmelin Handbuch der Anorganischen Chemie, Springer Verlag, Berlin, Heidelberg, New York 1976
4. Haire, R. G.: in: Actinides in Perspective (ed. Edelstein, N.), Pergamon Press, Oxford, p. 309 (1982)
5. Spirlet, J. C.: in: Actinide in Perspective (ed. Edelstein, N.), Pergamon Press, Oxford, p. 361 (1982)
6. Spirlet, J. C., Müller, W.: J. Less-Common Metals *31,* 35 (1973)
7. Haire, R. G., Baybarz, R. D.: J. Inorg. Nucl. Chem. *36,* 1295 (1974)
8. Baybarz, R. D. et al.: in: Transplutonium 1975 (eds. Müller, W., Lindner, R.), North-Holland Publishing Comp., p. 69, Amsterdam 1976
9. Spirlet, J. C.: J. Phys. (Paris), Coll., C4, *40,* 87 (1979)
10. Erdmann, B., Keller, C.: Inor. Nucl. Chem. Letters *7,* 675 (1971)
11. Müller, W., Reul, J., Spirlet, J. C.: Atomwirtschaft *17,* 415 (1972)
12. Lorenz, R. et al.: J. Nucl. Mat. *37,* 203 (1970)
13. Brown, D., Tso, T. C., Whittacker, B.: AERE-R 8638 (1976)
14. Bohet, J., Müller, W.: J. Less-Common Metals *57,* 185 (1978)
15. Jones, D. W. et al.: in: Rare Earths and Actinides 1977, Internat. Conf. on Rare Earths and Actinides, Durham, Inst. Phys. Conf. Series *37,* 11 (1978)
16. Müller, W., Reul, J., Spirlet, J. C.: Revue de Chimie Minérale *14,* 212 (1977)
17. Chauvin, G., Corion, H., Hure, J.: Industries *439,* 1 (1962)
18. Mullins, C., Leary, J. A.: Nuclear Applications *6,* 287 (1969)
19. Schmidt, F. A., Outlaw, R. A., Lunde, B. K.: Report 15-M 171 (1979)
20. Müller, W.: Chem. Ztg. *160,* 105 (1982)
21. Brown, D.: Halides of the Lanthanides and Actinides, Wiley Interscience Publishers, New York 1968

22. Brown, D.: in: Transurane (ed. Koch, G.), Teil C, Verbindungen, Gmelins Handbuch der Anorganischen Chemie, Ergänzungswerk zur achten Auflage, Band 4, Verlag Chemie, Weinheim 1972
23. Baybarz, R. D. et al.: J. Inorg. Nucl. Chem. *34,* 3427 (1972)
24. Peterson, J. R. et al.: J. Phys. (Paris) Coll., C4, *40,* 111 (1979)
25. Hussonois, M. et al.: J. Crystal Growth *51,* 11 (1981)
26. Akimoto, Y.: J. Inorg. Nucl. Chem. *29,* 2650 (1969)
27. Leger, J. M., Yacoubi, N., Loriers, J.: J. Solid State Chem. *36,* 261 (1981)
28. Henrick, C. C., Behrens, R. G.: J. Crystal Growth *51,* 183 (1981)
29. Finch, C. B., Clark, G. W.: J. Appl. Phys. *36,* 2143 (1965)
30. Finch, C. B., Clark, G. W.: J. Crystal Growth *6,* 246 (1970)
31. Martinot, L. et al.: Bull. Soc. Chim. Belges *79,* 250 (1962)
32. Van Lierde, W. et al.: J. Nucl. Mat. *5,* 250 (1962)
33. Spirlet, J. C. et al.: J. Phys. (Paris) Coll., C4, *40,* 108 (1979)
34. Spirlet, J. C. et al.: J. Crystal Growth *49,* 171 (1980)
35. Nomura, Y., Kamegashira, N., Naito, K.: ibid. *52,* 279 (1981)
36. Suski, W.: in: Uranium Chalcogenides, Wroclwaska Drukania Naukowa 1976
37. Damien, D.: Synthèse et Cristallochimie des Chalcogénures de Transuraniens, Dissertation, Clermont-Ferrand 1976
38. Hery, Y.: Synthèse, Cristallochimie et Propriétés Magnétique des Pnictures de Protactinium, Dissertation, Strasbourg 1978
39. Damien, D., Haire, R. G., Peterson, J. R.: J. Less-Common Metals *68,* 159 (1979)
40. Damien, D., Haire, R. G., Peterson, J. R.: J. Inorg. Nucl. Chem. *42,* 955 (1980)
41. Damien, D., Haire, R. G., Peterson, J. R.: Inorg. Nucl. Chem. Lett. *16,* 537 (1980)
42. Endebrock, R. W., Forster, E. L., Keller, D. L.: Compounds of Interest in Nuclear Reactor Technology, Proc. of a Symp., AIME, Nucl. Mat. *10,* 557 (1964)
43. Henkie, Z.: Technologia Monocrystalow, Polish Scientific Publishers, Warszawa *2,* 56 (1974)
44. Henkie, Z., Misiuk, A.: Kristall und Technik *14,* 539 (1979)
45. Charvillat, J. P., Damien, D., Wojakowski, A.: Revue de Chimie Minérale *14,* 178 (1977)
46. Vogt, O.: in: Actinides in Perspective (ed. Edelstein, N.), Pergamon Press, Oxford, p. 289 (1982)
47. Spirlet, J. C. et al.: Journée des Actinides, Orsay 1982
48. Eberle, A. R. et al.: NBC-252 (1970)
49. Charvillat, J. P.: Cristallochimie des Pnictures de Transuraniens, Dissertation, Clermont-Ferrand 1977
50. Henkie, Z., Markowski, P. J.: J. Crystal Growth *41,* 303 (1977)
51. Calestani, G., Spirlet, J. C., Müller, W.: J. Phys. (Paris), Coll., C4, *40,* 106 (1979)
52. Calestani, G.: Herstellung, Kristallzüchtung und Charakterisierung von Thorium-, Uran- und Protactiniumpnictiden, Dissertation, Saarbrücken 1980
53. Haschke, J. M., Hodges, A. E.: Actinides 1981, LBL 12441, p. 229 (1981)
54. Ward, J. W. et al.: Actinides 1981, LBL-12441, p. 86 (1981)
55. Ward, J. W., Bartscher, W.: 12ème Journée des Actinides, Orsay 1982
56. Arko, A. J., Weber, L. W.: Actinides 1981, LBL-12441, p. 57 (1981)
57. Spirlet, J. C.: 1980 World Conf. of the Internat. Target Development Society, Gatlinburg, Tenn. 1980
58. Leider, F.: Thèse de Doctorat, University of Liège (Belgium) 1983
59. Naegele, J. R.: J. Phys. (Paris) *C2,* 841 (1984)
60. Gandolfi, G.: Miner. Petrog. Acta *10,* 149 (1964)

Chapter C

Structural and Thermodynamic Properties of Actinide Solids and Their Relation to Bonding

L. Manes and U. Benedict

Commission of the European Communities, Joint Research Centre, Karlsruhe Establishment, European Institute for Transuranium Elements, Postfach 2266, 7500 Karlsruhe, F. R. G.

The problem of bonding in actinide solids has been approached through the establishment of semi-theoretical correlations between thermodynamic and structural properties and the degree of participation to the bond of the different electronic orbitals of the actinide atom.

The first part of the chapter is devoted to an analysis of these correlations, as well as to the presentation of the most important experimental results. In a second part the following stage of development is reviewed, i.e. the introduction of more quantitative theories mostly based on bond structure calculations. These theories are given a thermodynamic form (equation of states at zero temperature), and explain the typical behaviour of such ground state properties as cohesive energies, atomic volumes, and bulk moduli across the series. They employ in their simplest form the Friedel model extended from the d- to the 5f-itinerant state. The Mott transition (between plutonium and americium metals) finds a good justification within this frame.

In actinide binary compounds an equation of state can also be developed on the same lines. The difference in electronegativity of the actinide and the non-actinide element plays an important role, determining the degree of mixing between the actinide orbitals (5f and 6d) and the orbitals of the ligand. A mixture of metallic, ionic and covalent bond is then encountered. In the chapter, two classes of actinide compounds are reviewed: NaCl-structure pnictides or chalcogenides, and oxides.

In oxides a large departure from the stoichiometric composition is usually met. The importance of this phenomenon, also for the understanding of the oxide bond, is highlighted at the end of the chapter.

I. Introduction

1. Thermodynamic and Structural Studies as First Approach to Bonding

a. The "Idiosyncratic Structures"[1] of Lighter Actinide Metals

The lower actinides, uranium, neptunium and plutonium, present crystallographic structures which may be called "idiosyncratic". The simplest qualitative model ("electron sea" model) for a metal is one in which ions resulting from the loss of outer electrons are "glued together" by the "sea" of free electrons providing the special "glue" called the metallic bonding. Except in the immediate neighbourhood of each ion, the metallic bonding has no directionality. "Simple" metals, therefore, (e.g. the monovalent alkali metals) have structures of the highest symmetry, as appropriate to the packing together of (ionic) spheres, a feature similar to the crystallographic behaviour of other non-directional types of bonding, i.e. van der Waals and ionic bonding.

For the lighter actinide metals, however, we encounter, at normal temperature and pressure such asymmetric structures as monoclinic for plutonium. There is a complete "idiosyncracy" to the description of a "simple" metal: strong directionality sets in. In the light of the discussion of Chap. A on the 5 f orbitals, it is easy to attribute this directionality to the 5 f participation to the bonding, in competition with the 7 s and 6 d electrons. Thus, the degree of their participation becomes a central issue for the understanding of the conduction band of these "idiosyncratic" metals.

b. Bond Indicators for Actinide Metals

The cohesive energy of a metal (and the related measurable thermodynamic properties) is relatively easy to understand in a simplified picture of the metallic bond.

Thus, the first attempts to understanding of chemical and physical properties in the actinide series dealt with the systematic inspection, across the series, of the thermodynamic properties influenced by the cohesive energy. As well as for the structure, the variations encountered can be attributed to the participation of outer electrons in setting up the metallic bond, with the peculiar behaviour of the 5 f orbitals among them.

These attempts may be called "thermodynamic semi-theoretical approaches". They concern mostly the simplest kind of bonding, namely the metallic bond. The underlying hypothesis is that the contributions of different outer orbitals (7 s, 6 d, 5 f) in some chosen thermodynamic or structural property can be linearly combined, the coefficients of this linear combination being related to the "degree of participation" of the different orbitals in the bonding: an approach clearly related to the molecular orbital approach of quantum chemistry and to the hybridization concept, and which had been previously employed in other transition metals and to the rare-earth metallic systems[2–5] (for a criticism of this approach, see Ref. 6). The chosen thermodynamic and structural properties are, therefore, "bonding indicators", since they will reflect contributions introduced by the fact that the wavefunctions of bonding electrons have mixed orbital characters.

c. *Thermodynamic Equations for Bond Related Properties*

In recent times, the "bond indicators", which are the ground state properties of the solid related to its cohesion (metallic radii, cohesive energy, bulk moduli), have been interpreted in the light of band calculations. The bond in metals and in compounds has been described by an easily understandable and convincing thermodynamic formalism, which we shall illustrate in this chapter. Essentially, narrow bands, as the 5f electrons form, are considered to be "resonant" with the wider (spd) conduction band. The 5f electronic population is seen as a "fluid" the partial (bonding) pressure of which assists in cohesion along with the partial pressure of another "fluid" constituted by the conduction electrons of (s and d) character[7)].

II. The Crystal Structure of Actinide Metals

1. *General*

Most of the metallic elements of the Periodic Table crystallize in one or more of the highly symmetric structure types A 1 (cubic close packed, ccp[1]), A 2 (body-centered cubic, bcc) and A 3 (hexagonal close packed, hcp)[8)].

A few other structures encountered in metallic elements can be considered as distortions or slight modifications of one of these simple structure types. The structures of mercury fall into this category. The body-centered tetragonal structure observed for indium, protactinium, and plutonium is a distortion of the ccp structure type. The so-called double hexagonal close packed (dhcp) structure first found for americium, then also assigned to other actinide and lanthanide metals, is obtained by stacking hcp units along their c axis, with the condition that every second unit is rotated by 180° around that axis. This requires doubling of the c parameter with respect to the hcp structure, with the consequence that the dhcp unit contains four atoms in comparison to the two atoms contained in the hcp unit.

Departures from this orderly behaviour of the metals are only found in a few regions of the Periodic Table. Among the A groups, these exceptions are found in the 3rd, 4th and 5th column and concern gallium, white tin, and the pnictogens arsenic, antimony and bismuth. Among the transition metals, there is (with the exception of a metastable primitive cubic modification of tungsten, which forms under certain conditions) only one element which departs from the three basic metallic structures: manganese which in its beta form has a complicated primitive cubic structure. The exceptional crystallographic character of manganese was explained by Brewer through the existence of different electronic configurations with approximately equal energies. It is interesting to note in this context that plots of different physical properties versus atomic number in the 3d series[9)] show an anomaly for this metal.

1 The term ccp (*c*ubic *c*lose *p*acked) is here employed rather than the more familiar fcc (*f*ace *c*entered *c*ubic) in analogy with the term hcp (*h*exagonal *c*lose *p*acked)

The remaining exceptions concern: the lanthanide series, where samarium at room temperature has a particular hexagonal structure; and especially the lower actinides uranium, neptunium, and plutonium. Here the departure from simple symmetry is particularly pronounced. Comparing these three elements with other metals having partly filled inner shells (transition elements and lanthanides), U, Pu, Np have the lowest symmetry at room temperature, normal pressure. This particular crystallographic character is the reason why Pearson did not succeed to fit the alpha forms of U, Pu, and Np, as well as gamma-Pu into his comprehensive classification of metallic structures[1] and treated them as "idiosyncratic structures". Recent theoretical considerations[10] reveal that the appearance of low symmetries in the actinide series is intimately linked to the behaviour of the 5f electrons.

The higher actinide metals americium, curium, berkelium and californium have – at normal pressure – again the "common" structure dhcp and are in this respect similar to some of the lanthanide metals. In fact, the theoretical calculations and certain experimental observations show that in these actinide metals, 5f electrons are localized, as are the 4f electrons in the lanthanide metals. More detailed considerations on the possible correlations between electronic and crystal structure are found in[11].

Table 1 gives the known structures of the actinide metals.

2. *Structural Changes With Temperature*

The allotropic behaviour of the actinides with change in temperature is summarised in Fig. 1a.

3. *Structural Changes With Pressure*

a. *Trends for Actinide Metals Under Compression*

The actinide metals up to californium were studied under pressure at room temperature (Fig. 1b). For the lighter actinides up to plutonium, the (densest) room temperature allotrope remained in general stable under compression to the highest pressure attained (68 GPa for Th, 53 GPa for Pa, 21 GPa for Pu). α-U was stable up to about 50 GPa; preliminary results indicate that at 71 GPa a different structure of uranium may exist. Np was only studied to 3.5 GPa by a piezometric technique.

In contrast, americium undergoes three phase transitions in the pressure range below 16 GPa. The first of these (to the cubic close-packed structure, which is not found under pressure in the lighter actinides), underlines – as does the structural sequence with temperature at normal pressure – the resemblance of americium with the lanthanide metals. But the appearance of low symmetry structures Am III and Am IV above 10 GPa can be reasonably described as a consequence of the participation of the 5f electrons in bonding which is caused by pressure. Correspondingly, in the lighter actinides, 5f's already participate in the bond at normal pressure; thus pressure will not change the 5f configuration, and very high pressures are required (see uranium) to induce structural changes by other mechanisms.

Table 1. Crystallographic structures of actinium and the actinide elements. **a)** Temperature and pressure ranges of existence, and lattice parameters. **b)** Space group, number Z of atoms in the unit cell, atomic volume, and number of nearest neighbours. Most data are taken from Ref. 12. Recent high pressure data are from Refs. 13, 14 (Th), Ref. 15 (Pa), Ref. 16 (U), Ref. 17 (Np), Ref. 18 (Pu), Refs. 19–22 (Am), Ref. 23 (Cm), Ref. 24 (Bk, Cf). bct: body-centered tetragonal, or: orthorhombic, d.bc.: double body-centered. Number of 5f electrons according to Johansson[4, 5]

Table 1 a

Element, number of 5f electrons and allotrope			Structure	Lattice parameters				
				a, pm	b, pm	c, pm	β	T or p
89	*Ac* $5f^0$	RT	ccp, A 1	531.1				RT
90	*Th* $5f^0$							
	I	50–1673 K(10^5Pa) ≤68 GPa(RT)	ccp, A 1	508.51(7)				RT
	II	>1673 K	bcc, A 2	411				1723 K
91	*Pa* $5f^1$							
	I	50–1450 K(10^5Pa) ≤53 GPa(RT)	bct	394.5(2)		324.0(1)		291 K
	II	>1450 K	bcc, A 2	381				1473 K
	III	metastable at RT	ccp, A 1	501.9(3)				RT
92	*U* $5f^2$							
	I(α)	43–935 K(10^5Pa) ≤50 GPa(RT)	or. (α-U)	285.38(4)	586.8(2)	495.57(13)		RT
	α_1, α_2, α_3	4–43 K(10^5Pa)	similar to α-U					
	II(β)	935–1045 K(10^5Pa) 1020–1070 K(2 GPa)	tetragonal	1075.9		665.6		993 K
	III(γ)	1045–1400 K(10^5Pa) 1090–1420 K(7.5 GPa)	bcc	353.4				1078 K
93	*Np* $5f^4$							
	I(α)	RT-533 K(10^5Pa) RT-680 K(7.5GPa)	or.	472.3	488.7	666.3		RT
	II(β)	533–810 K(10^5Pa) 680–1020 K(4GPa)	tetragonal	489.7		338.8		590 K
	III(γ)	850–910 K(10^5Pa) 920–970 K(2GPa)	bcc, A2	352				873 k

94	*Pu* $5f^5$							
	I(α)	77–395 K(10^5Pa) RT-673 K(12 GPa) <21 GPa(RT)	monoclinic	618.3	488.2	1096.3	101.78°	294 K
	II(β)	395–479 K(10^5Pa) 673–853 K(12 GPa)	monoclinic	928.4	1046.3	785.9	92.13°	463 K
	III(γ)	479–592 K(10^5Pa) 720–790 K(2 GPa)	or. fc.	315.87	576.82	1016.2		508 K
	IV(δ)	592–724 K(10^5Pa)	ccp, A 1	463.71				592 K
	V(δ′)	724–749 K(10^5Pa)	bct	332.61		446.30		723 K
	VI(ε)	749–913 K(10^5Pa) 753–850 K(1 GPa)	bcc, A 2	363.43				749 K
95	*Am* $5f^6$							
	I	RT-878 K(10^5Pa) <6 GPa(RT)	dhcp	346.81		1124.1		RT
	II	~900–1100 K(10^5Pa) 5–10 GPa(RT)	ccp, A 1	489.4				295 K
	III LLNL:		monoclinic, α″-U	310.1(9)	621.2(19)	462.5(14)	γ = 93.5(1)°	11 GPa
	LANL:	11–14 GPa(RT)	monoclinic, d.bc.	303.3(6)	1190.0(23)	283.3(6)	105.78(16)°	11.1 GPa
	IV	15–18 GPa(RT)	or. (α-U)	306.0(5)	596.2(11)	515.5(8)		16.1 GPa
96	*Cm* $5f^7$							
	I	<18 GPa(RT)	dhcp	349.6		1133.1		293 K
	II	HT phase	ccp, A 1	503.9				RT
	III	12–18 GPa(RT)	hcp, A 3					
	IV	18–20 GPa(RT)	unkown					
97	*Bk* $5f^8$							
	I	RT phase	dhcp	341.6		1106.9		RT
	II	HT phase 8–24 GPa(RT)	ccp, A 1	499.7				RT
	III	25–57 GPa(RT)	or. (α-U)	231.7	558.4	448.0		45.9 GPa
98	*Cf* $5f^9$							
	I	RT-870 K(10^5Pa)	dhcp	339(1)		1101(5)		RT
	II	870–1000 K(10^5Pa) 17–30 GPa(RT)	ccp, A 1	494				RT
	III	30–40 GPa(RT)	distorted ccp					
	IV	41–48 GPa(RT)	or (α-U)	231.3	552.6	447.2		46.6 GPa
99	*Es* $5f^{11}$		fcc	541; 575				RT

Table 1b

Element and Allotrope		Space group Nr.	Symbol	equiv. pos.	Z	$V_{at} \cdot 10^3$, nm^3	T or p	Number of nearest neighbours (distance, pm)
89	*Ac* $5f^0$	225	Fm3m	(4a)	4	37.45	RT	12 (375.5)
90	*Th* $5f^0$							
	I	225	Fm3m	(4a)	4	32.87	RT	12 (359.6)
	II	229	Im3m	(2a)	2	34.7	1723 K	8 (356), 6 (411)
91	*Pa* $5f^1$							
	I	139	I4/mmm		2	25.21	291 K	8 (321.4), 2 (324.0)
	II	229	Im3m	(2a)	2	27.65	1443 K	8 (330.0), 6 (381)
	III	225	Fm3m	(4a)	4	31.61	RT	12 (354.9)
92	*U* $5f^2$							
	I (α)	63	Cmcm	(4c)	4	20.75	RT	2 (275.3), 2 (285.4), 4 (326.3), 4 (334.3)
	II (β)		P4nm, P4/mnm, or P4n2		30	21.82	993 K	67 (282–353)
	III (γ)	229	Im3m	(2a)	2	22.07	1078 K	8 (306.1), 6 (353.4)
93	*Np* $5f^4$							
	I (α)	62	Pmcn	(4c)	8	19.22	RT	4 (260–264), 7 or 9 (300–353)
	II (β)	90	$P42_12$	(2a, 2c)	4	20.31	590 K	4 (276), 10 (324–356)
	III (γ)	229	Im3m	(2a)	2	21.81	873 K	8 (305), 6 (352)
94	*Pu* $5f^5$							
	I (α)	11	$P2_1/m$	(2e)	16	20.00	294 K	3–5 (257–278)
	II (β)	12	I2/m or B2/m	not standard	34	22.43	463 K	11–13 (259–362)
	III (γ)	70	Fddd	(8a)	8	23.14	508 K	4 (302.6), 2 (315.9), 4 (328.8)
	IV (δ)	225	Fm3m	(4a)	4	24.93	592 K	12 (327.9)
	V (δ′)	139	I4/mmm		2	24.69	723 K	8 (324.2), 4 (332.6)
	VI (ε)	229	Im3m	(2a)	2	24.00	749 K	8 (314.7), 6 (363.4)

95	*Am* $5f^6$							
	I	194	$P6_3/mmc$	(2a, 2d)	4	29.27	RT	6 (346.8), 6 (345.1)
	II	225	Fm3m	(4a)	4	29.30	295 K	12 (346.1)
	III LLNL:		c-centered		4	22.3	11 GPa	
	LANL:	11	$P2_1/m$	(2a, 2e)	4	24.44	11.1 GPa	2 (283), 2 (303), 2 (321)
	IV	63	Cmcm	(4c)	4	23.51	16.1 GPa	
96	*Cm* $5f^7$							
	I	194	$P6_3/mmc$	(2a, 2d)	4	29.98	293 K	6 (349.6), 6 (347.9)
	II	225	Fm3m	(4a)	4	31.99	RT	12 (356.3)
97	*Bk* $5f^8$							
	I	194	$P6_3/mmc$	(2a, 2d)	4	27.96	RT	6 (341.6), 6 (339.8)
	II	225	Fm3m	(4a)	4	31.19	RT	12 (353.3)
	III	63	Cmcm	(4c)	4	14.49	45.9 GPa	
98	*Cf* $5f^9$							
	I	194	$P6_3/mmc$	(2a, 2d)	4	27.4	RT	
	II	225	Fm3m	(4a)	4	30.13	RT	12 (349.3)
	IV	63	Cmcm	(4c)	4	14.29	46.6 GPa	
99	*Es* $5f^{11}$					39.59; 47.53	RT	

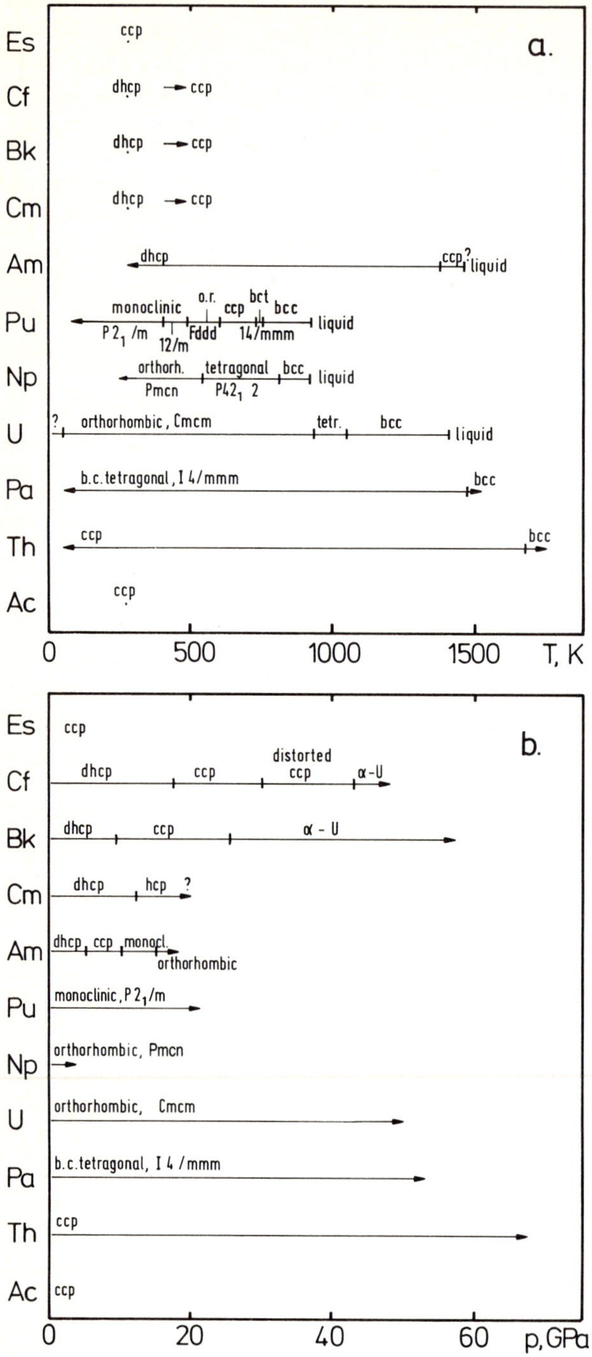

Fig. 1 a, b. Phase transitions in actinide metals, based on the data of Table 1; **a)** with temperature at ambient pressure, **b)** with pressure at room temperature. *Arrows* mark the highest (or lowest) temperatures and pressures attained up to now

The structural sequence dhcp $\xrightarrow{p}$ ccp was also expected to occur in the next element, curium. But the ccp phase was not observed under pressure. In contrast, the simple hexagonal close-packed (hcp) and, at still higher pressure, an as yet undetermined structure is formed. If these results are confirmed by further study, curium structures will have to be considered as another intermediate stage between the lighter and the heavier actinides[2].

The next two elements, berkelium and californium, were recently found to have identical structural sequences under pressure (Fig. 2b, c). The first high pressure transition for both Bk and Cf is dhcp → ccp as in the lanthanides. Thus the "lanthanide character" of heavy actinides again seems confirmed. But a second transition to the low symmetry α-uranium type structure follows in both metals. This transition reflects the start of 5f participation in bonding. The transition pressures increase monotonically on going from Am to Bk and Cf: 5, 7 and 17 GPa for the dhcp → ccp transition, 10, 25, 30 GPa for the ccp → An III (low symmetry phase) transition. The second transition in Cm occurs at 18 GPa; this transition pressure fits well into the sequence of delocalization pressures. But the dhcp → hcp transition in Cm occurs at 12 GPa and thus does not fit into the increasing Z sequence with respect to both structure type formed and transition pressure[2].

The lanthanide metals should also be investigated to higher pressures than previously applied. It is not excluded that their 4f electrons also participate in bonding as do the 5f's of Bk and Cf, after the dhcp, ccp and, possibly, "distorted fcc" phases have been reached. An indication of this possibility can be seen in the recent discovery of the α-uranium structure type in praseodymium (Pr IV)[25, 26]. This structure type was previously observed for cerium, but was thought to be restricted to that metal which has an exceptional position among the lanthanide elements.

b. *The Case of Americium*

The behaviour of americium under pressure provides important information for the understanding of the properties of actinides. It merits thus a more detailed consideration.

Published records of high pressure work on americium in the Lawrence Livermore and Los Alamos National Laboratories (LLNL and LANL)[19–22] cover the range up to 18 GPa. Figure 2a gives a summary of results for this pressure range. Both laboratories agree that above the range of the normal pressure, room temperature, allotrope of double hexagonal close-packed (dhcp) structure the structure is face-centered cubic up to about 10 GPa. LANL and LLNL report a monoclinic structure for pressures between 10 and 15 GPa. Space group No. 11 ($P2_1/m$) is assigned by LANL, but LLNL gives a c-centered structure with different lattice parameters. Roof[27] showed that both structures are related. A similar situation exists for the allotrope described for the pressure range 15 to 18 GPa. LANL reports it to be of the α-uranium type structure, LLNL describes it to be similar to that structure type.

Further study of the high pressure allotropes of americium seems to be required under these circumstances[3]. An additional argument for further invesigation is the discrepancy in volume change between experimental results and calculations. Figure 2a

---

2 See, however, the "Note Added in Proof" at the end of the Chapter

3 New structural data for Am up to 52 GPa[28] were obtained after this manuscript was completed

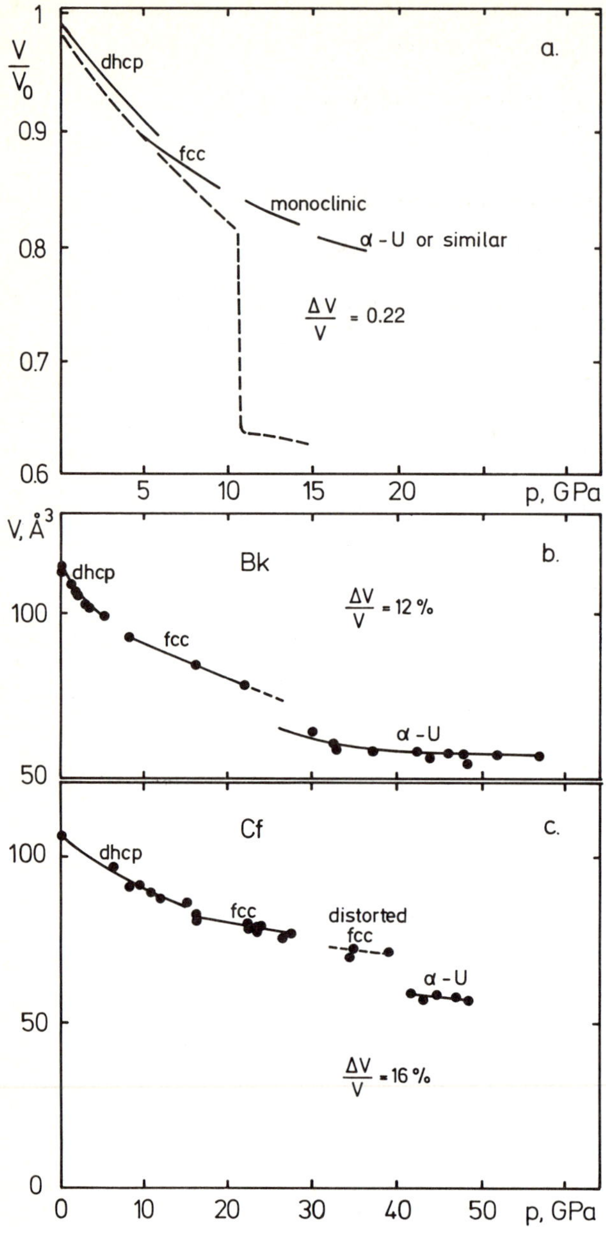

Fig. 2.a Calculated (*dashed line*[10, 29]) and experimental (*full line*[19–22]) volume change of americium under pressure. **b, c:** Experimental volume and structural changes for Bk and Cf

shows that only small, if any, volume discontinuities occur at the phase transitions. The dotted line gives the volume calculated by ASA-LMTO[10), the relativistic, non spin-polarized volume calculated by Brooks[29) being used as a lower limit of the volume collapse. A volume change of – 22% is thus evaluated, while – 35% was obtained in the original ASA-LMTO calculations. Though the volume decrease is smaller on the basis of Brook's calculations, its order of magnitude and discrepancy with experiment remain the same.

4. Phase Diagrams

Only for U, Np and Pu sufficient experimental work was carried out so that a pressure-temperature phase diagram can be constructed.

Figure 3 shows that the diagram simplifies when relatively low pressures (3 GPa) are applied. In each case, only 2 of the allotropes are left at four GPa and above. The explanation is probably that phases of relatively low density can no longer exist under pressure for these actinide metals.

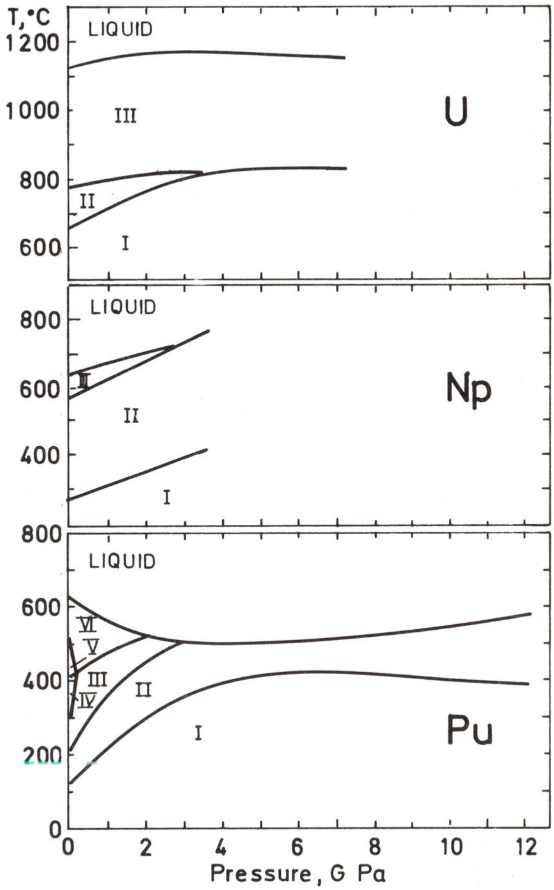

Fig. 3. Phase diagrams of U, Np and Pu metals

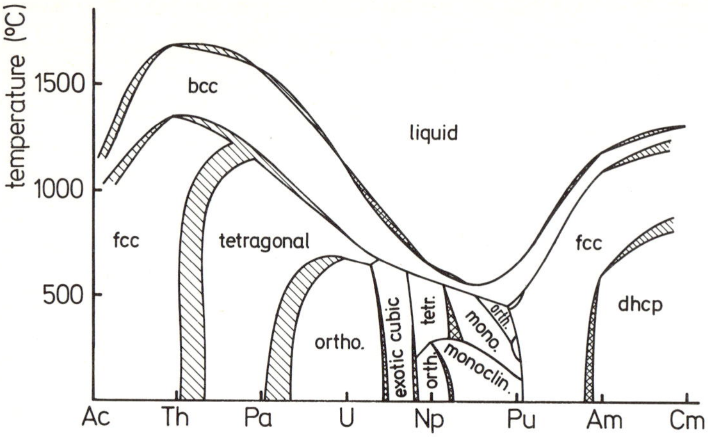

Fig. 4. Interconnected binary phase diagrams of actinide metals (after Smith and Kmetko[30]). Two phase regions: ▨ experimentally observed, ▧ assumed

Binary interactinide normal pressure phase diagrams are known in more or less detail for U-Np, Np-Pu, Pu-Am, Th-U, U-Pu.

An interesting way of simultaneously visualizing the structural relation in the actinide metal series are the interconnected binary phase diagrams of actinides given by J. L. Smith and E. A. Kmetko[30] (Fig. 4). When using these it must be kept in mind that some of the reported regions are based on assumptions made by interpolating between the constituents. Some simplification is also used: the (different) tetragonal structures of Pa I and U II are given the same phase field: they should in fact have two separate phase fields. However, these phase diagrams achieve their goal, which is a description of f-f overlapping (see Chap. A) in a simple and convincing way.

5. *Atomic Volume of the Actinide Metals*

Figure 5 gives the variation of the atomic volume in the actinide series, for the room temperature crystal structures as well as for the ccp and bcc high temperature allotropes, which exist for a number of actinides. The graph is based on the lattice parameters of Table 1, which includes also recent results. The marked dip in the curve from Th to Am illustrates the shrinkage of interactinide distance which is linked to the itinerancy of the 5f electrons in this part of the actinide series.

III. Thermodynamic Semi-Theoretical Treatments of Bonding in Actinides

1. *The General Description*

The following procedures are common feature to the thermodynamic semitheoretical approach across the An-metal series:

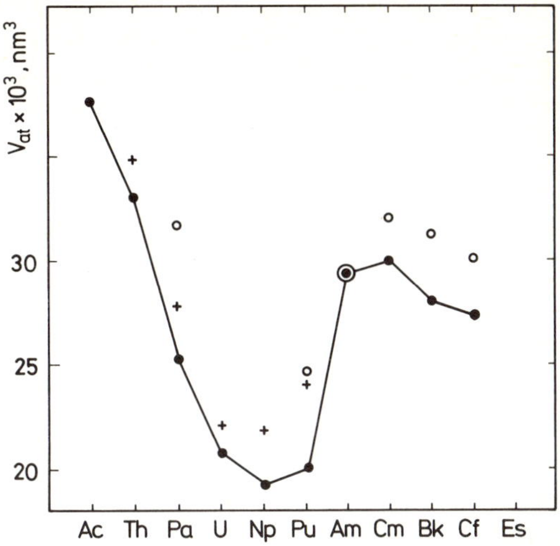

Fig. 5. Atomic volume of the actinide metals: ● room temperature crystal structures; ○ high temperature ccp allotropes at room temperature; + high temperature bcc allotropes at high temperature

- the definition of the metallic valence state for the An element in the metal (choice of a "valence indicator");
- the choice of "bonding indicators", i.e. of experimental properties, usually connected to the cohesive energy of the An-metal, in which a linear contribution from the outer electronic orbitals is postulated;
- the inspection of the Periodic Table to assess the most likely contribution from each electronic orbital to the different bonding indicators;
- the set up of a system of linear equations in which the degree of participation of the outer electrons in the bonding constitute the unknown.

Among bonding indicators, the most usually employed are:

a) the sublimation enthalpy ΔH_s, as obtained from vapour pressure measurements;
b) the melting point T_m;
c) the boiling point T_b.

Call $(BI)_j$ the bonding indicator. A system of linear equations is then set up taking the form:

$$(BI)_j = \sum_i \frac{\partial (BI)_j}{\partial b_i} b_i \quad (1)$$

$$v = \sum_i b_i \quad (2)$$

where by i we have indicated the different orbitals taken into consideration (typically, for Actinides, 5f, 6d, (6p), 7s), the b_i's are the degrees of participation in the bonding and v is the metallic valence of the elements.

We first observe that:

a) the valence state of the elements is chosen on the basis of the structural properties of the metal; typically[2)]
   - the metal radius
   - the symmetry of the crystal structures

   are taken as valence indicators.

b) the sublimation enthalpy ΔH_s is usually not the "bonding indicator" employed. ΔH_s is the difference in enthalpy between a collection of gaseous atoms at infinite distance between each other and in their (electronic) ground state configuration (see Chap. A, Table 1) and the same collection in the condensed phase (in which their (electronic) ground state configuration may be different). Usually, a quantity called cohesive energy E_{coh} connected to ΔH_s by the following thermodynamic cycle:

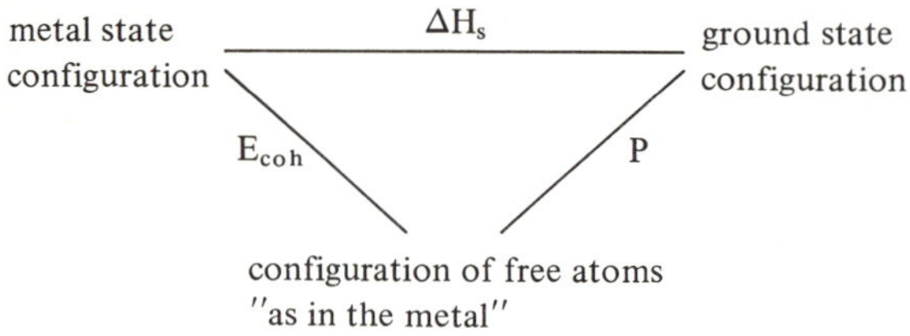

P is called promotion energy. The configuration of the free atoms "as in the metal" is usually an excited state of the atom: hence, the promotion energy can be usually obtained by atomic spectroscopy results.

c) The melting point T_m is usually not directly connected to the cohesive energy; however, in the treatment of critical phenomena it can be related in a simple way to the variation of the internal energy (and, although usually neglected, entropy) between the solid and the liquid state *when a particular model of the liquid is assumed.* Usually, a model of the liquid state is taken, according to which *just one* of the bonds between atoms in the solid state is broken during melting.

The $\partial(BI)/\partial b_i$ contributions of the different orbitals are computed through a systematic inspection of the Periodic Table, i.e. by recording the characteristic variation in the bonding indicator when a new outer orbital appears in the atomic electronic configuration. In this way, the s and p contributions are usually assessed. In the transition series', however, the further complication exists that the *unsaturated* shells across the series may give different contributions to the bonding, i.e. the contribution $\partial(BI)/\partial b_i$ depends on the occupancy number $n_{v,i}$ in the atomic configuration v.

This problem is handled differently by different authors, on the basis, usually, of reasonable assumptions suggested by the inspection of the plots of the (BI)'s across the series. This is done also, in the case of actinides, for the 5f unsaturated shell.

The results of the treatments are usually given in the following form:

$$An(5f^x, 6d^y, 6p^w, 7s^z)$$

where x, y, w, z *may be* fractional numbers. It is clear, as stated in the introduction of this chapter, that the interest of this kind of treatments lies in the assumption that it gives an insight into the orbital composition of the wave-functions of the binding electrons, i.e. that they will describe in some way the state resulting from hybridization occurring between the outer electron. In fact, x, y, w, z are often called "hybridization factors".

When applied to actinides, especially the lighter ones, these treatments encounter, however, typical difficulties:

a) since, as indicated in Chap. A, very low promotion energies separate already in the neutral atom, different ($d^m f^n$) configurations, it *may happen* that the electronic configuration chosen for the "metal state" and employed to extract from the sublimation enthalpies ΔH_s the cohesive energy E_{coh} would vary considerably along the series, *hence* requiring an ad hoc hypothesis for different elements in the use of these bonding indicators; this is equivalent to say that it is difficult to assume a common "valence state" for all elements in the actinide series;

b) the spatial extension of the 5f wave-functions changing with atomic number Z, the amount of f-f overlapping *and* the type of hybridization with other outer electrons change too. This is an effect which is not found in lanthanides, where the 4f electrons may be considered localized, and therefore with no f-f overlapping.

This requires that the result of the simple treatments described be sustained by a critical survey of many other physical properties[3)]. Unfortunately, and especially for heavier actinides, one is hampered many times by the lack of suitable experimental information. One of the merits, on the other hand, of these approaches is that they have often led to suggestions for particular decisive measurements.

We shell now examine in more details two particularly successful treatments of the actinide series.

2. *Examples of Semi-Theoretical Thermodynamic Treatments for the Actinide Metal Series*

a. *The Thermodynamic Treatment by Fournier*

This treatment[3)], aiming to evaluate thermodynamically the orbital character of the bond in actinide metals, follows closely the general features illustrated above and has a particular value inasmuch as it is accompanied by a fairly comprehensive survey of the chemical and physical properties of actinide metals known at that time. In it, the metallic radius and the crystal structures are taken as "valence indicators"; ΔH_s and T_M as the "bonding indicators". The metallic valence, however, is not taken as constant throughout the actinide series, but rather allowed to vary. The particular choices are justified by physical *and* chemical arguments, which are taken in support of the hypothesis chosen.

The outer electrons contributing to the bond are: 5f, 6d, 6p, 7s.

A system of linear equations as in Eq. (1) and (2) is employed. Rather than the value B of the bonding indicators in each actinide metal, ΔB variations are calculated with respect to the configuration of a reference state. The reference state configuration is inspired by the Engel-Brewer correlations, amply used for transition metals and alloys[2, 6)]. It is seen that the system of equations contains one equation less than the number of unknowns, so that only a range of the Δb_i solutions can be determined. However, this range can be shown, by a simple iterative procedure, to be limited.

The importance of the treatment lies in the physical considerations used for the choice of the valence and reference states as much as on the final results.

The first consideration deals with probable metal valences across the series. Correlations of the type introduced by Zachariasen have been discussed in Chap. A.

In this treatment, the line connecting the metallic radii of actinium and curium is considered as representative of an actinide contraction, analogous to the lanthanide contraction. This actinide contraction line may be considered as a trivalent basis line as for lanthanides; therefore, the considerable departure to a lower value for the metallic radii of light actinides is indicative of a higher metallic valence (or of a larger participation to the metallic bond of 5f electrons). Am and Cm should be considered as roughly trivalent whereas Cf and Es should be considered divalent at least in their hcp structures. Therefore, the following choices for valence are made:

Tetravalent metals:	Th, Pa, U, Np, Pu
Trivalent metals:	Ac, Am, Cm, Bk
Divalent metals:	Cf, Es

This classification is assumed to be justified by the examination of the crystal structures. In particular:

a) the low symmetry low temperature structures of Pa (tetragonal), U, Np (orthorhombic), Pu (monoclinic) are compatible with the hypothesis of a significant 5f contribution to the bonding, hence to the loss of (almost) one 5f electron with respect to the trivalent state;
b) the c/a ratio in h.c.p. Cf and Es is much larger than that expected for trivalent h.c.p. metals (1.73 rather than 1.63) and found in lanthanides; this fact is consistent with the divalent hypothesis.

The evaluated partial orbital characters of the bond are given in Table 2.

b. The Thermodynamic Treatment by Johansson and Rosengren

A somewhat different but very fruitful thermodynamic approch is offered by Johansson and Rosengren ([4], see also[5]). The merit of this approach is that an exhaustive and

Table 2. Comparison of the electronic configurations of actinide metals derived by Fournier[3] and by Johansson[4]

		Fournier[3]			Johansson[4, 5]	
An	Valence v	(f) Non-bonding state	Bonding conduction band	Valence v	(f) Non-bonding state	Bonding conduction band
Th	tetravalent	$(5f)^0$	$(s^{1.8}p^{0.2}d^{1.8}f^{0.2})$	tetravalent	$(5f)^0$	$(sd)^4$
Pa	tetravalent	$(5f)^1$	$(s^{1.9}p^{0.1}d^{1.6}f^{0.4})$	tetravalent	$(5f)^1$	$(sd)^4$
	pentavalent	$(5f)^0$	$(s^{1.9}p^{0.1}d^{1.6}f^{1.4})$			
U	tetravalent	$(5f)^2$	$(s^2\ p^{0.4}d^{0.6}f^1)$	tetravalent	$(5f)^2$	$(sd)^4$
	pentavalent	$(5f)^1$	$(s^2\ p^{0.4}d^{0.6}f^2)$	pentavalent	$(5f)^1$	$(sd)^5$
Np	tetravalent	$(5f)^3$	$(s^2\ d^{0.3}f^{1.7})$	trivalent	$(5f)^4$	$(sd)^3$
	pentavalent	$(5f)^2$	$(s^2\ d^{0.3}f^{2.7})$			
Pu	tetravalent	$(5f)^4$	$(s^2\ p^{0.2}d^{0.2}f^{2.6})$	trivalent	$(5f)^5$	$(sd)^3$
	pentavalent	$(5f)^3$	$(s^2\ p^{0.2}d^{0.2}f^{2.6})$			
Am	trivalent	$(5f)^6$	$(s^{1.3}p^{0.1}d^{1.3}f^{0.3})$	trivalent	$(5f)^6$	$(sd)^3$
Cm	trivalent	$(5f)^7$	$(s^{1.3}p^{0.4}d^{1.2}f^{0.1})$	trivalent	$(5f)^7$	$(sd)^3$
Bk	trivalent	$(5f)^8$	$(s^{1.5}p^{0.4}d^{0.9}f^{0.2})$	trivalent	$(5f)^8$	$(sd)^3$
Cf	divalent	$(5f)^{10}$	$(s^{0.7}p^{0.7}d^{0.6})$	trivalent	$(5f)^9$	$(sd)^3$
				divalent	$(5f)^{10}$	$(sd)^2$
Es	divalent	$(5f)^{11}$	$(s^1\ p^{0.5}d^{0.5})$	divalent	$(5f)^{11}$	$(sd)^2$

accurate comparison is made between the lanthanide and the actinide metals series. In this way, essential differences between the two series are highlighted.

As pointed out before, the main differences between the two series are the following:

- for the *lanthanides,* a non-bonding character of the 4f states due to their localization, *and* (with the exception of Eu and Yb) a trivalent state in the metal may be safely assumed
- for the *actinides:*
  i. at least for light actinides, the non-bonding character hypothesis cannot be any longer retained;
  ii. the metal valence state (as already shown in the precedent treatment) is no longer constantly trivalent, but varies across the series, assuming "chemical" values (i.e. the most common oxidation states for the different actinide elements in compounds) from pentavalent (in uranium) to divalent in the heavier actinides.

In the treatment of Johansson and Rosengren, no attempt is made, as in the previous treatment, to deduce fractional degrees of participation of different orbitals in the bonding of actinide metals, but, rather, attention is focussed on the correct attribution of metallic valences and on the discrepancy found between the thermodynamically calculated E_{coh} in a certain valence state and the experimental one. This difference is attributed to the bonding character of the 5f orbitals due to itineracy.

The method consists essentially of:

α) the building up of an interpolation scheme for cohesive energies of the trivalent state in lanthanides and a correspondent (ideal) trivalent state in the actinides;

β) the evaluation of the differences in cohesive energy when from the ideal trivalent state one goes to other valences, and the consequent assignment of a valence state to the different elements;

γ) the evaluation of the differences in cohesive energy in the metallic bonding due to the 5f electrons.

In α) and β) the non bonding-hypothesis for 5f electrons is retained, differences in cohesive energy being only due to promotion of outer electrons from one to another orbital state and ionization energies (or electron affinities) due to the different valence states attained. Therefore, any further discrepancy found with experimental values, is indicative of the metallic bonding introduced by delocalization of the 5f electrons (point γ).

We illustrate briefly the three steps.

α) The interpolation scheme for the E_{coh} of the trivalent state is built up by considering the quantity E_b (bonding energy) defined as the energy gained when free ions and free electrons are brought together to form a solid:

$$E_b = E_{coh} + I(III)$$

where I(III) is the ionization energy for the process: $R \rightarrow R^{+3}$ (R indicates the atom and R^{+3} the metal ion). Ce (trivalent), Gd, Lu, among the lanthanides, undergo the process: $f^n d^1 s^2 \rightarrow f^n$ from the atomic to the metal ion state, and their I(III) are known. Therefore, they are considered "pivot" elements, through which an E_b vs. Z smooth curve is passed across the lanthanide series. E_{coh} is then back calculated from interpolated values for all others lanthanides and found to agree well with experimental values (sublimation enthalpies ΔH_s).

For actinides, the ideally trivalent E_b vs. Z curve is then simulated by passing a smooth curve parallel to the lanthanides' one through the *only truly trivalent* value known at the time of the publication[4]: Cm, with $\Delta H_s = 89 \pm 4$ Kcal/mol. I(III)'s, also unknown, are taken from atomic spectroscopy and calculations[32, 33].

In this way, E_{coh} (trivalent) for all actinide metals are derived and for light actinides found in large discrepancy with the experimental ones. For them, a valence higher than three is suggested by the discrepancies.

β) The true valence states (and their relative cohesive energies) are then obtained by another thermodynamic consideration.

Consider the standard enthalpy for formation ΔH_f° of an ion in aqueous solution from a metal, corresponding to the process:

M^{+3} (metal) $\rightarrow$ M^{+3} (aqueous ion).

These quantities have been measured for most lanthanide and actinide metals (see Nugent[32]). For trivalent lanthanide metals, they correspond, invariably, to the destruction of the (s, d) metallic bond, without "damage" to the $4f^n$ shell. Therefore, ΔH_f° vs. Z is a slowly varying curve across the lanthanide series.

It is possible to pass the same curve through the experimental ΔH_f° values of two *truly trivalent* actinide elements, Am and Cm, and interpolate from the curve ΔH_f° (trivalent) for all other actinide metals.

An energy level scheme (or thermodynamic cycle) between the free atom and the aqueous M^{+3} ion can then be built as follows:

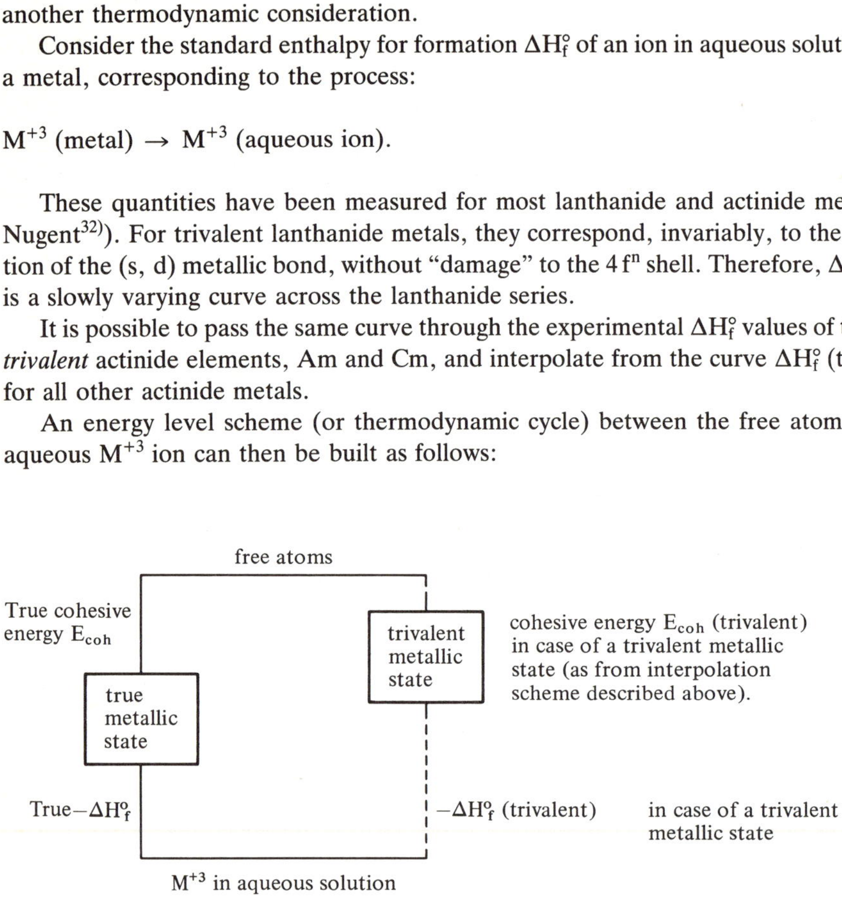

The (dashed) right part corresponds to the case in which both the metal and the aqueous ions are trivalent; this case is the unrealistic case for the light actinides.

From this cycle, the interpolation scheme of point a), giving, for all actinides, E_{coh} (trivalent), and the known ΔH_f° values for actinide metals, "true" cohesive energies E_{coh} may be evaluated. They are found in much better agreement with the experimental ΔH_s, although some deviation still occurs.

The difference between E_{coh} and E_{coh} (trivalent) may be due to higher valence as well as to 5f metallic bonding effects.

It is clear that the difference between the two effects in *not* clearcut. In fact, one may even doubt whether in the case of strong itineracy of 5 f's, the metallic valence concept retains any usefulness at all.

Nevertheless, the inspection of other transition metal series shows that, just as atomic volumes, there are regular variations of cohesive energies when the metal valence changes. Thus, a general increase of about 45 Kcal/mol is found when a metal transforms from a trivalent to a tetravalent state.

Applying these considerations, one may obtain the results of Table 2 in which a metallic valence is attributed to the light actinides.

For actinides heavier than Cm, a very similar scheme is worked out consisting in a comparison with a) trivalent lanthanides; b) surely divalent lanthanides Eu and Yb. In it, E_{coh} (trivalent) calculated with the above interpolation scheme, are compared with E_{coh} for divalent metals, as obtained by assuming a behaviour across the actinide series, similar to the one found in divalent lanthanides. The divalency of the heavier actinides (and the trivalency of Am and Cm) is concluded.

γ) The difference between the experimental E_{coh} and E_{coh} for the most stable valence chosen can be attributed to 5 f-bonding effects. The results are summarized in Table 3.

3. Conclusions

It is interesting to examine comparatively the results of the two treatments in term of composition of the bonding (participation of outer orbitals to the metallic bonding). This is done in Table 2, in which the comparison is performed up to Es. The bonding composition is given in the same formalism employed by the two authors in[3,4] respectively.

In the case of Fournier's approach, the partial factors of composition, of the conduction bonding band are certainly remindful of the hybridization concept. These factors may be understood as representing the percentage of each orbital character to be found in the hybridized conduction band (or, in a more chemical sense, the participation of different outer electrons to the bond). Thus, for instance, in the case of thorium, the small amount of 5 f participation seems to be in line with X-ray absorption spectroscopic result[34], which can be interpreted (because of the application of atomic selection rules) only when keeping in mind this concept. Recent high pressure X-ray diffraction experiments on thorium metal, however, fail to evidence this f-participation: it might be thought, in fact, that the 5 f character in the bond would have given some peculiar

Table 3. Contributions of valence and 5 f electrons to the cohesive energy of the lighter actinides[4]

	Valence v	$E_{coh}(v)$	5 f-Bonding
Th	tetravalent	143	
Pa	tetravalent	(130)	(29)
U	tetra- or pentavalent	103	25 kcal/mol
Np	trivalent	87	20 kcal/mol
Pu	trivalent	72	10 kcal/mol

behaviour to the high pressure compressibility, since, when approaching thorium atoms by means of pressure, the (fsd) hybridization should increase. On the other hand, band calculations[35)] yield a certain amount of f-character in the conduction band of thorium metal.

In the case of Johansson and Rosengren[4)], the conduction band has an (sd) character, and is occupied by an integral number of electrons, the atomic orbital character of which is unspecified. The 5f state, occupied by an integral number of electrons, may be a localized state or a band, this information being contained in the part of E_b due to f-itineracy (see Table 3). In the treatment, the concept of hybridization is consciously not used[5)]. In fact, in[5)], Johansson explicitly criticizes the point of view, predominant in the work of Jullien et al.[36)] who consider the 5f electrons as localized but *strongly* hybridized with the (sd) band. Johansson focusses mostly on the localization-itineracy problem of 5f bands (hybridization becoming important only between delocalized 5f and 6d, 7s states).

Both treatments were effective in predicting essential thermodynamic quantities, related to the cohesive energy E_{coh}, for elements that, at the time, had not been investigated. As shown above this was connected with the correct evaluation of their metallic valence. In this respect (Table 2), a difference exists between the two authors as for the valence of plutonium (and neptunium). Whereas tetravalency may be justified for plutonium on the basis of solid state oxidation states in compounds (e.g. PuO_2), pentavalency (as found by Fournier) seems rather awkward, whereas Johansson's trivalency is in line with the existence of solid state trivalent compounds (e.g. Pu_2O_3) as well as with the general tendency to lanthanide-like behaviour when increasing the actinide atomic number Z. However, the prediction of divalency for heavier actinides common to both authors, is now amply accepted and confirmed also by the easy existence of divalent compounds. This inspires nowadays chemical preparative research to synthesize such compounds as actinide monoxides the properties of which should reveal metallic character (e.g. in electric conductivity) instead of the refractory character of tetravalent or trivalent oxides.

IV. The Thermodynamics of the Metallic Bond

We shall now examine recent theoretical results concerning those ground state properties of metals and compounds which in the previous section, we have called "bond" and "valence" indicators. In this section, those properties are not taken as starting points for correlations aiming at the composition of the bond. They are instead the final point of electronic structure theories in which the different contributions to cohesion are analysed.

The results are conveniently and clearly expressed in a thermodynamic formalism: this is why they find their place in this chapter. They depend however on parameters which are drawn from band-theory, especially from the LMTO-ASA (Linear Muffin-Tin Orbitals-Atomic Sphere Approximation) method.

1. *The Composition of the Cohesive Energy and the Equation of State at 0 K for the Light Actinide Metals*

We define the cohesive energy E_{coh} (Johansson, Skriver[37]) as the difference between the energy of an assembly of free atoms in their ground state (see Table 1 of Chap. A) and the energy of the same assembly in the condensed phase (the solid at 0 °K), (this definition yields a positive number for E_{coh}). It coincides with the enthalpy of sublimation ΔH_s (see Chap. A) (which is usually extrapolated at room temperature).

As we have seen in the previous section, it is a property not only of the solid, but also of the atom. For instance, since the ground state configurations of the atom are not the same in the isolated state and in the solid, it contains a quantity – the promotion energy P – which is the difference in energy between the two configurations. We shall see that other atomic quantities will enter E_{coh} as well.

By this definition, it is clear also that E_{coh} is not the same as another quantity often employed, the bonding energy E_b (see the preceding section). The latter may be defined as the difference between the total energy of the atoms of the condensed phase (hereafter "metal atoms") at the equilibrium distance and the total energy of the *same atoms* at infinite distance. From the definition, E_b is a *negative* quantity, insuring cohesion in the condensed phase. If we take a usual picture of a metal, i.e. ionic cores and "valence" (or "conduction") electrons, and assuming that the ionic cores are "frozen" (i.e. do not change during this ideal expansion process) E_b coincides with the total energy of the valence electrons. Notice that E_b depends on the total volume of the solid, i.e. on the interatomic distance (in the expansion process we are keeping the symmetry of the crystal structure intact).

Let us specify better the model of the actinide metals we are discussing. With[37], we assume:

i. the metal atoms have a ground state configuration:

   $$(\text{radon})\,5f^n\,(s, p, d)^v$$

   which is different from the ground state configuration of the free atoms (see Table 1 of Chap. A).

ii. the v outer electrons are in a conduction band of mixed (s, p, d) character (see conclusions and Fig. 20 of Chap. A). v is the metallic valence of the light actinide metal. Johansson and Skriver[37] assume v = 4 in Th, Pa, U and v = 3 in Np, Pu, Am, a reasonable choice which we have discussed in chapter A and in the last section of this chapter. The occupation numbers of the 5f state are accordingly:

   $$n(\text{Th}) = 0,\ n(\text{Pa}) = 1,\ n(\text{U}) = 2,\ n(\text{Np}) = 4,\ n(\text{Pu}) = 5,\ n(\text{Am}) = 6.$$

   Complete band calculations[38, 39] would yield non-integral numbers for n, though close to these values.

iii. The 5f electrons are itinerant and bonding, and from narrow 5f-bands (eventually hybridized). As discussed however in chapter A, they are bound to be spin-polarized, so that:

iiii. The energy of the 5f electrons consists of their band-energy E_f and of a spin polarization term Δ_{sp}. Δ_{sp} has an "atomic" origin, and tends to localize the itinerant 5f electron within the core. There are other terms of "atomic" origin, as discussed by Brooks[40], namely a spin-orbit coupling and an orbital interaction term.

We come therefore to the following cycle which is valid for light actinide metals up to Pu (the case of Am being discussed separately):

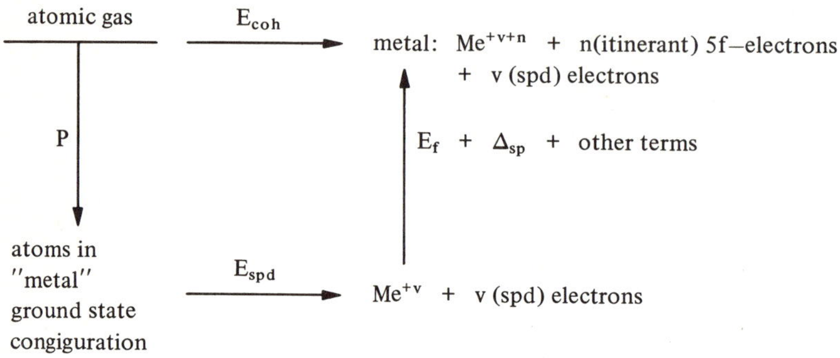

and:

$$E_{coh} = -E_{spd} - E_f - P - \Delta_{sp} - \text{other terms} \tag{3}$$

bonding energy of electrons E_b: $E_{spd} + E_f$ (4)

atomic terms: P, Δ_{sp}, other terms (5)

It is possible to estimate the metallic radii (or volumes) and the bulk moduli of the light actinide metals by the well known thermodynamic equations:

$$-\frac{dE_b}{d_V} = p_{electrons} = 0 \tag{6}$$

$$B = -V\frac{dp}{dV} \tag{7}$$

Equation (7) gives the necessary condition for equilibrium at 0 K. Notice that it may be written in the form of an equation of state for the crystal at 0 K:

$$p_{electrons}\, V = f(V) = 0 \tag{8}$$

The total electronic pressure $p_{electrons}$ can be expressed in terms of partial pressures due to the different electronic waves forming the itinerant electrons cloud (partial waves analysis):

$$p_{electrons} = p_s + p_p + p_d + p_f = p_{spd} + p_f \tag{9}$$

(Actually this is more rigorous than the energy analysis of (4)). Notice that the system looks like a mixture of two electronic fluids, one composed of the (spd) electrons and the other of the 5f electrons.

The considerations leading to (9) or to the equation of state (8) are quite general, in the sense that in all solids we may write similar expressions to describe the bond.

2. *Friedel's Model for E_f and the Evaluation of Atomic Volumes for the Light Actinides*

The contributions E_f to the bonding energy E_b of the itinerant 5f electrons can be evaluated in a simple way following the lines of Friedel's model[41]. Essentially the density of state peak $N_f(E)$ of the 5f band is substituted by a square function (see Fig. 16 in Chap. A):

$$N_f(E) = \frac{14}{W_f} \text{ per atom} \qquad \text{for } C - \frac{1}{2}W_f < E < C + \frac{1}{2}W_f$$

$$N_f(E) = 0 \qquad \text{otherwise} \tag{10}$$

where C is the center of the 5f-band, and W_f its bandwidth. The 5f band contains 14 states per atom, and W_f depends, as a first approximation, only on the interatomic distance (hence on the atomic volumes) determining the degree of f-f overlapping.

Then, the E_f contribution is obtained by the use of the expression:

$$E_f(n) = \int^{\mu_F} (E - C)\, N(E)\, dE \tag{11}$$

the occupation numbers n of the 5f band having been given above, and μ_F being the Fermi level of the metal (dependent on n).

Since the $N_{spd}(E)$ of the (spd) band, to which the 5f band is superimposed, is very small compared with $N_f(E)$, the use of conditions (10) and the evaluation of $(E - C)$ in the rectangular band of the model (Fig. 16 of Chap. A) yield the simple expression:

$$E_f = \frac{1}{2} n \left(1 - \frac{n}{14}\right) W_f \tag{12}$$

In order to apply Eq. (12) for the evaluation of equilibrium metallic volumes V_{at}, it is necessary to know the dependence on V of W_f in (12) and E_{spd} in (4). The LMTO-ASA theory yields the R-dependence: $W_f \sim R^{-(2l+1)}$. However, Johansson and Skriver chose in[37]:

$$W_f = W_f^0(R_0/R)^{-6} = W_g^0(V_0/V)^{-2} \tag{13}$$

where $V_0 = R_0^3$ (R_0 is the atomic sphere or Wigner-Seitz sphere radius – see Chap. A).

E_{spd} is the energy of the electrons in a f-non-bonding solid; e.g., thorium or the lanthanide or the heavy actinide metals. Hence, an expression valid for these less complicated cases may be chosen. Johansson and Skriver chose to integrate Murnaghan's[42] first-order equation of state:

$$P_{spd} = \frac{B^0_{spd}}{B'_{spd}} \left[\left(\frac{V_{spd}}{V} \right)^{B'_{spd}} - 1 \right] \tag{14}$$

where B^0_{spd} is the bulk-modulus at zero pressure and B'_{spd} its derivative with respect to pressure, and V_{spd} is the equilibrium volume if the 5f are non-bonding. The three quantities are then chosen from Th metal for the tetravalent case, from heavy actinides or lanthanides for the trivalent case (in particular $B'_{spd} = 5$). (Notice that Eq. (14), when integrated gives an R^{-2} dependence for E_{spd}, as usually assumed for conduction electrons in a broad band[43].)

With (12) and (14), an equation of state can be written:

$$p/B^0_{spd} = \frac{1}{5} \left[\left(\frac{V_{spd}}{V} \right)^5 - 1 \right] - n\left(1 - \frac{n}{14}\right) W^0_f \left(\frac{V_{spd}}{V} \right)^3 \left(\frac{V_0}{V_{spd}} \right)^2 \frac{1}{B_{spd} V_{spd}} \tag{15}$$

which, set to zero, gives the equilibrium volumes (see Fig. 20 of Chap. A) for the light actinides.

Figure 6 shows the results of this simple model and compares it with the experimental values and the results of full band calculations. Notice that:

i. The results are fairly good for the light actinides, yielding the parabolic trend with respect to Z which has already discussed in Chap. A[4].

   This parabolic trend can be surmised from Eq. (15), where the occupation numbers n are proportional to Z (the parabola should have a minimum at Cm, i.e. or the half-filling of the 5f shell). The same parabolic trend exists in d-transition metals, and is explained in Friedel's model in a similar way. This fact had seemed to early theorists (see, in Chap. A, the discussion of Zachariasen's model) to suggest that the actinides were 6d-transition metals. In reality, it means that the light actinides are 5f-transition metals, with the 5f wavefunctions playing the role of d-wavefunctions.

ii. A discrepancy is encountered already for Np and Pu. This discrepancy (which is however of the same order as those encountered in d-transition metals) is due to the atomic effects, which are totally neglected in the model, and which tend to localize 5f's. (In particular, Brooks[40] has shown, that for Np and Pu, spin-orbit coupling plays an overwhelming role. Introducing this effect gives a somewhat higher equilibrium volume for these two elements.)

iii. The discrepancy becomes important at americium which lies completely on the trivalent, 5f non-bonding curve of Fig. 6. Heavier actinides do not follow any longer the parabolic trend yielded by the model.

   This is the Mott-like transition in the actinide metals series, introduced by Johansson and already discussed in Chap. A.

---

4 Notice that in the whole treatment, as presented here, there is the simplifying implicit assumption that W^0_f is constant throughout the actinide series. The same assumption, when used for d-metals, could not explain that the V_{at} vs. Z curve departed from a parabola (to a curve showing skewer behaviour) from the minimum at the half filling on. In fact, W^0_f narrows with increasing occupation of the d shell. In light actinides, however, it is shown[37] that the constancy of W_f is a good hypothesis, as can be seen from band calculations' values for this quantity

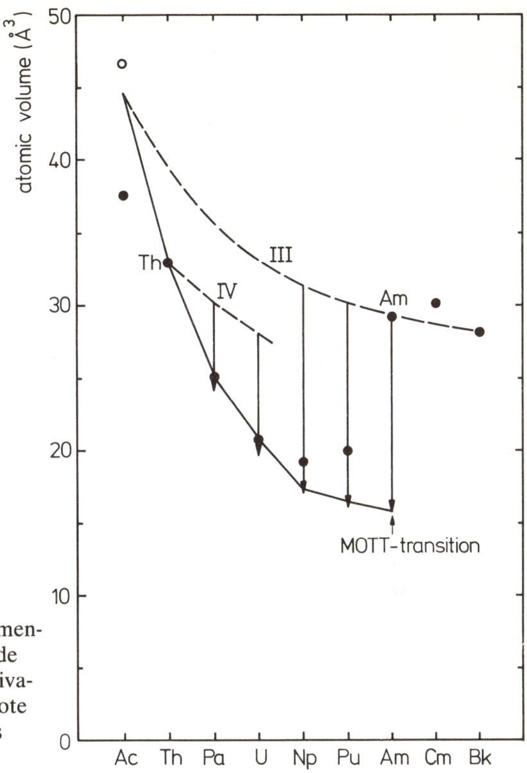

Fig. 6. Calculated[38] (*full lines*) and experimental values for the atomic volumes for actinide metals. The labels III and IV refer to the trivalent and tetravalent states. The *arrows* denote the estimation from the simplified Friedel's model. The figure is from[37]

A well known qualitative argument, often employed in discussions of the actinide series, takes the departure of metal radii from a simple actinide-contraction curve as a measure of 5 f-itinerant bonding (see Figs. 3 and 5 of Chap. A). The model presented here justifies and gives a quantitative basis to this argument.

3. The Cohesive Energy and the Bulk Modulus of Light Actinides in Friedel's Model

Equation (4) together with Eq. (15), integrated for volume, can now be used for the calculation of the cohesive energy, within this model, of the light actinides. Differentiation with respect to volume of Eq. (15) yields the bulk moduli.

For the evaluation of these quantities, however, the relevant atomic terms, and especially Δ_{sp}, must be added. We must therefore consider this quantitaty.

The promotion energies P in Eq. (3) refer usually to Hartree-Fock solutions of the atomic Schrödinger equations, and refer then to the so called "grand barycentre" of the configurations. The atomic maximisation of spin in unfilled shells, leading to Hund's rules, and already qualitatively discussed in Chap. A, is a correlation which is missing in the Hartree-Fock theory. Jørgensen[44] has discussed this stabilising atomic energy term within his spin-pairing theory. Nugent et al.[45, 46], based on the work by Jørgensen, have

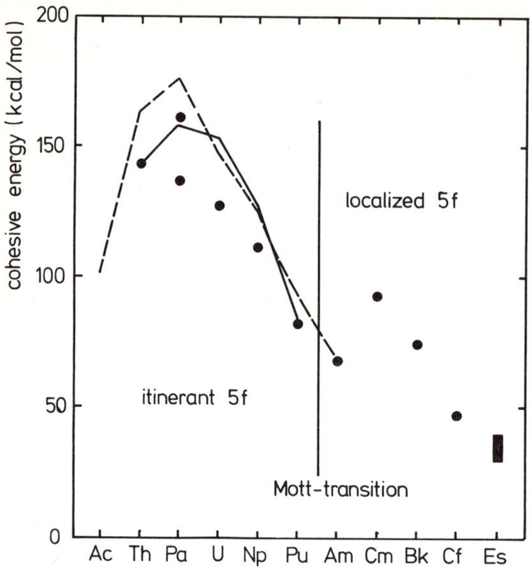

Fig. 7. Calculated[40], and estimated (from Friedel's model[37]) cohesive energies of actinide metals compared with experimental values ([37] and [47]). The figure is from[37]

extracted the energy difference between the atomic spin-stabilized $5f^n$ state and its grand barycentre from spectroscopic data. This term may be taken as Δ_{sp} in Eq. (3).

Conversely, Brooks and Johansson[7] have calculated the spin-polarization term Δ_{sp} from a spin-polarized local-density approximation in atoms. These values can also be used in Eq. (3). The two evaluations, though not identical, do not differ greatly.

In Fig. 7 the results of the model for the cohesive energy are given, and compared with the experimental values and with the results of band calculations. The agreement is satisfactory (at least of the same order as for similar models for d-transition metals). For americium, the simple model yields too low a value, and one needs spin-polarized full band calculations (dashed curve in Fig. 7) to have agreement with the experimental value.

The evaluation of bulk moduli from the model agrees with the trend of the experimental values, with one major difficulty (see Fig. 15 of Chap. A): the model would predict a maximum for the bulk moduli at U; a recent experimental determination of this quantity by Benedict et al.[15] finds a very high bulk modulus for Pa and the bulk modulus vs. Z curve peaks at this element.

4. *The Case of Americium: the Mott-Hubbard Transition and the Effects of Pressure*

In Fig. 6, it is seen that the metallic radius of americium metal is not obtained when applying the simplified Friedel-type model which, on the contrary, explains well the metallic radii of Pa, U, and, to some extent, of Np and Pu. Between Pu and Am we have indicated the Mott-like transition proposed by Johansson[5], which was discussed in Chapt. A. This transition from itinerant to localized behaviour occurs because of the stabilizing effect in the atom of spin polarization: the bonding effect of itineracy is overwhelmed by the potential energy gain due to the alignment of spins within the atom.

From a thermodynamic viewpoint, we may imagine that, in an actinide metal, the model of the solid in which completely itinerant and bonding 5 f electrons exist and that in which the same electrons are localized, constitute the descriptions of two thermodynamic phases. The 5 f-itinerant and the 5 f-localized phases may therefore have different crystal properties: a different metallic volume, a different crystal structure. The system will choose that phase which, at a particular T and p (since we are dealing with metals, the system will have only one component) has the lower Gibbs free-energy. A phase transition will occur then the "fugacity" in the two possible phases is equal: e.g. the pressure. To treat the transition, therefore, the free energies and the pressures of the two phases have to be compared. We recall that:

Gibbs free-energy $$G = E + PV - TS = H - TS \qquad (16)$$

$$p = -\frac{\partial G}{\partial V} \qquad (16\,a)$$

where H is the enthalpy of the system, S its entropy, E is its internal energy, that we have learnt to calculate, in the itinerant case, in the preceding subchapter within Friedel's model (in what follows, at T = 0, the term TS vanishes, so that enthalpies, rather than Gibbs free energies, are compared).

We have divided both E and more rigorously, p in the preceeding subchapter in their partial components (for the itinerant case): p_{spd} and p_f (see Eq. (9)). In the case of spin-polarization of the 5 f-electrons, it is necessary therefore, to have a new way of writing p_f: it is this "fugacity" which will determine the existence of one or the other phase.

All the pressures are strongly dependent on the interactinide distance, i.e. metallic volume dependent. Therefore, we shall have one or the other phase if we change one or the other of the thermodynamic variables p and T. If we confine ourselves to T = 0, we may induce the transition by compressing the solid: hence the great importance of the pressure studies reported previously in this chapter.

It is perhaps useful to distinguish the two ways in which the concept of a Mott-Hubbard transition is introduced in the discussion of actinide metals.

i. Mott-transition *within* the actinide series at T = 0, p = 0;
ii. Mott-transition occuring in one metal system *when compressed* (T = 0, p = p*).

We have seen (preceding subsections and Chap. A) that, in both cases, there is a change in two properties: the metallic volume (interactinide distance) *and* the crystal structure (low-symmetry for the itinerant case and compact high symmetry for the localized case).

In Fig. 6, it is seen that the americium metallic volume at T = 0, p = 0, is well reproduced by spin-polarized band calculations[37)]. The same band-calculations predict the pressure p* in which the transition from localized to itinerant behaviour occurs for americium. Here, we want to present the simplified Friedel-type model (Skriver, Anderson and Johansson[37)]) by which the spin-polarized americium system is described and both transitions, i and ii, predicted. One word of caution: the model, as well as the band calculations, do not account for the change in structure accompanying the transition, and which is an important fact in actinide metals, which will be discussed in the following subsection. What changes at the transition is the metallic volume, and, as was seen in Fig. 2, the predicted "volume collapse" is much larger than that experimentally found. The pressure p* is however, very well predicted. The full validity of the model presented here for the transition under compression is, therefore still somewhat a matter of controversy.

We shall now proceed to derive a suitable expression for p_f in the spin-polarized case.

In Chap. A, we have seen that, in the Stoner model, (ferromagnetic) spin-polarization of electrons originates two electron states E_+ and E_- from each electron state E of a non-spin-polarized electron band, the difference between the two being $(E_+ - E_-) = Im$, where I is the Stoner parameter and $m = n_+ - n_-$ is the magnetization density.

We may therefore assume that the 5f non-spin-polarized band splits into two sub-bands because of spin-polarization. Approximation of the two sub-bands, according to Friedel's model, by two rectangular ones, having densities of state $N_+(E) = N_-(E) = 7/W_f$, and occupation numbers n_+ and n_-, leads to the following expression for the total p_{spd} pressure:

$$p_f^{(pol)} = -\frac{2l+1}{2}\left[n_+\left(1-\frac{n_+}{7}\right)+n_-\left(1-\frac{n_-}{7}\right)\right]\frac{W_f}{3V} \tag{17}$$

in which we used the $R_0^{-(2l+1)}$ dependence for W_f. (Equation (17) is obtained, as in the preceding chapter, by differentiating an $E_f^{(pol)}$ energy obtained with conditions such as in Eq. (10) for the two sub-bands.)

The integration within the sub-bands is performed to a *common* Fermi level for the two spin-populations, which is the usual Stoner thermodynamic condition (an neglecting the N_{spd} density of state). The occupation numbers $n^{\pm}$ are therefore:

$$n_{\pm} = \frac{1}{2}(n \pm m) \tag{18}$$

By substitution in (17), one obtains:

$$p_f^{(pol)} = -\frac{2l+1}{2}\left[n\left(1-\frac{n}{14}\right)-\frac{m^2}{14}\right]\frac{W_f}{3V} \tag{19}$$

in which the first term in brackets is the usual 5f itinerant non-spin-polarized bonding partial pressure that can be obtained by differentiating with respect to volume Eq. (13).

Let us analyse Eq. (19) in general, since it may be applied for *all* actinide metals:

i. if no or little spin-polarization is present, then $m = 0$, and $p_f^{(pol)} = p_f$ as in Eq. (13);
ii. if spin-polarization is complete, then $m = n$ $(n_- = 0)$. Equation (18) can be re-written as:

$$p_f^{(pol)} = \frac{2l+1}{2}n\left(1-2\frac{n}{14}\right)\frac{W_f}{3V} \tag{20}$$

One sees that, at $n = 7$ (i.e. at Cm, for a half-filling of the 5f shell) $p_f^{(pol)} = 0$. If one considers the total electronic pressure $p_{electronic}$ of Eq. (9), with p_f substituted by $p_f^{(pol)}$, at the half-filling of the electronic series this is given totally by p_{spd} as in a lanthanide.

In americium, the occupation number is n = 6 (actually, n = 6.4 from band calculations[38)]). If we assume full spin-polarization for this metal, then the 5f bonding partial pressure is very small, and we may be reconducted to the lanthanide case for this metal also.

The problem is whether this happens or not. In order to decide, we have to know the volume dependence of m, and substitute to p_f the full Eq. (19) in (10) and (16), setting then p = 0 as in Eq. (7). The volume dependence of m can be obtained by band calculations, and is shown, for Am, in[37)].

When the procedure illustrated above is followed, Eq. (7) has two *stable* solutions for the equilibrium volume V:

α) one very near to the itinerant bonding 5f solution;

β) one very near to the localized, fully spin-polarized, non-bonding 5f solution, which would be obtained by disregarding completely any 5f bonding contribution in the solid.

The situation is well illustrated qualitatively by Fig. 8. The problem, now, whether to choose α or β as the metallic volume for americium metal at T = 0 and p = 0 is brought to the evaluation of the enthalpies of the two phases in this thermodynamic condition. This is illustrated in Fig. 8 (from[37)]), where the two enthalpies are evaluated.

From Fig. 8, one notices that the localized enthalpy is lower by about 0.9 eV than the itinerant one, thus classifying americium metal among the lanthanide-like, heavier actinides. The picture is consistent with the results of magnetic measurements, which explain magnetism in americium metal as derived from a $5f^6$, J ≅ 0 atomic ground state.

In Fig. 2, it is seen also that, when increasing the pressure on the metal, the two enthalpies meet at p* ≅ 140 kbar (another estimate[37)], gives p* ≅ 100 kbars). At that pressure p* a transition from the full spin-polarized to a no spin, itinerant phase is predicted to take place. Figure 2 shows that at p^*_{exp} = 100 kbar, a transition from ccp to a monoclinic structure takes place, the relative change in volume being however much smaller than that predicted by the theory.

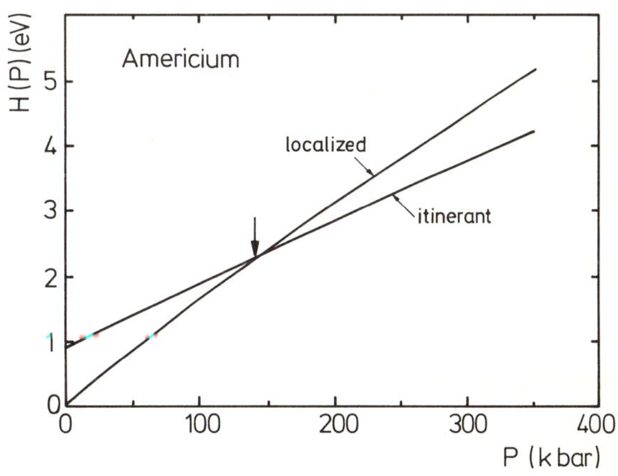

Fig. 8. Enthalpy vs. pressure curves for localized and itinerant 5f electrons in americium metal (from[37)])

5. Transformations of Crystal Structures Under Pressure

The monoclinic structure encountered in Am at high pressures is a low symmetry crystallographic form: this is a clear indication that the 5f orbitals have become bonding in the solid. The formation in compression of lower symmetry crystallographic modifications occurs also in the heavier actinides, as we may see from Figs. 1 and 2.

For the lanthanide-like heavy actinides, this trend is not surprising, since the same trend is followed in lanthanides as well. From the data available to date we have, on compression:

lanthanides: hcp → samarium type structure → dhcp → ccp → α-U (for Pr[25, 26)])

heavy actinides: dhcp → ccp → monoclinic or α-U
(starting from Am)

We may, thus, formulate a qualitative rule (which looks like a modification of the Le-Chatelier principle):
"the effect of *pressure* in f-bonding systems is to transform crystal structures to the *low-temperature* (low-symmetry) phases of light actinides".

If one accepts that the loss of symmetry is originated by the high directionality of bonding and hybridized f-orbitals, as we have discussed above and in Chap. A, the rule seems to describe rather well the thermodynamics of the f-bond in its angular aspect in 6f as well as in 5f systems (notice that all model considerations developed in the previous subchapters focused rather on the radial characteristics of the 4f-wavefunction than on angular ones). The fact that the transition pressure to the low-symmetry form increases with increasing Z in the actinide series (Figs. 1 and 2) is also consistent with this explanation, since the 5f wavefunctions contract progressively to the core with Z. Apparently, pressure liberates the 5f "fluid" for bonding purposes[5].

When the itinerant state is formed, a volume collapse $\Delta V/V$ is always encountered, as predicted by the theory of the preceding sections. In one of the lanthanides, cerium, this volume collapse is particularly accentuated for its *isostructural* transition from the γ to the α form, possibly associated with a change in metallic valence from three to four (both oxidation numbers are stable in cerium chemistry) (see Fig. 1 of Chap. A).

The much weaker collapse encountered experimentally in the *non-isostructural* transition of americium might perhaps be associated with the fact that the 5f are already (slightly) bonding in this metal in its low pressure modification. As seen in Fig. 2, $\Delta V/V$ increases the more the 5f are localized.

V. Crystal Structures and Thermodynamic Properties of Simple Actinide Binary Compounds

The only compounds the structure of which will be discussed in some detail are those of the NaCl(B 1) and of the CaF_2(C 1) type. They are relatively easily accessible to theoreti-

5 See the "Note Added in Proof" at the end of the Chapter

Table 4. AnX compounds with the NaCl-type structure and their lattice parameters (in pm)

	borides (III A) and carbides (IV A)		pnictides (V A)					chalcogenides (VI A)			
	B	C	N	P	As	Sb	Bi	O	S	Se	Te
Th		533.8	520	581.8	597.2	631.8		(530.2)	568.2	587.5	
Pa		506.74	504.7		575.60			496.1			
U		495.98	488.99	558.9	577.9	619.1	636.4	(492)	548.4	575.0	616.3
Np		500.4	489.7	561.48	583.89	624.85	637.32	(501)	553.2	580.4	619.8
Pu	492	497.37	490.5	566.3	586.10	623.96	635	(495.9)	553.6	577	618.3
Am			499.5	571.1	587.42	623.97	636.7	(505)	559.2	582.1	617.1
Cm			504.1	574.3	590.1	624.8	623.8		557.5	579.8	615.0
Bk			495	566.9	583.3	618.8					
Cf					580.9	616.5					

cal calculation of properties because their lattices are highly symmetric (cubic) and because there are only two or three atoms in the formula unit.

Both structure types are face-centered cubic, space group Fm3m (Nr. 225) and have four formula units in the unit cell (Z = 4).

1. The NaCl (B1) Structure Type

The actinide compounds crystallizing in this structure type are listed in Table 4.

There are only very few actinide 1 : 1 (AnX) compounds which have structures other than the B 1 structure type. A review based on the 430 1 : 1 (MeX) compounds covered by Wyckoff[8] and on the 32 additional AnX compounds described more recently is given in Table 5.

Practically all 1 : 1 compounds of actinides and lanthanides have the B 1 structure, whereas only 31% of the other MeX compounds have that structure. Another comparison shows that, of the B 1 type compounds, 72% have an actinide or lanthanide as the metal component, while lanthanides and actinides constitute only 34% of all the metals of the Periodic Table.

This hints to a particularity of lanthanides and actinides that favors the B 1 structure as soon as they combine to another element in a 1 : 1 ratio. This trend seems to be less pronounced, among the other metals, as these have a much larger variety of structure types for their 1 : 1 compounds.

In principle, the B 1 structure should be strongly ionic in character. But in the actinide compounds having this structure, the ionic character is more or less weakened as their anions are less electronegative than the halogens which are the anions of the typical representatives of this structure type.

It was inferred from simple geometrical considerations (condition for contact between neighbouring X anions) that this structure could only be formed if the ratio of the cation radius to the anion radius is between 0.41 and 0.73. It is not easy to check this condition for the AnX compounds because there is some difficulty to define the radius to be used for this comparison. If the tabulated ionic radii are used, many of the AnX compounds are in the range mentioned, but some of them are outside, e.g. the monocarbides.

Table 5. Occurrence of B 1 type compounds in the periodic table. Me: metal

Me	number of MeX compounds		
	all structures, = 100%	B 1 structure type	
		absolute number	in %
Y, Ln (incomplete 4f shell)	89	89	100
Ac, An (incomplete 5f shell)	68	65	96
all other metals	314	98	31
Hf to Ra (4f shell completed)	53	11	21

Table 6. Range of hypostoichiometry in AnC_{1-x} systems

AnC_{1-x}	$1 - x$
ThC_{1-x}	0.67–1
PaC_{1-x}	
UC_{1-x}	0.985–1
NpC_{1-x}	0.82–0.96
PuC_{1-x}	0.78–0.89
AmC_{1-x}	

Most of the actinide monoxides described as having the B 1 structure have been put between brackets in Table 4 because they are nowadays thought of to be stable only if stabilized by additions of carbon or nitrogen.

2. *Non-Stoichiometry in the NaCl-Structure Compounds*

A system particularly studied from a thermodynamic point of view is the monocarbide (AnC_{1-x}) system. A range of stability for carbon deficient compositions in the monocarbides is present in many metal-carbon phase diagrams. Table 6 shows the composition range at room temperature for actinide monocarbides. The non-stoichiometry range is very limited for uranium monocarbide; for neptunium and plutonium monocarbides, the stoichiometric $AmC_{1.00}$ composition is not stable.

Figures 9 and 10 present the Th-C and PuC phase diagrams.

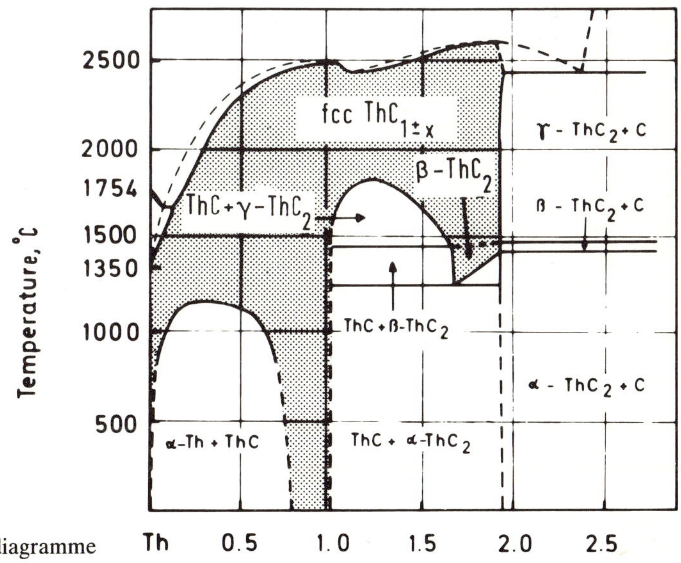

Fig. 9. The Th-C phase diagramme

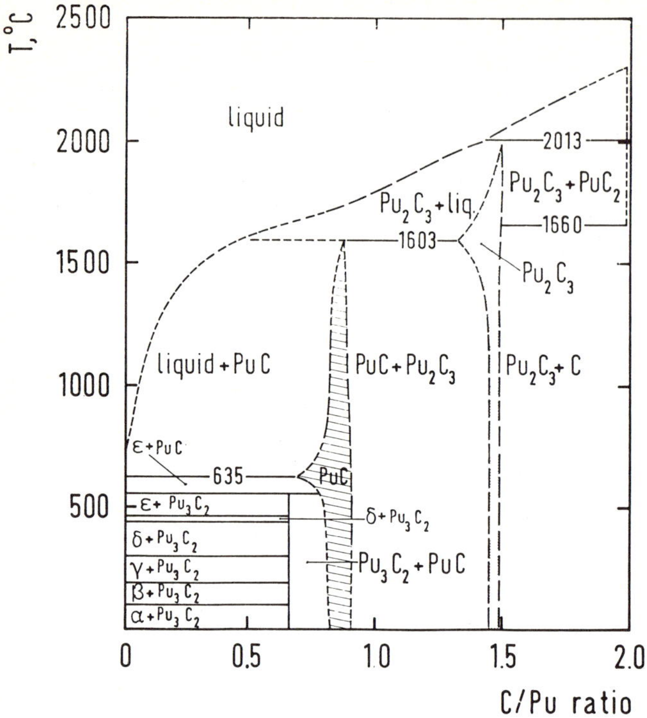

Fig. 10. The Pu-C phase diagramme

3. The CaF_2 (C1) Structure Type

The fluorite (CaF_2) type structure is a structure often encountered in ionic solids. It follows the same general principles as described above for the ionic AnX compounds: the packaging ensures that the chief contacts are between atoms of opposite sign and that each atom is surrounded by the maximum number of atoms of opposite sign. The cations in the fluorite structure are surrounded by eight equidistant anions at the corners of a cube. Inversely, each anion has around it four cations at the corners of a tetrahedron. As a rule, this structure is only formed if the ratio radius cation/radius anion is greater or equal to 0.73.

The An atom and the X atoms are in the special positions 4a and 8c, respectively, of space group Fm3m.

The largest, and by their stability in many respects very important, group of actinide compounds with the CaF_2 structure are the dioxides. These, and the remaining fluorite type compounds, are shown in Table 7, which gives the lattice parameters for the stoichiometric $AnX_{2.00}$ composition. Hyperstoichiometric (AnX_{2+x}) compositions have lower lattice parameters and hypostoichiometric (AnX_{2-x}) compositions have higher lattice parameters.

For PuOF, which in Table 7 is described with the CaF_2 type structure, a second structure, which is a slight distortion of the fluorite structure, was also reported: tetragonal, P4/nmm with a = 405, c = 572 pm, c/a = 1.412.

Table 7. Lattice parameters (in pm) of the fluorite-type AnX_2 compounds

	Dihydrides AnH_{2-x}	Dinitrides AnN_{2-x}	Dioxides $AnO_{2\pm x}$	Oxyfluorides AnOF
Ac	567.0			594.3
Th	548.9		559.97	
Pa			550.5	
U		531	547.1	
Np			543.41	
Pu	539.5		539.5	571
Am			537.6	
Cm			537.2	

In the CaF_2 structure, a typical feature is present: a large octahedral hole (sited at (1/2, 1/2, 1/2)) which corresponds to a favorable potential position in which other (interstitial) anions may be accommodated. It is the site in which oxygen excess is thought to be located in hyperstoichiometric oxides.

4. *Non-Stoichiometry in CaF_2-Structure Binary Compounds*

The most studied non-stoichiometric system in actinide CaF_2-structured compounds is the An-O system: all actinide dioxides (with the exception of ThO_2) present large departures from stoichiometry. Since uranium and plutonium dioxides (and their solid solutions) are employed as fuels in nuclear reactors, a very large effort has been dedicated to the study of their physical and physico-chemical properties. All these properties are affected by the oxygen composition of the compound.

Non-stoichiometric actinide dioxide can be classified in two groups:

a) oxygen-excess oxides MO_{2+x} (hyperstoichiometric)
b) oxygen-deficient oxides MO_{2-x} (hypostoichiometric)

a. Oxygen-Excess Dioxides: AnO_{2+x}

To this group belong PaO_{2+x} and UO_{2+x}, the latter one by far the most studied. Figure 11 shows the U-O phase diagram.

The fluorite-structure hyperstoichiometric range spans for high temperatures to compositions near to O/M = 2.25, which is the composition of the next oxide U_4O_9.

Between the two possible defects which may be responsible for hyperstoichiometry (i.e. uranium interstitials or oxygen vacancies) the latter is well evidenced by measurements of lattice parameter and density[48] and neutron diffraction[49, 50]. Oxygen interstitials order in U_4O_9 to provide a crystal structure which can be derived from the fluorite structure of UO_{2+x}.

The negative charge (with respect to the lattice) of oxygen interstitials (O_i^{-2}) is balanced by the oxidation of U^{+4} to either U^{+5} or (more unlikely) U^{+6}.

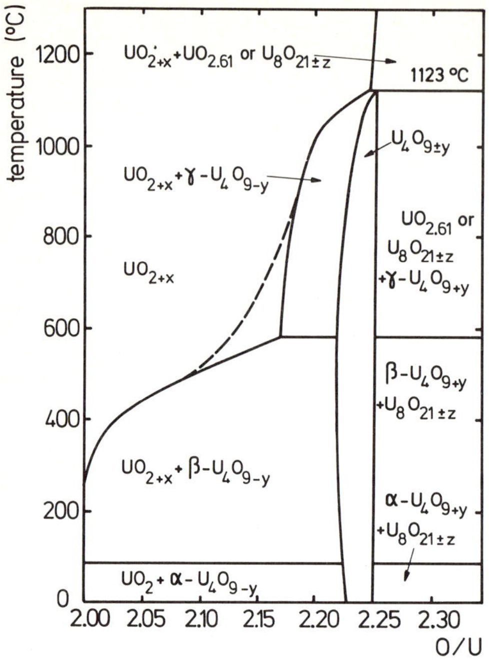

Fig. 11. The U-O phase diagramme

b. Oxygen-Deficient Dioxides: AnO_{2-x}

Actinide oxygen deficient dioxides are PuO_{2-x}, AmO_{2-x}, CmO_{2-x}, BkO_{2-x} and CfO_{2-x}. Fluorite structured oxygen deficient dioxides are encountered also among lanthanide oxygen systems, namely PrO_{2-x}, CeO_{2-x}, TbO_{2-x}.

It is usually assumed that the dominant defect species responsible for non-stoichiometry is constituted by oxygen vacancies (rather than metal interstitials). This assumption can be justified on the basis of X-ray and neutron diffraction and density measurements[51–53)].

Electroneutrality in the lattice is ensured by reduction of An^{+4} to An^{+3}, since for all the actinides in the hypostoichiometric oxide class, the trivalent state is stable (see Fig. 1 in Chap. A).

We shall underline a very interesting problem which emerges when comparing the systems Pu-O and Am-O with the following AnO systems in the series.

When inspecting the phase diagrams of the lanthanide oxides PrO_{2-x} and CeO_{2-x}, it is seen that the very wide non-stoichiometric range present at high temperature for the fluorite-structured MO_{2-x} gives rise at low temperatures to a series of phases, having a composition which can be represented by a general formula Ln_nO_{2n-2}[54)]. The crystal structures of these phases are sometimes very complicated, but can be understood as being distorted fluorite structures, in which ordering of oxygen vacancies, present at higher temperature, has been achieved. There is gradual transformation from the fluorite structure of the $MO_{2.00}$ stoichiometric composition to the structure of the sesquioxide M_2O_3. Quantitatively, it can be stated that essentially ionic MO_{2-x} fluorite structures tend to form intermediate phases of a general formula M_nO_{2n-2} at low temperature.

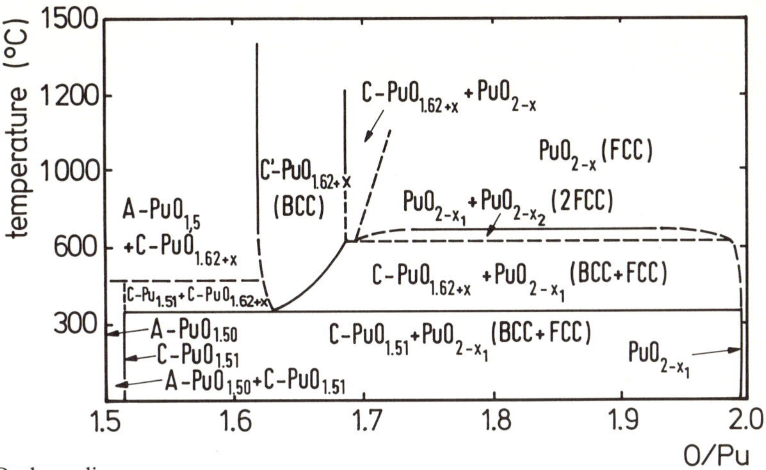

Fig. 12. The Pu-O phase diagramme

In the actinide oxygen deficient dioxides, the An_nO_{2n-2} subphases have not as yet been evidenced in PuO_{2-x} (see Fig. 12) and AmO_{2-x}. However, phases of composition $AnO_{1.82}$ and $AnO_{1.714}$ ($n = 11$ and 7 respectively) have on the contrary been identified in the Cm-O and Cf-O systems[55, 56)], i.e. in typical lanthanide like actinides.

The absence of the An_nO_{2n-2} subphases in the Pu-O and Am-O systems may therefore qualitatively be attributed (see Blank[51)]) to a somewhat different type of bond in these two oxides, e.g. a greater covalency, influencing the type of ordering process of defects and stabilizing the fluorite structure.

The Pu^{+4} and Am^{+4} ions have a $5f^4$ and $5f^5$ well localized f-shell: this is borne out clearly by the fact that their magnetic susceptibility is well explained in the atomic picture (see Chap. D). Nevertheless, this f-configuration is non-localized in metals, therefore, it might well be assumed to form some amount of covalent bonding by hybridization with the 2p electrons of the oxygen ion (see later, and Chap. E).

VI. The Bond in Binary Compounds

A formalism similar to that presented for actinide metals has been developed for the ground state properties of binary compounds by Andersen et al.[57, 58)], leading to a general form of equation of state (see Chap. F). However, this analysis of bonding contributions must draw from detailed results of band calculations more heavily than for the metals' case (where the explanation of the qualitative behaviour of ground state properties vs. atomic number needed only the hypothesis of a constant 5f-bandwidth and its volume dependence as predicted by the general theory). In fact, the bond is more complicated:

i. the electronegativity difference between the actinide and the non-actinide element induces an ionic part in the bonding;

ii. hybridization takes place not only with the other outer actinide electronic orbitals (especially the 6d) but also with the outer electronic orbitals of the non-actinide element, inducing a covalent part in the bonding.

Hence, for compounds, we have a composition of metallic, covalent, and ionic bond. A formula for the equation of state for the general case of metals and compounds is:

$$3pV = \sum_T n_T(2l+1)(E_T - C_T) + E_M \tag{21}$$

Here, T = (t, l) labels the orbital l for the component atom t of the solid, C_T is the centre, E_T the centre of gravity of the T band, given by

$$E_T = \frac{1}{n_T} \int^{\mu_F} E\,N_T(E)\,dE \tag{22}$$

where $N_T(E)$ is the density of state of the band T, μ_F is the Fermi level, and n_T the number of electrons occupying it. The factor $(2l + 1)$ arises from the differentiation of the bandwidth $W_T \alpha R_0^{-(2l+1)}$. E_M, finally, is the Madelung long range electrostatic energy between ions, arising from the change transfer between the differently electronegative components of the compound, and expresses the ionic part of the bond (For this simplified model, in (21), as well as in (11), a part has been neglected, arising from the volume dependence of the centers C_T and of exchange and correlation terms).

Equation (21) is perfectly comparable with Eq. (11), apart from the term E_M. Equation (21) can be simplified within a Friedel's model, by assuming square bands, as for Eq. (12) and following.

If no hybridization is considered between the bands, n_T is the number of electrons which can be evaluated from the oxidation numbers of the components of the compound. For instance, in the case of an actinide dioxide AnO_2, suppose An to have the ground state configuration: $f^m(sd)^v$, the An^{+4} ion will have a ground state configuration f^n with $n = m - q$, and the $q + v = 4$ electrons are filling the oxygen 2p band. Then, the corresponding n_T's are: $n = m - q$, $n_{O,p} = 6$. In this case, Eq. (21) describes a combination of *metallic* (the first term) and *ionic* (the second term E_M) parts.

The metallic part would vanish, of course, for empty or filled bands, and in the case of strong localization $((E_T - C_T) = 0)$. To the first term, a spin polarized form may also be given as for metals and a further source of vanishing of this term may be identified when full atomic spin polarization is achieved.

If hybridization is taken into account, n_T can be provided only by band calculations. This is true for metals (where T inicates only the orbital) e.g. when hybridization between the f state and the (s, p, d) band is considered. In the case of cation-anion hybridization (ii.), covalency sets up. The complete charge transfer of the ionic model is reduced; some charge sharing occurs between cation and anion.

1. NaCl-Structure AnX Compounds

The physical properties of these compounds have been already discussed in Chap. A. A further discussion is given in Chap. D and F. A strong 5f metallic bonding is present in these compounds (see e.g. Table 7 of Chap. A). Band calculations, however, evidence a very important covalent f-p contribution to the bond[59, 60]. The Madelung contribution (ionic bond) derives mostly from the transfer of the 7 s and 6 d electrons of the actinide atom ground state configuration into the p-band of the anion.

Equation (21) describes well the bond, as shown by Brooks and Glötzel[59, 60]. In the simplified square band model, the contribution to (21) from the f-f metallic and f-p covalent band can be written as[59, 60]:

$$(3\,pV)_{ff+fp} = (3\,pV)_{ff} + (3\,pV)_{fp} = -7(14 - n_{fp})F(1 - F)W_f/2 - 10\,n_{fp}(C_f - C_p)(1 - F) \tag{23}$$

where $F = n/14$ is the fraction of f-electron states occupied in the unhybridized band. If one compares the first term with Eq. (12), the number of 5f bonding electrons of the actinide has been reduced by a number $F\,n_{fp}$, transferred to the oxygen atom. The number n_{fp}, representing the number of f states introduced by hybridization on the p band and, symmetrically ($n_{fp} = n_{pf}$) the number of p states introduced in the f band, is a quantity which arises from hybridized band calculations, and, in LMTO-ASA formalism, can be written as[61]:

$$n_{fp} = \frac{2\,M_{fp}^2 \Delta_f \Delta_p}{(C_f - C_p)^2} \tag{24}$$

where Δ_f and Δ_p are the bandwiths of the unhybridized bands, which are changed by hybridization, and M_{fp}^2 is a matrix element depending only on the crystal structure.

Equation (22) governs, as Eq. (12) for metals, the atomic volume vs. Z curves for these compounds (see Fig. 5 of Chap. A). Although the f-f metallic part has a parabolic behaviour vs. Z, the covalent part introduces a skewness in the parabola. Also the position of the minimum is drawn from the centre of the series towards its beginning. The skewness and the distance of the minimum from the centre increase as the strength of the covalent term increases.

For the AnBi and AnSb systems, furthermore (see Fig. 5 of Chap. A), the metallic bond might be also different from a prevalently 5 f bond from the actinide as in Eq. (22), since the elemental Bi, and Sb solids have themselves a metallic character.

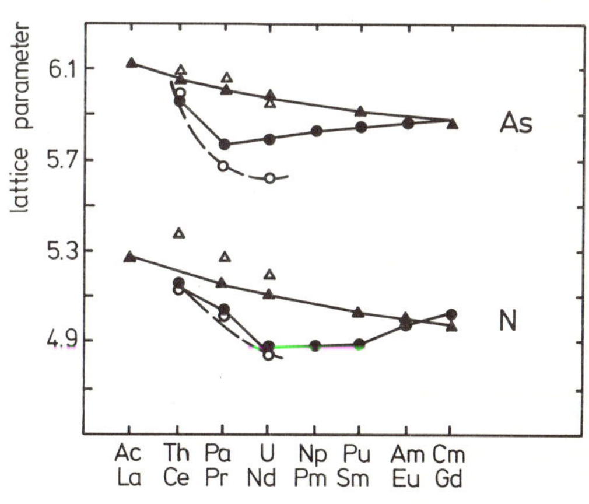

Fig. 13. Lattice parameters of the actinide and rare earth nitrides and arsenides. *Full circles* and *triangles* represent experimental values; *open circles* and *triangles*, the theoretical values, for actinides and rare earths (from[61])

Figure 13 compares the results of the calculations with Eq. (22), for the AnN and AnAs systems with the experimental lattice parameters and with the corresponding lanthanide systems. This latter comparison evidences, also, in these compounds, that the presence of a departure from a monotonous, almost linear curve, as found for lanthanides, is a clear sign of metallic 5f bonding.

Spin-polarization sets up in the second part of the series, as in the case of metals, and, correspondingly, Eq. (22) should be modified with the use of spin-polarized terms. This explains the onset of an actinide contraction trend in heavier actinide NaCl compounds, as shown in Fig. 5 of Chap. A for the AnN system.

2. CaF_2-Structure Actinide Dioxides

The analysis of the dioxide bond has been performed with band calculations by Brooks and Kelly[62–64] for UO_2. The method employed is essentially the same as explained above. Only, as suggested by the experimental (especially magnetic) evidence for strong localization of the $5f^n$ in the An^{+4} ion, band calculations cannot take into account conveniently these orbitals. Cluster molecular orbital calculations[65–69] have been also performed on actinide dioxides. They usually show a strong f-p hybridization, which increases first with increasing Z across the series, becoming strongest for NpO_2 PuO_2, CmO_2, and then decreasing when the 5f states are pushed to very high binding energies, becoming real atomic states of the actinides.

The band calculations of Brooks et al. have, however, evidenced a strong (uranium) 6d-(oxygen) 2p hybridization (Table 8), leading to a strong covalent part in the bond of these systems. Charge transfer is strongly reduced from that of a pure ionic model. In[64], Brooks and Kelly have shown (from calculating the oxygen charge density in the solid) that the oxygen ion in UO_2, usually assumed to be O^{-2}, is likely to be a O^{-1} ion: a fact

Table 8. A comparison of the electronic structures of CaF_2 and UO_2, after LMTO-ASA band calculations (energies in Ryd) (from[70])

	CaF_2	UO_2
Cation charge	1.86	3.4
Theoretical lattice parameter	3% Experiments	1/2% Experiments
No. of anion electrons		
s(core)	2.0	2.0
s(conduction)	0.0	0.0
p	5.9	5.7
No. of cation electrons		
s(core)	1.98	2.0
s(conduction)	0.0	0.13
p	5.87	5.61
d	0.29	0.82
d-bonding contribution to 3 pv	− 0.68	− 1.53
Madelung contribution to 3 pv	− 2.01	− 6.50

already evidenced by the analysis of the structural and mechanical properties (Blank[54]) and optical properties[71,72] of these oxides. The charge density distribution of the uranium ion has also been calculated by Brooks and Kelly and is shown in[64]: although it is very similar to that of a $U^{+4}(5f^2)$ ion, it presents a higher value at higher radial distances. This is also in the direction of a covalent part in the oxide bond.

In the same paper[64], Brooks and Kelly have considered the possible contributions of 5f orbitals to the bonding of UO_2. While the hypothesis of an itinerant picture for these orbitals in the solids leads to a 35% higher atomic volume than the observed one, the assumption of a 5f Mott-Hubbard spin-localized band, comprising seven states per atom (instead of 14) (see Chap. A) yields the correct value for this quantity. A certain amount of f-p hybridization is found as a weak and diffuse percentage of 5f character in the predominantly 2p-6d valence band of this oxide.

3. *Large Non-Stoichiometry in Oxides and its Relation to Bond*

Non-stoichiometry is a very important property of actinide dioxides. Small departures from stoichiometric compositions, are due to point-defects in anion sublattice (vacancies for AnO_{2-x} and interstitials for AnO_{2+x}[73]). A lattice defect is a point perturbation of the periodicity of the perfect solid and, in an ionic picture, it constitutes a point charge with respect to the lattice, since it is a point of accumulation of electrons or electron holes. This point charge must be compensated, in order to preserve electroneutrality of the total lattice. Actinide ions having usually two or more oxidation states within a narrow range of stability, the neutralization of the point charges is achieved through a Redox process, i.e. oxidation or reduction of the cation. This is in fact the main reason for the existence of non-stoichiometry. In this respect, actinide compounds are similar to transition metals oxides and to some lanthanide dioxides.

For a very low departure from the stoichiometric composition, there is in principle no reason why point lattice defects and the compensating cationic charges (constituting also a defect in the cationic lattice) should be grouped together in special regions of the lattice or occupy special lattice sites: usually, they are considered randomly distributed in the lattice.

However, if the concentration of these two defect species becomes high (large deviations from stoichiometry), then both the point-defect hypothesis and that of a random distribution become questionable. A lattice defect causes a redistribution of charge density in certain regions, affecting strongly the surrounding lattice. Relaxation of the lattice may be achieved through small displacements of the ions surrounding the imperfection, causing a polarization of the lattice which may help trapping other defects, thus giving use to clustering processes, which contradict the random hypothesis. In covalent solids[74], the charge density redistribution in the zone perturbed by one or more defects can be described sometimes, in a good approximation, as the formation of a new molecular species in the solid matrix. Since, usually, there are difficulties in defining well this molecule (the size and chemical composition of which depends upon the size of the perturbed lattice region) the term "molecularity" is often employed.

The thermodynamic functions of non-stoichiometric solids at very high deviations from stoichiometry are strongly affected by defect clusters and molecularities. The detailed theoretical description of the interactions between defects and the lattice as well as of defects between each other should be known if the thermodynamic functions are to be interpreted. This is usually far from being an easy task. Then, modeling of the defect structure is used for their interpretation[73].

a. Interaction Thermodynamics and Cluster Formation for Very Large Deviations From Stoichiometry

Introduction of defects (vacancies or interstitials) in a perfect lattice determines a variation in the free energy G of the lattice which may be written as:

$$\Delta G = G - G_{\text{perfect lattice}} = \{N\Delta H_{form} + \Delta H_{inter}(N,N')\} - T\{\Delta S_{conf}(N,N') + \Delta S_{vib}(N,N')\} = \Delta H - T\Delta S \quad (25)$$

N is here the number of lattice defects (vacancies or interstitials) which are responsible for non-stoichiometry. ΔH_{form} is the variation of lattice enthalpy when one non-interacting lattice defect is introduced in the perfect lattice. Since two types of point-defects are always present (lattice defect and altervalent cations (electronic disorder)), the ΔH_{form} takes into account not only the enthalpy change due to the process of introduction of the lattice defect in the lattice, but also that occurring in the Redox reaction creating the electronic disorder.

If defects are not independent but interact between each other, the total enthalpy of the lattice is affected by this interaction. ΔH_{inter} takes care of the gain or loss of total enthalpy due to the interaction fields. ΔH_{inter} is dependent on both N and N′ (where N′ is the number of cationic compensating charges, usually related through stoichiometry to N), since both lattice defects and electronic disorder are electrical charges with respect to the lattice.

If the defects can be considered point charges, localized on their own lattice sites, the polar electrostatic interaction between them is usually written as a long range monopolar Coulomb energy. If, on the other hand, for large concentrations of defects, local charge effects, as described in the introduction, are present, then ΔH_{inter} is much more difficult to write.

In fact, ΔH_{inter} determines the shape of the total ΔH for very concentrated defect solutions. Simple statistical descriptions putting $\Delta H_{inter} = 0$ (independent defects) fail completely to describe the experimentally determined behaviour of the thermodynamic functions[73].

We turn now to ΔS in (20), which describes the total change of entropy in the non-stoichiometric lattice with respect to the perfect one. The term ΔS_{vib} represents the change in entropy due to the change in the phonon spectrum of the lattice due to the presence of defects[73]. This part can be particularly important when the lattice contains a large number of defects. Information on the change of the phonon spectrum can be obtained from vibrational spectroscopy; infrared or Raman optical spectroscopy and, above all, inelastic neutron scattering (see Chap. D). Due to the experimental difficulties with actinide oxides (their high radioactivity; lack of large single crystal adequately conditioned to keep a well defined stoichiometry) unfortunately very few results are available in this field (except for ThO_2 and UO_2). Perhaps for this, very little care has been given to the extremely important ΔS_{vib} in statistical interpretations of the thermodynamic functions of grossly non-stoichiometric compounds.

The other term in ΔS of (25) is ΔS_{conf}, representing the change in entropy when the perfect stoichiometric lattice is disordered by the presence of defects. ΔS_{conf} is expressed by:

$$\Delta S_{conf} = k \ln \Omega(N,N') \tag{26}$$

where $\Omega(N,N')$ represents the number of ways (configurations) by which the N,N′ defects defined above dispose themselves in the perfect lattice.

If defects are non-interacting, Ω is given by combinatorial terms and mass action equilibrium equations are obtained.

In the case of concentrated solutions of defects, i.e. large departures from stoichiometry, a fundamental fact in encountered[73]: ΔH_{inter} and ΔS_{conf} are strongly correlated, i.e.

$$\Delta S_{conf} = f(\Delta H_{inter}) \tag{27}$$

Clusters or molecularities' models trying to reproduce the correlation (27) are useful tools for gaining insight into the bonding of the oxide.

b. Clusters in Oxygen-Excess Dioxides

The only thoroughly studied grossly overstoichiometric actinide oxide is UO_{2+x}.

Willis[48, 75–77], using Bragg neutron scattering techniques, performed an investigation on the defect structure of $UO_{2.13}$ and U_4O_9. According to these studies, interstitial oxygen O_i is incorporated in the lattice, causing non-stoichiometry. At large deviations from stoichiometry, the O_i's do not occupy the center of the "octahedral hole" of the fluorite structure (see Part V). Rather, they occupy two other sites, termed O′ and O″. O′ is displaced from the center (1/2 1/2 1/2) of the octahedral hole along the $\langle 110 \rangle$ axis, the second along the $\langle 111 \rangle$ axis. One or two adjacent oxygen atoms are displaced from their perfect-lattice equilibrium position leaving vacancies at their place. This finding is confirmed by the application of a deuteron channeling technique (Matzke[51, 52]), which also yields the important result that the heavy uranium atoms are also displaced somewhat (about 0.25 Å) from their normal position.

Oxygen interstitials of the two types as well as oxygen vacancies are associated in different ways, forming clusters. Proposed clusters are (V_0 = oxygen vacancy):

1 : 2 : 2 i.e. (1 O′ − 2 V_0 − 2 O″)
2 : 2 : 2 i.e. (2 O′ − 2 V_0 − 2 O″)
4 : 3 : 2 i.e. (4 O′ − 3 V_0 − 2 O″)

Catlow and Lidiard[53] calculated, by computer assisted cluster calculations in an ionic model, that the 2 : 2 : 2 and 4 : 3 : 2 clusters are particularly stable. Similar clusters are reported to exist in other ionic fluorite-structure solids, e.g. CaF_2 + YF_3[78, 79], indicating that they are a feature of anion-excess fluorite compounds.

Table 9. Possible clusters in different composition ranges of UO_{2+x}, as deduced from ln p_{O_2} vs. ln x experimental curves (from[80])

composition range	cluster
....0–0.026	$\{O_i' \cdot 2\,U^{+5}\}$
....0.026–0.115	$\{2(O_i'' \cdot V_0 \cdot O_i'' \cdot 2\,U^{+5}\}$ i.e. a 2:2:2 cluster associated with 2 U^{+5}
....0.115–0.18	unknown type of cluster, but certainly *not* the 4:3:2
....0.18–0.24	long-range ordered, almost U_4O_{9-y} phase

Sørensen[80] analysed the ln O_2 vs. lnx experimentally determined functions at different temperatures. The slope of such curves can be shown to be simply related, by equilibrium equations, to the number of atoms and defects composing a cluster. By this method, clusters and clusters' aggregations as given in Table 9 are proposed.

It may be observed that

i) even at very low deviation from stoichiometry, it seems possible to postulate a trapping of 2 electron holes (2 U^{+5}) in the neighbourhood of an interstitial oxygen atom;

ii) the observed 2:2:2 cluster also traps 2 electron holes;

iii) there is possibly a building up and progressive ordering of the Willis' clusters, leading to the formation, at very high deviations of stoichiometry, of other oxide phases.

The theoretical interpretation of the Willis' clusters has been attempted by Catlow[81, 82]. These authors employed the HADES program, a code used successfully for many ionic compounds.

In this method, the respective positions of about 100 ions (inner zone) are calculated under various assumptions for the inter-ionic potentials, assuming the rest of the lattice as a quasi-continuum. A polarization per unit cell is attributed from measured values of the static dielectric constant. The configuration within the inner zone (which may generate a "cluster") is varied until the total energy of the lattice is minimized[82].

The problem of this method is to simulate correctly the inter-ionic potential. Not dissimilarly to other types of calculations in the ionic picture (e.g. application of Born-Haber cycles), the most important part of this potential is the monopole Coulombic potential (radial dependence: R^{-1}) between the ionic charges. This is corrected either by assuming "effective point charges" for the ions or by adding repulsive and attractive short-range terms, having a different radial dependence and representing either higher poles, or simulating the charge density distribution in the inter-ionic space (i.e. covalency effects).

The calculation shows that the 2:2:2 and the 4:3:2 Willis' clusters are essentially interstitial dimers stabilized by a coupled interstitial-lattice relaxation mechanism[82]. Oxidized (U^{+6}) cations localized at cation sites adjacent the cluster are also needed[82].

This affirmed necessity of positioning around the cluster, U^{+5} (U^{+6}) cations is a very relevant fact, especially if coupled with the result of the channeling experiment reported above, which shows a displacement of uranium ions from their equilibrium positions. The Willis' clusters can then be described also as molecularities in which the electronic charge transfer between the oxygen interstitials and the surrounding metal ions is incomplete. A covalent picture may then be invoked.

The recent band calculations by Brooks et al.[64] show indeed from charge density calculations that in stoichiometric UO_2 a purely ionic picture is incorrect: the oxygen ion is singly charged, the second electron not being retained by the Madelung potential. Simple considerations based on mechanical properties[54] and spectroscopic evidence[72] had already pointed at these conclusions.

c. Clusters in Oxygen-Deficient Dioxides

In fluorite-structure oxygen-deficient oxides MO_{2-x} there is general agreement that the oxygen vacancy is the point-defect responsible for non-stoichiometry. Unfortunately, no *direct* observation is available of a basic cluster species for MO_{2-x} such as the Willis cluster in UO_{2+x}.

The problem, instead, which polarizes attention is constituted by the difference, illustrated before, between the phase diagrams of fluorite-structure lanthanide oxides PrO_{2-x} and CeO_{2-x} as well as of heavier actinide oxides CmO_{2-x} and CfO_{2-x}, in which M_nO_{2n-2} subphases have been observed, and those of PuO_{2-x} and AmO_{2-x}, which do not display these subphases.

For the lanthanide dioxides PrO_{2-x} and CeO_{2-x}, many authors[83-86] have proposed, as the basic structural entity generating the M_nO_{2n-2} series, a linear infinite string of MO_6 octahedra – along the $\langle 111 \rangle$ direction-surrounded by a contiguous sheet (of different "thickness in different M_nO_{2n-2} species") of MO_7 cubes (in the normal fluorite-structure, one metal is coordinated to 8 oxygen ions, i.e. MO_8). Catlow[82] attributes these structures to the existence of an even more basic cluster unit constituted by a vacancy composed of two oxygen vacancies at the opposite ends of an anion cube of the fluorite structure. This vacancy needs, however, to be stabilized by the association of neighbouring trivalent M^{+3} cations.

The important point for bonding is not so much the "structural description" of the strongly disturbed oxygen sublattice, but (as discussed for UO_{2+x}), the required association postulated between M^{+3} ions and oxygen vacancies. Once again, molecularities are formed, strongly hinting at covalent effects.

"Local" association of the M^{+3} reduced cation and the oxygen vacancy is clearly suggested by the thermodynamics of the hypo-stoichiometric mixed oxides $(U_{1-y}Pu_y)O_{2-x}$, where the thermodynamic functions do not depend on x and y separately, but rather on a quantity, called "plutonium valence", which contains the ratio x/y[73, 87]. Clusters consisting of this association have been proposed[88, 90] in order to explain the thermodynamics of actinide hypostoichiometric dioxides.

Manes and Manes-Pozzi[87] have suggested a cluster of the type $(V_0 \cdot 2\ Me^{+3})$, which has been taken as the basis species for a statistical treatment aimed at the interpretation of the thermodynamic data on $(U_{1-y}Pu_y)O_{2-x}$ and PuO_{2-x}. This cluster has later been called by Manes, Sørensen et al. the "tetrahedral defect"[91]. The reason of this name lies in the fact that the local bond is supposed to occur in a coordination tetrahedron of an oxygen ion in the fluorite structure: in this tetrahedron, one oxygen vacancy is formed, and the two electrons are shared with the four surrounding cations, giving rise (formally) to $2(Me^{+3})$ locally bonded with the vacancy. In[91], Manes, Sørensen et al. showed that by the periodic packing of tetrahedral effects in the fluorite lattice, some of the M_nO_{2n-2} substructures could be generated: an interesting structural observation coupling thermodynamics and crystal structure.

The tetrahedral defect[73]:

a) is no longer a point defect but has its own volume in the lattice (the tetrahedral volume);
b) constitutes a neutral entity with respect to the lattice, but has a dipole moment since the centers of positive and negative charge do not coincide when a charge transfer has taken place between the vacancy and the cations (the center of the positive charge being in the vacancy and that of the negative charge being on one of the tetrahedron's sides, halfway between the reduced cations);
c) represents local short-range ordering with respect to the "free" defects and thus has an important effect on the total energy and entropy of the crystals;
d) may be packed in different ways, thus extending the range of ordering of the defect population.

When packing progressively tetrahedral defects in the fluorite lattice, it is assumed that they interact with the lattice (elastic strains) and with one another (dipolar interaction). Both mechanisms of interaction are strongly dependent on the "local" charge distribution within the coordination tetrahedron, in which the defect is formed.

The elastic strain imposed on the lattice may be reconducted to an increase in volume of the tetrahedron when the tetrahedral defect is formed, due to the local presence of M^{+3} ions having a much larger ionic volume. In this case, also the dipole within the tetrahedron is at its maximum. If, however, a certain amount of covalency is assumed (already in the perfect oxide), the formation of the tetrahedral defect is predicted to occur without great increase in the volume of the original tetrahedron, and with a lower dipole. In a more covalent fluorite structure oxide, therefore, a more relaxed situation exists and a greater disorder of defects is preserved even at high deviations from stoichiometry. In this case the fluorite lattice is not submitted to the strains due to defects packing, and this even at the low temperatures at which, from a more ionic lattice, M_nO_{2n-2} subphases would precipitate. This may well explain the difference between the phase diagrams of PuO_{2-x} and AmO_{2-x} and those of PrO_{2-x}, CeO_{2-x} and the heavier actinide oxides. Cluster molecular orbital calculations (already reported), for the perfect oxides indicate that the covalent f-p contribution to the bond is at a maximum for PuO_2 and AmO_2 in the actinide dioxides' series[67].

This argument had been anticipated by Blank[92] on the bais of a critical examination of structural properties. In later publications, the same author showed[56, 93] that many properties (especially, mechanical and structural properties) of fluorite-structured dioxides are better explained when assuming a considerable amount of covalency in their bond.

VII. Conclusions

In this chapter, we have examined relevant structural and thermodynamic properties of metals and simple binary compounds, and which information they may convey on the bonding. Crystal structures and chosen thermodynamic quantities, such as the heat of vaporization and heat of solution of metals, and, especially, their correlations and trends across the actinide series, contain a description of the bonding, provided scientists are able to display enough ingenuity to read and interpret it with the help of adequate models.

Given the characteristics of spatial extension of 5f wavefunctions, the important parameter for the 5f overlap phenomena, hence the metallic bond, is interactinide distance. If hybridization sets in, either between the actinide atom or with the non-actinide atom in compounds, the type of crystal structures can change due to the angular and spatial characteristics of the hybrid orbitals. Studying structural changes under pressure, therefore, has shown itself to be of paramount importance in actinide physics, and will for certain keep this importance in the future. The pressure parameter is in fact able to change the nature of the valence band considerably. Applied pressure competes in a solid with the 5f-pressure p_f, which as we have seen, decides in metallic bond between localization and itineracy of these states. It seems that pressure will decide whether an element (be it actinide or lanthanide) will behave, as far as the f-bonding properties are concerned, as an f-transition metal or a lanthanide-like element. Therefore measurements under pressure are becoming more and more favoured: crystallography under pressure, electrical resistivity measurements under pressure, neutron diffraction under pressure and optical spectroscopy under pressure.

Temperature effects are not, however, to be overlooked: the phase diagramme with its numerous allotropes of Pu metal is there for us to remember. Usually crystallographic studies are made from room to higher temperatures but low temperature, cryogenic studies would be also indicated. One reason is that the effect of phonons have a very strong influence in determining interactinide distances and crystal symmetry. Recent theoretical work on the high temperature f.c.c. structure of plutonium metal, in which, as shown by photoemission, 5f localization is not present, although the crystal structure is typical of a 5f localized behaviour, suggests that the softening of the lattice due to numerous phonon excitations is responsible for the anomaly.

As for ionic solids, e.g. oxides, attention is drawn in the chapter to phenomena, such as the huge non-stoichiometry, which hint to "local" bonding of defects and their aggregation. In this respect, covalency (f-p or d-p type) may play a major role. Unfortunately, although a growing interest in this problem exists, opinions are controversial. A way to study defect clustering effects in PuO_{2-x}, where they may be very important, is by elastic, inelastic and diffuse neutron scattering. Unfortunately, 239-Pu has an unfavourable cross section: the much more costly and rare 242-Pu would be ideal (single crystals of the oxide with this isotope would also be highly desirable).

Also, more accurate thermodynamic measurements, aiming at a precise determination of the slopes of the oxygen potential curves vs. oxygen deficiency, which are related strongly to the nature of the defect clusters, would be helpful. On the theoretical side, cluster molecular orbital calculations on defects or defect clusters, with their ability to provide information of short-range electron charge distribution and its orbital character, are being already performed and will certainly aid to the understanding of this difficult matter.

VIII. Note Added in Proof

Recent Results for Curium Metal Under Pressure

In Part II.3.a it was pointed out that the structures proposed for curium metal and reported[23)] in Table 1 and Fig. 1b represented an anomaly in the trend encountered when applying pressure on lanthanide and heavier actinide metals. Thus, curium metal should have represented a somewhat unexplained intermediate stage between lighter and heavier actinides.

However, very recent results obtained in a collaborative study by one of the authors[94, 95)], eliminate this anomaly, and correct Table 1 and Fig. 1b. Starting with the dhcp phase (Cm I) at normal temperature and pressure, a dhcp $\xrightarrow{p}$ ccp (Cm II) transition occurs at 23 GPa, and a ccp $\xrightarrow{p}$ Cm III (α-uranium type structure) transition at 43 GPa. The $\Delta V/V$ (see Fig. 2) from the ccp to the low symmetry phase is, for Cm metal, 21%. Therefore, curium metal behaves as the heavier Cf and Bk and in a different way from the lighter actinides up to plutonium. The case of americium metal stands apart as a particular transition stage as underlined in parts II.3.b and IV.5.

These recent results for Cm metal, however, show that for Cm the "delocalization" pressure and volume change $\Delta V/V$ (relative to the ccp $\xrightarrow{p}$ α-uranium transition) are higher than for the higher Z Cf and Bk. This is an apparent anomaly, which is however in line with the spin-polarization model for 5f-localization, illustrated in this chapter. Cm is in fact to be considered as the most localized system, since its $5f^7$ half-filled configuration has the largest number of unpaired spins, hence the largest m in Eq. (19) (see the discussion of Eq. (19)). Also, the spin polarization energy Δ_{sp} should be the largest. Therefore, the necessity of a higher pressure to induce the transformation and the largest volume collapse are justified.

IX. References

1. Pearson, W. B.: The Crystal Chemistry and Physics of Metals and Alloys, Wiley 1972
2. Brewer, L.: Science *161,* 115–122 (1968)
3. Fournier, J. M.: J. Phys. Chem. Solids *37,* 235–244 (1976)
4. Johansson, B., Rosengren, A.: Phys. Rev. *B11,* 1367 (1975)
5. Johansson, B.: ibid. *B11,* 2740 (1975)
6. Hume-Rothery, W.: The Engel-Brewer Theories of Metals and Alloys, Progress in Materials Science, *Vol. 13* (eds. Chalmers, B., Hume-Rothery, W.), Pergamon Press, Oxford, 231–264 (1968)
7. Brooks, M. S. S., Johansson, B.: J. Phys. F: Met. Phys. *13,* L197 (1983)
8. Wyckoff, R. W. G.: Crystal Structures, 2nd edition, Interscience Publishers 1965
9. Gschneidner, K. A., jr.: p. 275–430 of: Solid State Physics, *Vol. 16* (eds. Seitz, F., Turnbull, D.), Academic Press, New York 1964
10. Johansson, B., Skriver, H. L., Anderson, O. K.: p. 245–262 of: Physics of Solids under High Pressure (eds. Schilling, J. S., Shelton, R. N.) North-Holland Publ. Co. 1981

11. Fournier, J. M.: These Dr.-ès-Sc., Grenoble 1975
12. Donohue, J.: The Structures of the Elements, reprint edition 1982, Robert E. Krieger Publ. Co. Malabar, Florida
13. Bellussi, G., Benedict, U., Holzapfel, W. B.: J. Less-Common Metals *78,* 147 (1981)
14. Benjamin, T. M. et al.: Carnegie Institution of Washington Year Book *80,* 280 (1981)
15. Benedict, U. et al.: J. Magn. and Magn. Mat. *29,* 287 (1982)
16. Akella, J., G. S. Smith, H. Weed: J. Phys. Chem. Solids, submitted
17. Stephens, D. R.: J. Phys. Chem. Solids *27,* 1201 (1966)
18. Roof, R. B.: p. 221–230 of: Advances in X-ray Analysis *24,* Plenum Press 1981
19. Akella, J. et al.: J. Geophys. Research *85,* 7056 (1980)
20. Smith, G. S. et al.: paper presented at Actinides 1981, Asilomar, CA, Sept. 1981
21. Roof, R. B. et al.: Science *27,* 1353 (1980)
22. Roof, R. B.: J. Appl. Cryst. *14,* 447 (1981)
23. Akella, J.: Report on the Workshop "Actinides under Pressure", April 21–22, 1983, Commission of the European Communities, Joint Research Centre, Karlsruhe, 1983, p. 5
24. Benedict, U. et al.: J. Phys. F: Metal Phys. *14,* L 43 (1984)
25. Smith, G. S., Akella, J.: J. Appl. Phys. *53,* 9212 (1982)
26. Grosshans, W. A., Vohra, Y. K., Holzapfel, W. B.: J. Phys. F: Metal Phys. *13,* L 147 (1983)
27. Roof, R. B.: Z. Krist. *158,* 307 (1982)
28. Benedict, U. et al., in: Americium and Curium Chemistry, N. Edelstein, J. D. Navratil, W. W. Schulz, eds., Reidel Publ. Co., Dordrecht, in press.
29. Brooks, M. S. S.: J. Phys. F: Metal Phys. *13,* 103 (1983)
30. Smith, J. L., Kmetko, E. A.: J. Less-Common Metals *90,* 83 (1983)
31. Zachariasen, W. H.: in: The Metal Plutonium (ed. Coffinberry, A. S., Miner, W. N.), Univ. of Chicago Press, p. 99, Chicago 1961
32. Nugent, L. J., Burnett, J. L., Morss, L. R.: J. Chem. Thermodyn. *5,* 665 (1973)
33. Carlsson, T. A. et al.: Atomic Data *2,* 63 (1970)
34. Bonnelle, C., Lachere, G.: J. de Physique *35,* 295 (1974)
35. Skriver, H. L., Andersen, O. K., Johansson, B.: Phys. Rev. Letters *44,* 1230 (1980)
36. Jullien, R., Galleani d'Agliano, E., Coqblin, B.: Phys. Rev. *B 6,* 2139 (1972)
37. Johansson, B., Skriver, H. L.: J. of Magn. and Magn. Mat. *29,* 217 (1982)
38. Skriver, H. L., Andersen, E. K., Johansson, B.: Phys. Rev. Lett. *41,* 42 (1978)
39. Brooks, M. S. S.: J. Phys. *F 13,* 103 (1982)
40. Brooks, M. S. S.: J. Phys. F: Metl. Phys. *14,* 1157 (1984)
41. Friedel, J.: in: The Physics of Metals (ed. Ziman, J.), Cambridge University Press, London, Ch. 8 (1969)
42. Murnaghan, F. D.: American J. of Math. *L IX,* 235 (1937)
43. Harrison, W. A.: in: Electronic Structure and the Properties of Solids. The physics of the chemical bond (eds. Freeman, W. H., Co.), San Francisco 1980
44. Jorgensen, Ch. K.: in: Orbitals in Atoms and Molecules, Academic Press, London and New York 1962
45. Nugent, J. L.: J. Inorg. Nucl. Chem. *32,* 3485 (1970)
46. Nugent, J. L. et al.: ibid. *33,* 2503 (1971)
47. Bradbury, M. J.: J. Less-Common Met. *78,* 207 (1981)
48. Belle, J.: J. Nucl. Mats. *30,* 3 (1969)
49. Willis, B. T. M.: Proc. Roy. Soc. (London) *A 274,* 133 (1963)
50. Markin, T. L., Wheeler, V. J., Brones, R. J.: J. Inorg. Nucl. Chem. *30,* 807 (1968)
51. Matzke, Hj., Davies, J. A., Johansson, G. E.: Can. J. Phys. *49,* 2215 (1971)
52. Matzke, Hj.: Diffusion in non-stoichiometric oxides, Chapt. 4 of Non-stoichiometric Oxides (ed. Sørensen, O. J.), Academic Press, New York 1981
53. Faber, J. Jr., Seitz, M. A., Mueller, M. H.: J. Phys. Chem. Solids *37,* 903 (1976) ibid. 909 (1976)
54. Blank, H.: in: Thermodynamics of Nuclear Materials, 1974, *Vol. II,* IAEA-Vienna, 45 (1975)
55. Turcotte, R. P., Chikalla, T. D.: J. Inorg. Nucl. Chem. *35,* 3739 (1971)
56. Turcotte, R. P., Chikalla, T. D., Eyring, L.: ibid. *35,* 809 (1973)
57. Andersen, O. K., Klose, W., Nohl, H.: Phys. Rev. *B 17,* 1209 (1978)
58. Andersen, O. K. et al.: Pure and Appl. Chem. *52,* 93 (1979)

59. Brooks, M. S. S., Glötzel, D.: Physica *102 B,* 51 (1980)
60. Brooks, M. S. S.: J. Phys. F: Metal Phys. *14,* 653 (1984)
61. Brooks, M. S. S. et al.: Physica *102 B,* 84 (1980)
62. Kelly, P. J., Brooks, M. S. S., Allen, R.: J. de Phys. *C4,* C4-184 (1979)
63. Kelly, P. J., Brooks, M. S. S.: Physica *102 B,* 81 (1980)
64. Brooks, M. S. S., Kelly, P. J.: Solid State Comm. *45,* 689 (1983)
65. Gubanov, V. A., Ellis, D. E., Freeman, A. J.: Fiz. Tverd. Tela *19,* 409 (1977)
66. Gubanov, V. A., Rosén, A., Ellis, D. E.: Solid State Comm. *22,* 219 (1977)
67. Ellis, D. E., Gubanov, V. A., Rosén, A.: J. de Phys. *C4,* Ca-187 (1979)
68. Gubanov, V. A., Rosén, A., Ellis, D. E.: J. Phys. Chem. Solids *40,* 17 (1979)
69. Gubanov, V. A., Rosén, A., Ellis, D. E.: J. Inorg. Nucl. Chem. *41,* 975 (1979)
70. Programme Progress Report "Plutonium Fuels and Actinide Research", TUSR 27, Joint Research Center of the European Community, p. 102 (1979)
71. Manes, L., Barisich, A.: Phys. Stat. Sol. (a) *3,* 971 (1970)
72. Manes, L., Naegele, J.: in: Plutonium 1975 and Other Actinides (eds. Blank, H., Lindner, R.), North-Holland, Amsterdam, 361 (1976)
73. Manes, L.: A new method of statistical thermodynamics and its application to oxides of the lanthanide and actinide series, in "Non-stoichiometric Oxides" (ed. Sørensen, O. T.), Academic Press, New York, Chapt. 3, p. 99 (1981)
74. Stoneham, A. M.: Theory of Defects in Solids, Clarendon Press, Oxford 1975
75. Willis, B. T. M.: Proc. Br. Ceram. Soc. (London) *1,* 9 (1964)
76. Willis, B. T. M.: J. de Phys. *25,* 431 (1964)
77. Willis, B. T. M.: Acta Crystallogr. *18,* 75 (1965)
78. Cheetham, A. K. et al.: Solid State Commun. *8,* 171 (1970)
79. Cheetham, A. K.: in: Chemical applications of thermal neutron scattering (ed. Willis, B. T. M.), Oxford University Press, London and New York 1973
80. Sørensen, O. T.: Thermodynamics and defect structure of non-stoichiometric oxides, in "Non-stoichiometric oxides" (ed. Sørensen, O. T.), Academic Press, New York, Chapt. 1, p. 1 (1981)
81. Catlow, C. R. A.: J. de Phys. *C6,* L64 (1973)
82. Catlow, C. R. A.: Defect clustering in non-stoichiometric oxides, in "Non-stoichiometric oxides" (ed. Sørensen, O. T.), Academic Press, New York, Chapt. 2, p. 60 (1981)
83. Hyde, B. G., Eyring, L.: Proc. Conf. Rare Earth Res. 4th, 623 (1964)
84. Sawyer, J. O., Hyde, B. G., Eyring, L.: Bull. Soc. Chim. Fr. 1190 (1965)
85. Kunzmann, P., Eyring, L.: J. Solid State Chem. *14,* 229 (1975)
86. Summerville, E., Tuenge, R. T., Eyring, L.: ibid. *24,* 21 (1978)
87. Manes, L., Manes-Pozzi, B.: in: Plutonium 1975 and other actinides (eds. Blank, H., Linder, R.), North-Holland, Amsterdam, 145 (1976)
88. Schmitz, F.: J. Nucl. Mater. *58,* 357 (1975)
89. Catlow, C. R. A.: ibid. *67,* 336 (1977)
90. De Franco, A., Gatesoupe, J. P.: in: Plutonium 1975 and other actinides (eds. Blank, H., Lindner, R.), North-Holland, Amsterdam, 135 (1971)
91. Manes, L. et al.: in: Thermodynamics of Nuclear Materials, 1979, *Vol. I,* IAEA-Vienna, 405 (1980)
92. Blank, H.: EUR 3653e, EURATOM, European Atomic Energy Community, Bruxelles 1965
93. Blank, H.: J. Nucl. Mater. *51,* 269 (1974)
94. Benedict, U. et al., J. Phys. F: Metal Phys. *15,* 29 (1985)
95. Haire, R. G. et al., J. Less Common Metals, submitted

Chapter D

Magnetic Properties of Actinide Solids

J. M. Fournier

University of Grenoble and Centre d'Etudes Nucléaires de Grenoble, Grenoble, France

The various magnetic properties observed in actinide solids are largely determined by the degree the unsaturated 5 f shells retain an atomic localized character or lose it in an itinerant mode. Hence the investigation of magnetism (and of related properties such as low temperature electrical resistivity) has played a central role in understanding bonding in actinide solids. Between clear cut cases of "localization" or "itineracy" intermediate situations are observed (leading for instance to spin fluctuating behaviour) which have made actinide magnetic research a rich but also controversial field.

A short exposition of the basic concepts of actinide magnetic research and of the different aspects of magnetic ordering in actinide solids is followed by a discussion of experimental and interpretational aspects, in particular the difficulty always encountered of separating the different contributions.

A number of examples have been chosen to illustrate the variety of magnetic behaviour and its meaning in establishing a unified picture of actinide solid state physics.

One of the aims of the chapter is to inspire new research in actinide magnetism. To this purpose, tables collecting the main magnetic data for a large number of actinide binary compounds are provided.

I. Introduction

In this chapter our aim is to stress the main physical features governing the unique magnetic properties of the actinides and to illustrate these features by clear examples when available.

The most important factor governing the physical, and in particular magnetic, properties of the actinides, is the extended nature of the 5f wave functions. Depending on:
1) the number of 5f electrons involved (occupation of the 5f shell)
2) the interactinide spacing
3) the actinide environment in a compound

we find properties ranging from that of itinerant transition-like systems to that of localized lanthanide-like systems.

In Sect. II, we show how the localization of 5f electrons is related to the magnetic behaviour of pure metals and compounds.

In Sect. III, we illustrate the different kinds of magnetic behaviour displayed by actinides.

In Sect. IV, we present the main macroscopic properties and discuss the physical information which can be extracted from the experimental data.

In Sect. V, the same procedure is used in the case of microscopic properties.

The most practical use of such a chapter is a collection of data. Thus we tried to compile all magnetic data available up to end 1983 in the form of data tables which are the object of Sect. VII.

II. Localization and Magnetic Properties

1. *Stoner and Mott-Hubbard Models*

In this chapter we are concerned with the magnetic properties of the actinides. How the localization of electrons belonging to an incomplete shell is related to their magnetic properties? This is an old question to which it is possible to answer qualitatively if not quantitatively. There are 2 extreme points of view to approach this crucial problem.
- The one-particle band point of view (Stoner theory)
- The many-particle local point of view (Mott-Hubbard theory)

In the Stoner theory for magnetism, it is shown that magnetic ordering is possible in a band scheme, when spin-polarization is energetically favorable; the full discussion (see Chap. A) for the simpler case of ferromagnetism leads to a simple criterion known as the Stoner criterion[1)]

$$I \times N(E_F) \geq 1$$

where I is the so-called Stoner parameter and $N(E_F)$ is the density of electron states at the Fermi level. It has been found[2)] that the Stoner parameter is fairly constant for different actinides compounds at about 3 eV. We are thus led to a very simple and interesting result, namely that possible magnetic ordering is only governed by the 5f electronic density of states at the Fermi level.

In the Mott-Hubbard theory[3] on the other hand, it is shown that there exists an instability in the narrow-band electronic structure (Peierls instability[4]) and if the bandwidth decreases below a critical value, a sudden transition (Mott transition)[5] takes place toward a complete localized situation. In this approach, it is assumed, in fact, that band magnetism does not exist and one has to deal only with 2 classes of materials

- non-magnetic systems with 5f bands
- magnetic systems with 5f local states

A simple criterion for the occurence of magnetism within this scheme for non-degenerate electron-state is[3] (see Chap. A)

$$W_f \leq U_{eff}$$

where W_f is the 5f bandwidth and U_{eff} the effective intra-atomic Coulomb correlation.

2. *Pure Metals and Mott-Like Transition*

In Chap. A and in particular in Sect. IV, the formation of 5f bands in actinides metals was described. We shall recall here the points which directly affect the magnetic properties. In Table 1 we report W_f, U^{at}, and ($I \times N(E_F)$) for the actinide metals. This tabulation focusses on the transition that occurs between plutonium and americium. Americium metal behaves like a lanthanide in many physical properties; from the magnetic point of view its $J = 0$ ground state is non-magnetic ($5f^6$ configuration) and thus allows even superconductivity to occur. Since plutonium metal does not order magnetically, this has led Johansson to view this sudden change as a Mott-like transition of the 5f electrons along the actinide series[6]. Thus, it appears that in the case of the pure metals, the Mott-Hubbard approach is appropriate to describe the evolution of 5f behaviour along the actinide series: a Mott-like transition is the simplest and clearest description of what happens between plutonium and americium. Up to plutonium, i.e. in the band regime – the Stoner criterion is not fulfilled and this is more or less correctly predicted from the calculated $I \times N(E_F)$ reported in Table 1. Given their "critical" situation, plutonium metal – and also neptunium – have an abnormal behaviour.

Band calculations[8] show that 6d-5f hybridization reduces considerably the occurence of ferromagnetism (Stoner criterion) in the 5f band.

Then, curium metal is antiferromagnetic and its paramagnetic effective moment definitively supports the picture of a localized $5f^7$ configuration[10]; the same is true for what is known about berkelium and californium metals[11, 12].

Table 1. 5f bandwidth W_f intra-atomic Coulomb correlation U^{at} and Stoner parameter time the density of states $I \times N(E_F)$ for light actinide metals

	Pa	U	Np	Pu	Am
W_f(eV)	4	4	3	2	0.1
U^{at}(eV)	1.5	2.3	2.6	3.5 ± 1	5 ± 1
$IN(E_f)$	~ 0.3	0.9	0.9	0.8	>1

As a remark, it should be noted that a single electron ground state from a filled band (which would be the case of americium within a band description because the 5f band is spin-orbit splitted into a filled 5/2 sub-band and an empty 7/2 sub-band) is equivalent to a localized state[7)] and thus a spin-polarized band description leads to the same conclusion as a simple Mott description.

3. *Metallic Compounds and Hill Plots*

In 1970, Hill[13)] pointed out that one may classify uranium, neptunium and plutonium compounds into a magnetic and non magnetic group depending on interactinide distance $d_{An\text{-}An}$ (see Chap. A) with a critical value of about 3.4 Å. This systematics clearly evidences a localization process related to direct 5f-5f overlap. As we said at the beginning of this section, the critical distance may be viewed from the 2 different points of view
– it corresponds to the fulfilment of a Stoner criterion
or
– it corresponds to a Mott transition.

From band structure calculations[2, 14)] as well as from experimental results (photoemission, specific heat...), it is now clear that, at least for magnetic compounds near the critical spacing, 5f electrons are in bands. *This means that contrary to the pure metals, the Stoner criterion may be fulfilled before the band description breaks down.*

As noted by Hill himself[13)], the weakness of the Hill plots comes from the crude assumption that direct 5f-5f overlap is the only parameter governing the 5f bandwidth. That such plots are so successful is, however, in favour of such an assumption. Since that time, considerable new information has been obtained, showing that many other parameters influence the appearance of magnetism in early actinides compounds. As already pointed out in the case of pure metals, the most important correction comes from hybridization between 5f states and p or d states of the compound "partner". In the case of pnictides and chalcogenides, p states are well below the Fermi energy so that hybridization will not result in a f-p conduction band and is less pronounced than for other metallic compounds. Thus the main effect of hybridization for magnetism will be in the kind of exchange mechanism involved. Experimentally it is found that for pnictides, where f-p hybridization is strongest, exchange is very anisotropic – From the degree of experimental knowledge we have today, the main interest of Hill plots resides in anomalies such as for example $AnSn_3$, $AnRh_3$, on one side, for which magnetism is delayed to much higher $d_{An\text{-}An}$ values (USn_3 is not magnetic although $d_{U\text{-}U}$ = 4.63 Å)[1] and, on the other side, a compound like UNi_2, which is ferromagnetic, although $d_{U\text{-}U} \cong 3.0$ Å. If we compare with non-magnetic UCo_2, and by analogies with $LnNi_2$ intermetallics, we have good reasons to assume that magnetic ordering is due to uranium. Band structure calculations[16)] seem to support this hypothesis but direct proof will only be obtained through polarized neutron diffraction.

1 From the band point of view, this means a very strong hybridization to keep the density of state low enough although the direct 5f-5f overlap is very much reduced. This has been nicely verified for example in the case of paramagnetic URh_3, where the bandlike nature of 5f electrons was shown by de Haas van Alphen oscillation measurements which agreed with calculated band structure[15)]: strong hybridization with 4d rhodium states is the dominant factor

III. Different Aspects of Magnetic Ordering

1. Localized vs. Itinerant Picture in Actinide Compounds

From what has been presented in Sect. II, it is clear that the magnetic properties of actinide compounds are most difficult to interpret, since we encounter the situation of non-fully localized 5 f electron states and thus are faced by the very severe question: does a *localized* picture or an *itinerant* picture provide a better description of the magnetic properties? The answer is by no means unequivocal. Let us use a very simple scheme already discussed by Chan et al.[17)]. Apart from their kinetic energy, f electrons see their own centrifugal potential, due to their high angular momentum (see Chap. A, Sect. 2). Thus (see Fig. 12 b in Chap. A) the f electrons can tunnel the potential barrier to the next ion. If the barrier is high enough, this is the situation of f-resonance or critical bound state[18)]. Such hopping costs Coulomb energy U so that f electrons spend most of their time in the ground state: they appear localized and the number of electrons, n, will have only small fluctuations about a mean value $\bar{n}$ which is an integer. If the f-overlap or f-d/f-p hybridization increases the resonant picture gives way to a band.

From an experimental point of view, it appears that the resonant f level is the best starting hypothesis for most U, Np and Pu compounds. Only in some cases of strong hybridization (particularly for Laves phase and $AuCu_3$-type structure intermetallics) it will broaden into "true" bands and we shall try to give criteria for itinerant magnetism.

What has also emerged is that the tails of 5 f wave functions (see Chap. A), while not being much involved in making up the magnetic moments, are responsible for the f-bonding capacity: this is one important duality of 5 f states: compared with 4 f wavefunctions, tails are much more extended in real space. These tails will be greatly involved in the magnetic ordering process: anisotropy, RKKY and superexchange mechanisms, type of ordering in actinide compounds are often to be attributed to them.

2. Localized Magnetism

a) Classification of Interactions and Crystal Field Theories

Assuming 5 f localization, it is possible to classify the different interactions on an energy scale. Similar to lanthanides, the most impotant interaction is the intra-atomic Coulomb correlation U. Infrared spectroscopy results for solutions of trivalent actinide ions indicate that the Coulomb interactions are about 4×10^4 cm^{-1} and the spin-orbit constants are about 3×10^3 cm^{-1} [19)]. The 5f being much less screened than 4f, the crystal field interaction is, however, much stronger in actinides than in lanthanides. Magnetic and spectroscopic studies of ionic actinide compounds show that the crystal field interactions are about 10^3 cm^{-1} (it will be less for semimetallic compounds due to better screening). The exchange interactions may be estimated from the ordering temperatures of collective magnetism: they are about 10^2 cm^{-1}.

In Table 2 we indicate these results together with those of lanthanides for comparison.

Table 2. Comparison of Coulomb, spin-orbit, crystal field and exchange interactions for actinides and lanthanide ions, (U: Coulomb; H_{s-o}: spin-orbit; $H_{c.f.}$: crystal field, $H_{exch.}$ = exchange)

	Actinides		Lanthanides	
U	$\sim 4 \times 10^4$ cm^{-1}	or 5 eV	$\sim 10^5$ cm^{-1}	or 10 eV
H_{s-o}	$\sim 3 \times 10^3$ cm^{-1}	or 0.3 eV	$\sim 10^3$ cm^{-1}	or 0.1 eV
$H_{c.f.}$	10^2 to 10^3 cm^{-1}	or 0.01 to 0.1 eV	$< 10^2$ cm^{-1}	or $<$0.01 eV
$H_{exch.}$	$\sim 10^2$ cm^{-1}	or 0.01 eV	~ 10 cm^{-1}	or 0.001 eV

In the presence of Coulomb correlation only, the wave function is characterized by the total spin $S = \Sigma S_i$ and the total angular momentum $L = \Sigma_i l_i$ of the 5f electrons, and the total momentum J is given by Hund's rule ($J = L \pm S$). Important spin orbit coupling will mix LS multiplets and only J remains a good quantum number. The Russell-Saunders coupling scheme is no longer valid and an intermediate coupling scheme is more appropriate.

The inclusion of the crystal field destroys the rotational symmetry of the ion and lifts the degeneracy of J levels (except of course Kramer's degeneracy): the only good quantum numbers will be Γ's, the irreducible representations of the point-group symmetry operation. If the crystal field interaction is comparable to J-J splitting (and we see from Table 2 that this is the case of actinides) it will also cause an admixture of different J multiplets.

In Table 3 are given the ordered and effective moment obtained for the f^2 to f^5 configurations. In the case of free ions, we use the wave functions given by Chan and Lam in[17]. Up to the $5f^4$ configuration the departure from the pure Russell-Saunders ground state is small; the correction to the g_J Landé factor and thus to the magnetic moment is less than 3%. Only for the $5f^5$ configuration the correction becomes significant (20%).

The effective and ordered moments in a cubic crystal field were also calculated using the obtained g_J values and the wave functions given by Lea, Leask and Wolf[20]. In the case of the Γ_8 quadruplet the magnetic moment is anisotropic and depends upon the crystal field parameter x. The effective moment is space averaged and given as a function x[20]. The value of the ordered moment along $\langle 001 \rangle$ is given as a function of x also.

b) Examples of Localized Magnetism

We give here some examples of localized magnetism in actinide compounds.

– *UO_2:* in this semiconductor the 5f levels lie in the 6 eV gap between an oxygen 2p derived valence band and an uranium 6d derived conduction band[21]. This situation prevents hybridization insuring good localization with U^{4+} ions ($5f^2$ configuration). A first order magnetic transition towards a simple antiferromagnetic structure takes place at $T_N = 31$ K[22].

Within L-S coupling the ground state multiplet is $^3H_4$ leading to an effective moment $\mu_{eff} = 3.38\ \mu_B$ (see Table 3) which is slightly higher than the experimental value $\mu_{eff} = 3.2\ \mu_B$: thus J mixing is small and the ground state is essentially H_4 (Chan and Lam[17])

Table 3. Ordered (μ_0) and effective (μ_{eff}) moments for the f^2 to f^5 configurations in the Russell-Saunders (R.S.) and intermediate (Int.) coupling schemes for the full multiplet J or the crystal Γ states

	J	g_J R.S.	μ_{eff} R.S.	μ_0 R.S.	g_J [a] Int.	μ_{eff} Int.	μ_0 Int.	Γ (cubic symmetry)	$\mu_{eff}^{(\Gamma)}$ R.S.	$\mu_0^{(\Gamma)}$ R.S.	$\mu_{eff}^{(\Gamma)}$ Int.	$\mu_0^{(\Gamma)}$ Int.
$5f^2$	4	0.800	3.58	3.20	0.812	3.63	3.25	$\Gamma_4(3)$	0.56	0.39	0.57	0.40
								$\Gamma_5(3)$	2.80	2.00	2.84	2.03
$5f^3$	9/2	0.727	3.62	3.27	0.740	3.68	3.33	$\Gamma_6(2)$	1.48	1.33	1.51	1.38
								$\Gamma_8(4)$	1.30 to 3.10	1.05 to 2.89	1.32 to 3.16	1.07 to 2.43 ⟨001⟩
								$\Gamma_8(4)$	2.27 to 3.10	1.51 to 2.27	2.31 to 3.16	1.54 to 2.31 ⟨001⟩
$5f^4$	4	0.600	2.68	2.40	0.616	2.75	2.46	$\Gamma_4(3)$	0.42	0.30	0.43	0.31
								$\Gamma_5(3)$	2.10	1.50	2.15	1.54
$5f^5$	5/2	0.286	0.85	0.71	0.343 [b]	1.01	0.86	$\Gamma_7(2)$	0.41	0.24	0.49	0.29
								$\Gamma_8(4)$	0.67	0.52	0.80	0.63

[a] Calculated with the free-ion wave-functions given by Chan and Lam in[17)]

[b] Calculated with the Pu^{3+} wave function given in Lam, D.S. and Chan, S. K. Phys. Rev. *B6*, 307 (1972)

gives a composition 89% $^3H_4$, 10% $^1G_4$). Crystal field calculations[23)] lead to a Γ_5 triplet ground state with $\mu_{eff} = 2.9\ \mu_B$. Small amount of J mixing is sufficient to obtain the observed effective moment as well as the ordered moment $\mu_{ord.} = 1.8\ \mu_B$ which is slightly lower than the cubic Γ_5 value of 2 μ_B.

– *PuP:* amongst compounds of metallic character, plutonium pnictides PuP, PuAs, PuSb are ferromagnetic. PuP has been studied not only by the magnetization technique[24)] but also by neutron diffraction[25)], specific heat[26)] and NMR[24)]. Powder polarized neutron diffraction[25)] has confirmed that Pu is trivalent having a $5f^5$ configuration. The effective moment is $\mu_{eff} = \mu_B$ in good agreement with J = 5/2 ground state ($\mu_{eff} = 0.85\ \mu_B$ in the Russell-Saunders coupling scheme and $\mu_{eff} = 1\ \mu_B$ in the intermediate coupling scheme). The ordered moment obtained from neutron diffraction is $0.77 \pm 0.07\ \mu_B$; it is in agreement either with the J = 5/2 ordered moment or with the Γ_8 crystal field level ordered moment. Lander and Lam[25)] made a detailed discussion of the form factor result taking into account exchange mixing of the Γ_7 and Γ_8 levels which they find very closely spaced ($\sim 15\ cm^{-1}$). In agreement with this local approach is the low electronic specific heat coefficient $\gamma = 3$ mJ/mol K^2 as well as the value of the excess entropy at the transition temperature[26)].

3. *Itinerant Magnetism*

a) *Criteria for Itinerant Magnetism*

As in the case of transition metals magnetism, the question of localized magnetic moments has been debated for some time and is still open.

Fradin[27)] has considered the question recently and emphasized the relationship between particular measurements and the way they project out the localized and the band nature of the 5f electron states.

At the beginning of this section we said that a localized picture is very often the best starting point for U, Np and Pu compounds. Is there any clear-cut criterion to direct us in the way of itinerant magnetism? Good experimental evidence seems to be
– a reduced ordered moment (in particular when it is incompatible with the effective moment in the crystal field scheme[17)]) *and* – a large electronic specific heat (> 10 mJ/mol K^2) *and* – a reduced magnetic entropy at the transition ($\ll$ RLn 2)

b) *Examples of Itinerant Magnetism*

We present here some examples of itinerant magnetism.

– *UN:* uranium nitride orders antiferromagnetically at $T_N = 52$ K. While it is possible to explain its low order moment ($\mu_0 = 0.75\ \mu_B$) and its high effective moment ($\mu_{eff} = 3.1\ \mu_B$) assuming a $5f^2$ configuration with a Γ_4 ground state and a small crystal field splitting (see Table 3), it is *not* possible to explain concurrently the high electronic specific heat coefficient ($\gamma = 50$ mJ/mol K^2) and the small entropy at the transition (~ 0.1 Rln 2) as has been shown by De Novion[29]. These properties on the contrary strongly suggest itinerant antiferromagnetism. Strong support has been recently obtained from two kinds of measurements

- the rapid decrease under pressure of both T_N and μ_{ord} at the same rate[30]
- the observation of a structure in the photoemission spectrum corresponding to gap opening[31].

Both sets of results are in good agreement with the Fedder and Martin model of itinerant magnetism[32]. It should be noted that in the case of PuN, the absence of a detectable ordered moment by neutron diffraction[33] together with a high electronic specific heat coefficient (γ = 64 mJ/mol K^2 and a small entropy at the transition 0.004 Rln 2)[31] also strongly support the assumption of itinerant antiferromagnetism.

These experimental results received theoretical support: from band calculation results, Brooks[35] found that for UN and PuN the Stoner criterion was just satisfied and that these nitrides were itinerant magnetic systems.

– *$NpSn_3$:* in this antiferromagnet (T_N = 9 K) the ordered moment is reduced to μ_{ord} = 0.28 μ_B. The most elegant experimental evidence comes from specific heat measurements[36]: the electronic specific heat fits well a BCS (*B*ardeen-*C*ooper *S*chrieffer) shape predicted by the gap opening at the transition (Fig. 1). The large change in γ at the transition (γ_{ord} = 88 and γ_p = 242 mJ/mol K^2) is further evidence of the itinerant nature of magnetism. The γ_p value is the largest measured thus far in actinides.

– *$NpOs_2$:* this intermetallic Laves phase is ferromagnetic below T_c = 7.5 K with μ_{ord} = 0.4 μ_B while μ_{eff} = 3.3 μ_B[37]. The magnetic entropy is 0.2 RLn 2 while γ = 205 mJ/mol K^2[38]. Additional evidence for itinerant magnetism comes from the very large superimposed susceptibility in the ordered state

$$\chi_{ord} = 1.2 \pm 0.2 \times 10^{-2} \text{ emu/mol}$$

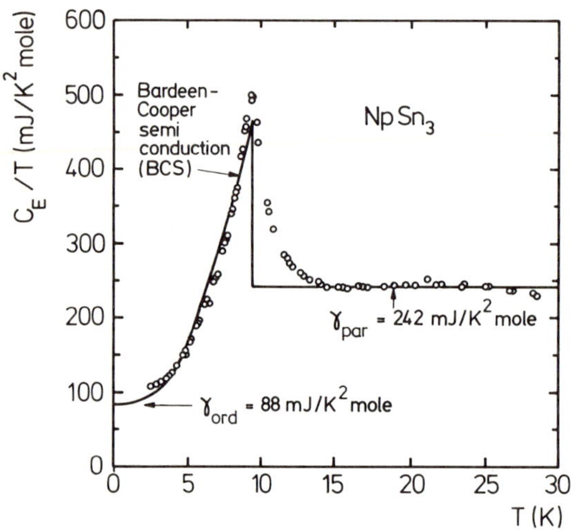

Fig. 1. C_E/T versus T for $NpSn_3$ where C_E is the electronic specific heat. The *solid curve* is the mean-field theoretical prediction. The electronic specific heat coefficients of the paramagnetic and ordered states are designated by γ_{par} = 242 and γ_{ord} = 88 mJ/(mol K^2), respectively. (Trainor et al.[36])

By using this low temperature superimposed susceptibility, the transition temperature and the Curie constant immediately above T_c, Dunlap et al.[39] found that $NpOs_2$ fulfilled the Murata-Doniach requirements for a weak itinerant ferromagnet[40].

4. Very Weak Itinerant Magnetism

A limiting case of itinerant magnetism is very weak ferromagnetism, a well known example of which is $ZrZn_2$ ($\mu_{ord} = 0.15\ \mu_B$). The Stoner-Wohlfarth-Edwards model of very weak magnetism[41] was applied to $NpOs_2$ by Brodsky et al.[38] although the ordered moment ($0.4\ \mu_B$) is not very low. A better example of very weak magnetism is UNi_2 (hexagonal Laves phase structure). While Brodsky et al.[42] give an ordered moment $\mu_{ord} = 0.13\ \mu_B$/f.u., two different studies by Sechovsky et al.[43] and by Fournier[44] indicate $0.045 < \mu_{ord}\ 0.06\ \mu_B$ with a Curie temperature $T_c = 21$ K. In each case these values of μ_{ord} and T_c were obtained by means of Arrot plots[45] (Fig. 2). An interesting question is: where is the magnetization located? In a simple minded band approach, by analogy with lanthanides the 3d band is filled by charge transfer from uranium[44]: as a matter of fact there is no moment on Ni in $LnNi_2$ Laves phases.

To support the argument is the fact that UCo_2 does not order but is a Pauli enhanced paramagnet. The very interesting conclusion is then that magnetic ordering is due to 5f electrons of uranium, although U-U distance (3.0 Å) is even smaller than in uranium metal (3.1 Å). We propose that in contrast with uranium metal, 5f electrons are *de-hybridized* because the uranium 6d band is almost empty and the nickel 3d band almost full. Then we are left with a pure, narrower, 5f band for which the Stoner criterion may be fulfilled. Above T_c, the reciprocal susceptibility is strongly curved as is usually the case for very weak ferromagnets[46]. Very recently, the specific heat of UNi_2 at low temperature has been measured[47]; the obtained γ value is very large ($\gamma = 85$ mJ/mol K^2).

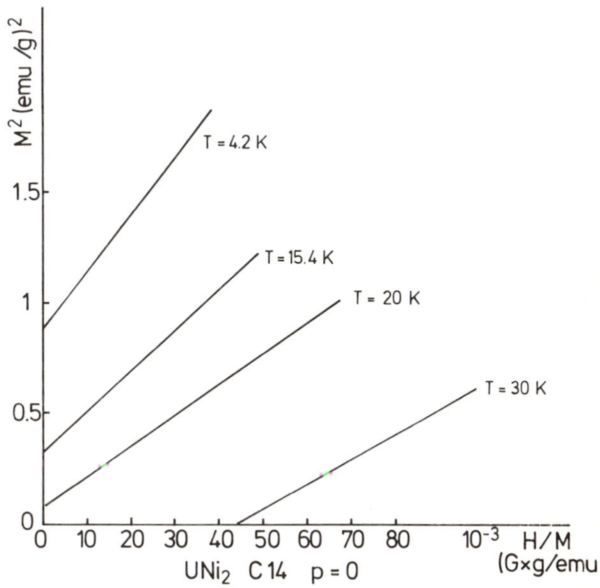

Fig. 2. Arrott plots (M^2 versus H/M) of the magnetization of UNi_2 (Fournier[44])

5. *Spin Fluctuations*

When, in metallic systems, the Stoner criterion is nearly fulfilled, a number of physical properties are quite anomalous. The explanation has been looked for either in spin fluctuation theory and/or in charge (valence) fluctuation theories[48]. The common physical basis is the existence of a narrow band – here 5f – pinned at the Fermi level; the increase of its width $1/\Delta$ can be identified as the fluctuation time τ_{vf}. If Δ is very narrow (as for example SmB_6) then τ_{vf} becomes of the order of experiment probing time: for example, in XPS measurements $\tau_{XPS} < \tau$ so that the incident photons "see" both f^n and f^{n-1} configuration[49]. On the contrary, lattice constant or isomer shift measurement have time scales larger than τ_{vf}. They will "see" only an average intermediate configuration[5]. If Δ is larger, which is surely the case of light actinides compared to lanthanides then all experiment time scales will be larger than τ_{vf}: whatever the probe, only an intermediate configuration is found, and no evidence for valence fluctuation will be found.

However, the intra-atomic Coulomb interaction $U_{f\text{-}f}$ affects the dynamics of f spin and f charge in different ways: while the spin fluctuation propagator $\chi(q, \omega)$ is enhanced by a factor $(1 - U_{ff}\chi^0(q, \omega))^{-1}$ which may exhibit a phase transition as $U_{f\text{-}f}$ is increased, the charge fluctuation propagator $C(q, \omega)$ is depressed by a factor $(1 + U_{ff}C^0(q, \omega))^{-1}$. In the case of light actinide materials no evidence of charge fluctuation has been found. Most of the theoretical effort for the concentrated case (by opposition to the dilute one-impurity limit) has been done within the Fermi liquid theory[51]. Main practical results are: a T^2 term in electrical resistivity, scaled to order T/T_{sf} where T_{sf} is the characteristic spin fluctuation temperature (which is of the order $- T_F/S$ where S is the Stoner enhancement factor ($S = 1/1 - IN(\mu F)$) and $T_F \sim \Delta/k_B$ is the Fermi temperature of the narrow band).

– a saturated susceptibility $\chi(0)$ at $T = 0$ K, enhanced by the Stoner factor and a $\chi(0)(1 - \alpha(T/T_{sf})^2)$ behaviour at low temperature

– a cross-over from enhanced Pauli paramagnetism to free spin behaviour for $T_{sf} < T < T_F$

– a $T^3 \ln T/T_{sf}$ term in the specific heat.

Many light actinide alloys which are not magnetic have a T^2 dependence of the resistivity at low temperature as well as a large electronic specific heat coefficient γ (Table 4). However, the archetype of a spin fluctuation system is UAl_2. The electrical resistivity is proportional to T^2 with a very large coefficient $a = 0.15\ \mu\Omega cm/K^2$ up to 5 K[52].

The magnetic susceptibility decreases as T^2 from the lowest temperature $T = 80$ mK with a coefficient $T^+ = 13$ K[53]. Above 10 K a modified Curie-Weiß law is followed. The specific heat is the only known case of a $T^3 \ln(T/T_{sf})$ behaviour with $T_{sf} = 25$ K and $\gamma =$

Table 4. Actinide intermetallics showing spin fluctuations' behaviour

Compound	γ(mJ/mol K^2)	a($\mu\Omega$ cm/K^2)
UGa_3	54	$\sim 20 > 10^{-3}$
USn_3	164	$\sim 300 \times 10^{-3}$
UAl_2	142	$\sim 200 \times 10^{-3}$
$NpRh_3$	85	$\sim 5 \times 10^{-3}$

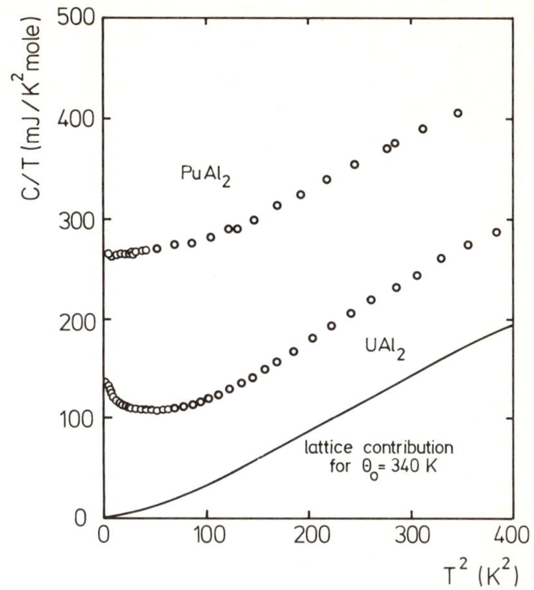

Fig. 3. C/T versus T^2 for UAl_2 and $PuAl_2$ up to 20 K. Also shown is the model lattice contribution for θ_D = 340 K. (Trainor et al.[54])

142 mJ/mol K^2 (Fig. 3)[54]. Photoemission measurements[55] have shown the existence of a narrow 5f band pinned at the Fermi energy. UAl_2 is even more interesting since it shows indeed two regimes[55]. While at low temperature $T < T_{sf} \sim 25$ K the behaviour is well described within the Fermi liquid phenomenology, at high temperature (T > 50 K) UAl_2 behaves rather like a collection of magnetic impurities ($\mu_{eff} \sim 3\ \mu_B$) with a Kondo temperature ~100 K. In fact, the $\chi(T)$ curve, for UAl_2 bears strong similarity with a liquid gas Clapeyron phase diagram and is illustrative of a Fermi liquid-gas phase diagram. The analogy may even be pushed further when looking to pressure effects[56] (Fig. 4).

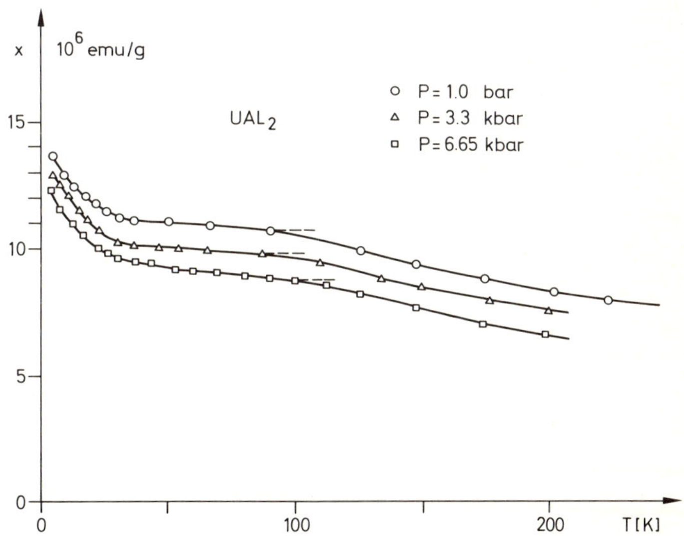

Fig. 4. Susceptibility of UAl_2 versus temperature for different pressure. (Fournier and Beille[56])

IV. The Macroscopic Properties

The experimental methods employed for the study of magnetism in solids are here generally described. Emphasis is given to the relevance of these methods for actinide solids. Therefore, the theory which is used in connection with the experiments is analyzed in some detail.

The complex magnetic behaviour of actinide solids, described in this chapter, as well as in Chap. A, can be better understood when a set of measurements is performed together. This is why we discuss here not only magnetic measurements (susceptibility and magnetization) but also transport properties and specific heat, which are sensitive to the band structure of the solid and to the form of its density of states (see Table 4 in Chap. A). We have called this section: "Macroscopic properties", in which thermodynamically averaged functions are measured, rather than microscopic, atomic ones. The latter (neutron, electron, and photon scattering and absorption experiments) will be discussed later in this chapter, with regard to neutrons, and in Chap. E with regard to the photoelectron spectroscopic method, of great relevance to our problem.

1. Paramagnetic Susceptibility

Magnetic susceptibility measurements are basic to the study of the magnetic properties of a compound: samples under powder form are sufficient to start with and, what is very important in the case of actinides, small amounts of material are satisfactory (typically ~100 mg). When working with transuranium compounds, safety requirements are fulfilled by working with sealed containers[57].

The magnetic susceptibility can be written as a sum of different contributions

$$\chi_{exp} = \chi_D + \chi_c + \chi_{loc.} \tag{1}$$

where χ_D is the core diamagnetism, χ_c is the conduction electron term and χ_{loc} is the susceptibility of 5f localized levels.

a) Core Diamagnetism

Is due to the orbital moments of the electrons in the complete Radon shell and is expressed as (per mole)

$$\chi_D = -\frac{e^2N}{6\,mc^2}\sum_n \langle r_n^2 \rangle \tag{2}$$

where $\langle r_n^2 \rangle$ is the mean square radial extension of the atomic orbitals. Van Vleck[58] had obtained an empirical formula giving $\langle r_n^2 \rangle$ as a function of the electron quantum numbers. While correct for hydrogen, this formula becomes less valid as the atomic number Z increases: progressive screening of the nuclear charge has to be taken into account. Better results are obtained using the Slater formula[59].

$$\chi_D = -\frac{e^2 N a_0^2}{3\,mc^2} \sum_n \sum_{l=0}^{n-1} \frac{n^{*2}(n^* + 1/2)(n^* + 1)}{(2 - \sigma_{n,l})^2} \tag{3}$$

where $\sigma_{n,l}$ is a screening constant and n* an "effective" quantum number. In the case of Rn, one obtains $\chi_D^{Rd} = 76 \times 10^{-6}$ uem/mole[57]. The diamagnetism of Rn may also be obtained empirically in the following way: if we calculate $(\chi_D^{exp})/Z$ for the rare gases, the obtained ratio is always close to 0.8×10^{-6} uem/mol in good agreement with the value calculated using the Slater formula. This proportionality to Z is also verified using the Slater formula. In the case of H, $X_D^H \approx e^2 N a_0^2 = -0.79 \times 10^{-6}$ uem/mol. It follows that in the case of rare gases we have the interesting following result:

$$\sum_n \langle r_n^2 \rangle \cong Z a_0^2 \tag{4}$$

The $\langle r_n^2 \rangle$ values have now been accurately calculated by means of Dirac-Fock atomic calculations[6]. Using Desclaux's data[60], we have calculated the rare gases diamagnetism; the values obtained agree very well with the experimental ones (Table 5). Thus the relation (4) fits well to calculated $\langle r_n^2 \rangle$ values.

We have calculated the Radon core diamagnetism for actinide ions up to americium; the results are reported in Table 6. These values are smaller than usually assumed in the literature. Diamagnetic contribution for localized 5f electrons are also given – From Table 6, we see that the core diamagnetism is large and has to be taken into account when a detailed analysis of the susceptibility is made. In the case of Th metal, for example, it amounts to 40% of the experimental susceptibility.

b) Conduction Electron's Susceptibility

The s electrons may be considered as free electrons. In such case, the orbital diamagnetism is the Landau term

$$\chi_L = -1/3\, \mu_B^2 N(E_F) \tag{5}$$

The orbital paramagnetism is 0 (l = 0) and the spin paramagnetism is the Pauli term (see Chap. A)

$$\chi_p = \mu_B^2 N(E_F) \tag{6}$$

Table 5. Rare gases diamagnetism

Rare gas	$\chi_D^{theor} \times 10^6$ uem/mol	$\chi_D^{exp} \times 10^6$ uem/mol	$\lvert\chi_D^{exp}\rvert/Z \times 10^6$ uem/mol
He	− 1.87	− 1.54	0.77
Ne	− 7.4	− 7.2	0.72
Ar	−20.5	−19.4	1.07
Kr	−31	−28	0.78
Xe	−48	−43	0.80
Rn	−60		

Table 6. Radon core and localized 5 f electrons diamagnetic correction for light actinide ions

Ion	$\chi_D^{Rn} \times 10^6$ uem/mol	$\chi_D^{5f} \times 10^6$ uem/mol
Th^{+4}	−38	
Pa^{+4}	−36	−2
Pa^{+5}	−35	
U^{+3}	−36	−5
U^{+4}	−34	−3
Np^{+3}	−34	−7
Np^{+4}	−33	−4
Pu^{+3}	−33	−8
Pu^{+4}	−31	−5
Am^{+3}	−31	−8
Am^{+4}	−30	−6

Then the s contribution, given by

$$\chi_s = 2/3\ \mu_B^2 N_s(E_F) \tag{7}$$

is of the order of 10×10^{-6} uem/mol.

In the case of narrow bands – and this will be the case of hybridized 5 f bands when 5 f electrons are itinerant – an approximate treatment has to be done. Kubo and Obata[61)] have studied the case of transition metals in the tight binding approximation. The narrow band susceptibility is the sum of 4 terms

$$\chi_{5f} = \chi_L + \chi_0 + \chi_{s-0} + S \cdot \chi_S \tag{8}$$

χ_L is the Landau diamagnetism: in a first approximation, formula (5) may be corrected by a factor m/m* where m is the effective mass of the considered electron; m/m* being much smaller than one, χ_L may be neglected.

χ_0, χ_{s-0} and χ_s are orbital, spin-orbit and spin contributions, S is the Stoner enhancement factor (see Sect. III of this chapter). χ_s, χ_0 and χ_{s-0} are given by[61)]:

$$\chi_s = \mu_B^2 \sum_n \sum_{n'} \frac{f(E_n) - f(E_{n'})}{E_{n'}(k) - E_n(k)} |\langle nk|2s|n'k\rangle|^2 \tag{9}$$

$$\chi_0 = \mu_B^2 \sum_n \sum_{n'} \frac{f(E_n) - f(E_{n'})}{E_{n'}(k) - E_n(k)} |\langle nk|L|n'k\rangle|^2 \tag{10}$$

$$\chi_{s-0} = \mu_B^2 \sum_n \sum_{n'} \frac{f(E_n) - f(E_{n'})}{E_{n'}(k) - E_n(k)} |\langle nk|L|n'k\rangle\langle n'k|2s|nk\rangle| \tag{11}$$

These susceptibilities have been worked out in the case of transition metals[62, 63)]. These results may be used to indicate the importance of these terms for actinide metals[57)]: while χ_{s-0} is negative and comparable to the Radon core diamagnetism, χ_0 appears large, up to 50% of the experimental susceptibility. A very important term in (8) is the Stoner parameter S which is a central quantity for spin fluctuation models[64] (see above). We may evaluate it by comparing the $S \cdot \chi_s$ term obtained by substracting from $\chi_{exp.}$ all the other terms with a bare Pauli susceptibility calculated from the 5f-6d density of states; this density of states may be evaluated in two ways

- using a covalent 5f-6d band model, as proposed by Friedel for uranium[65]
- from the electronic specific heat coefficient γ account being taken of 7s contribution and of electron-phonon enhancement. Results are reported in Table 7. It is quite surprising that using the bare χ_{exp} and $\gamma_{exp.}$, similar s values are found.

Thermal broadening effects of the density of states should also be taken into account yielding

$$\chi(T) = \chi(0)\left\{1 - (\pi^2 k_B^2 T^2/6)\left(\left(N^{-1}(E)\frac{dN(E)}{dE}\right)^2 - N^{-1}(E)\frac{d^2N(E)}{dE^2}\right)\right\}_{E=E_F} \tag{12}$$

c) 5f Localized Levels Susceptibility

The general formula giving the paramagnetic susceptibility for free ions has been established by Van Vleck[58]

$$\chi = \frac{\Sigma_J(b_J + (g_J\beta^2 J(J+1))/3K_BT)(2J+1)\exp(-E_J^0/k_BT)}{\Sigma_J(2J+1)\exp(-E_j^0/k_BT)} \tag{13}$$

Here, b_J is an atomic quantity characteristic of the particular state J, $\beta = e\,h/2\,mc$, g_J is the Lande factor.

In the case, where multiplet splitting is large compared with T – which may be assumed to be the case for actinides – only the ground multiplet intervenes and (13) reduces to the well known Curie law

$$\chi = \frac{C}{T} \tag{14}$$

Table 7. Stoner enhancement factor S determination for light actinide metals

	Th	U	Np	Pu
$S \times \chi_s$ (st/eV.at.)	1.1	8.6	15.6	12
χ_p (band model) (st/eV.at.)	1.5	2.2	3.4	4.1
$\chi_p(\gamma)$ (st/eV.at.)	1	2.75	4	2.9–6.4
S (band model)	0.7	3.9	4.6	2.9
$S(\gamma)$	1	2.5	3.9	4.1–1.9
S(bare)	1.5	2.7	3	3.6–1.7

where

$$C = \frac{N}{3k_B}\mu_{eff} \text{ and } \mu_{eff} = \mu_B g_J \sqrt{J(J+1)}$$

If excited states are slightly thermally populated, they give rise to a constant susceptibility χ_0 and the susceptibility takes the modified Curie law form

$$\chi = \chi_0 + \frac{C}{T} \quad (15)$$

When the ion is embedded in a matrix, the crystal field splits the ground multiplet and the crystal field levels give rise to the same problem as the free ion multiplets, to be solved in the same way. If several crystal field levels are involved, a complete treatment using Eq. (13) is needed.

In a magnetic compound we have to consider exchange interactions; in the molecular field approximation, the Curie law takes the Curie Weiss form

$$\chi = \frac{C}{T - \theta} \quad (16)$$

θ being negative for antiferromagnetic coupling.

Very often, experimental data are well fitted by the so called modified Curie-Weiss Law:

$$\chi = \chi_0^* + \frac{C^*}{T - \theta} \quad (17)$$

where the experimental χ_0 may be a Pauli or Van Vleck term. However, adding a constant term to the Curie law leads to a renormalization of the Curie-Weiss constant as well as of the constant term[66]: χ_0^* and C^* are different from the physical C and χ_0. This applies both to the ferromagnetic and to the antiferromagnetic case.

We take as an example the case of plutonium monopnictides PuP, PuAs, PuSb, the susceptibility of which is well fitted by Eq. (17). Results are given in Table 8: while bare data are in agreement with the full J multiplet value ($\sim 1\ \mu\beta$ in the intermediate coupling), the corrected value is substantially smaller and points towards a Γ_8 ground state (see Table 8) in agreement with more detailed discussions[25].

Table 8. Bare (*) and corrected values for the constant susceptibility and the effective moment of plutonium monopnictides

	$\chi_0^* \times 10^6$ uem/mol	μ_{eff}^* (in μ_B)	$\chi_0 \times 10^6$ uem/mol	μ_{eff} (in μ_B)
PuP	190	1.06	157	0.87
PuAs	330	0.98	213	0.63
PuSb	200	1.01	171	0.85

It should be noted that while $C < C^*$ and $\chi_0 < \chi_0^*$ for ferromagnetic coupling, the contrary holds in the case of antiferromagnetic coupling.

The analysis of the paramagnetic susceptibility makes it possible – in principle – to determine whether the full J ground multiplet (for $k_BT \gg$ crystal field splitting) or on the contrary the crystal field ground state (for $k_BT \ll$ crystal field splitting) are determining magnetism. The crystal field ground state is more important for comparison with the ordered magnetic state. Such analysis has been done, for example, in the case of UAs[67]. It was concluded, in agreement with the ordered magnetic moment value $\mu_0 = 2.2\ \mu_B$ that Γ_8 was the ground state, Γ_6 being the first excited state.

2. Magnetization

a) Low Field Magnetization and Anisotropy

Very large anisotropy seems to be the rule for actinide magnetic compounds (except weak ferromagnets). This raises the question of the use of data mainly obtained on powders: knowing the easy axis of magnetization, it is always possible to relate averaged powder saturation moment and the saturation moment along the easy axis, μ_{sat}. (Distinguish between the saturation moment μ_{sat} and its localized part μ_{ord} which is the true local 5f moment.)

In the case of cubic symmetry, one obtains:

– $\mu_{sat} = 1.27\ \mu_{powder}$ for $\langle 100 \rangle$ easy axis

– $\mu_{sat} = 1.1\ \mu_{powder}$ for $\langle 111 \rangle$ easy axis

In the case of tetragonal symmetry, the difference reaches

$\mu_{sat} = 2\ \mu_{powder}$ for $\langle 001 \rangle$ easy axis.

In the case of PuP, μ_{sat} is unkown, but from neutron diffraction[68] the easy axis is $\langle 100 \rangle$ thus $\mu_{sat} = 1.27\ \mu_{powder} = 0.6\ \mu_B$ and not $0.42\ \mu_B$. This correction is important because the difference between $\mu_{ord} = 0.77\ \mu_B$ obtained by neutron diffraction and $\mu_{powder} = 0.42\ \mu_B$ has been attributed to a huge conduction electron polarization of $-0.35\ \mu_B$[25]. Although important, this term should in reality reach only half this value ($-0.17\ \mu_B$).

In the case of tetragonal $NpAs_2$ the extreme discrepancy is observed with $\mu_{powder} = 0.66\ \mu_B$[68] and $\mu_{ord} = 1.45\ \mu_B$[69].

Detailed magnetization measurements have been done on US[70] (Fig. 5). An extremely large anisotropy restrains the magnetic moments to the $\langle 111 \rangle$ axis resulting in a near $\cos\theta$ angular dependence of the magnetization away from the $\langle 111 \rangle$ axis. The anisotropy constant K_1 ($T = 0$) is about $8.5 \times 10^8\ erg/cm^3$ from which an anisotropy field $H_A = 2\ K_1/M_{sat} = 3000$ kOe is deduced (M_{sat} is the saturation magnetization). This extremely large value is of the same order as estimated for other uranium compounds, for example UAsS[71].

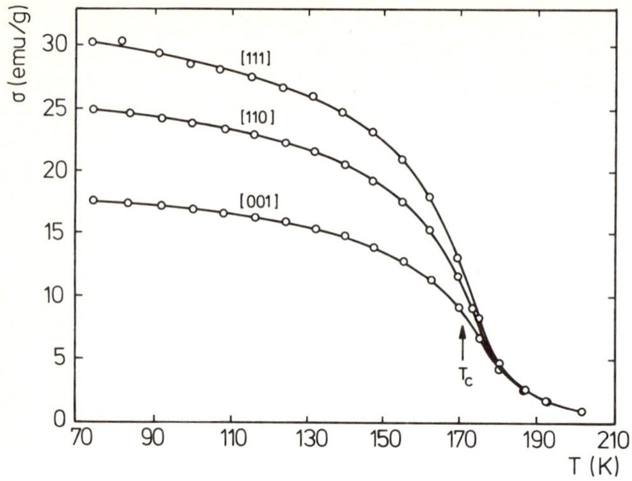

Fig. 5. Temperature dependence of the magnetization σ of US measured along the main crystallographic directions upon cooling from the paramagnetic region in 15 kOe. (Tillwick and Du Plessis[70)])

Ferromagnetic U_3As_4 and U_3P_4 have also been extensively investigated and show large magneto-crystalline anisotropy[71)]. For U_3As_4, measurements in pulsed fields up to 230 kOe[72)] give $K_1 = +2.7 \times 10^6$ erg/cm^3 and an extraordinary $K_2 = +3.5 \times 10^8$ erg/cm^3.

Intermetallic U_2Co_{11} is a highly anisotropic uniaxial ferromagnet ($H_A > 60$ KOe) which may be of practical interest[73)]: the high Curie temperature ($T_c = 360$ K) and the large magneto-optical Kerr effect suggest this material for thermal magnetic recording of information in magnetic domains (characteristics domain dimension: 0.2 μm).

In the case of very weak ferromagnets, like UNi_2[43, 44)] and UPt[74)], the anisotropy is very small but problems are connected with M(H). There is of course no saturation of the magnetization due to the very large superimposed susceptibility in the ordered state. Spontaneous magnetization may be conveniently obtained from Arrott plots (linear M^2 versus H/M for small M values) and extrapolation to $H = 0$. Such results for UNi_2 are shown on Fig. 2.

b) High Field Magnetization and Magnetic Phase Diagrams

In the preceding section, the magnetic field was mainly used to determine the saturation magnetization of ferromagnets. However, it is now well established that many actinide compounds have complex magnetic behaviour and the application of an external field may change or suppress antiferromagnetic structures.

Thus UAs is antiferromagnetic below $T_N \approx 127$ K and a change of antiferromagnetic order occurs at 66 K. Measurements as a function of temperature and crystal orientation in fields up to 150 kOe[75)] reveal first strong anisotropy, then the existence of a ferromagnetic phase with $\langle 110 \rangle$ easy axis in the temperature range 2 – 66 K and just around T_N (Fig. 6). Similar although not identical behaviour is found in UP[76)].

In the case of transuranium compounds, measurements on powder samples in fields up to 100 kOe show spin rearrangement in NpAs[77)] and $NpAs_2$[69)].

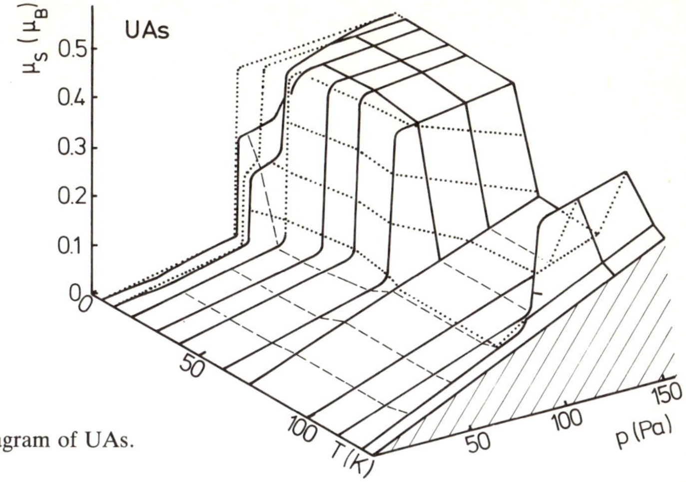

Fig. 6. Magnetic phase diagram of UAs. (Busch et al.[78])

NpAs is antiferromagnetic below T_N = 172 K; at 155 K occurs a transition toward another antiferromagnetic state and at 142 K a first order phase transition takes place toward still another antiferromagnetic state[78]. This very complex magnetic behaviour is much simplified under high field and above 50 kOe NpAs becomes a simple ferromagnet up to $T_c \simeq$ 175 K[77]. The understanding of the magnetic phase diagram (Fig. 7) needs urgently neutron diffraction studies.

$NpAs_2$ is antiferromagnetic below T_N = 52 K and becomes ferromagnetic below T_c = 18 K[69]. Under field, the Curie point rises rapidly, until the antiferromagnetic phase is suppressed by 30 KOe (Fig. 8).

Complicated magnetic phase diagrams also occur for solid solutions like for example $U As_x Se_{1-x}$ as a function of x (Fig. 9)[79]. The transition from the ferro to the antiferromagnetic region is accompanied by the occurrence of numerous phase changes. Under high fields, the ferromagnetic region is extended to higher x value.

All these complicated magnetic behaviours need further study by means of neutron diffraction and such results will be discussed later on.

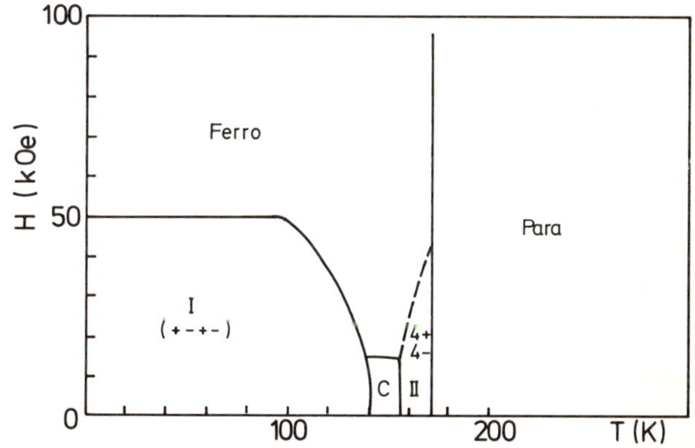

Fig. 7. Tentative magnetic phase diagram of NpAs. (Blaise et al.[77])

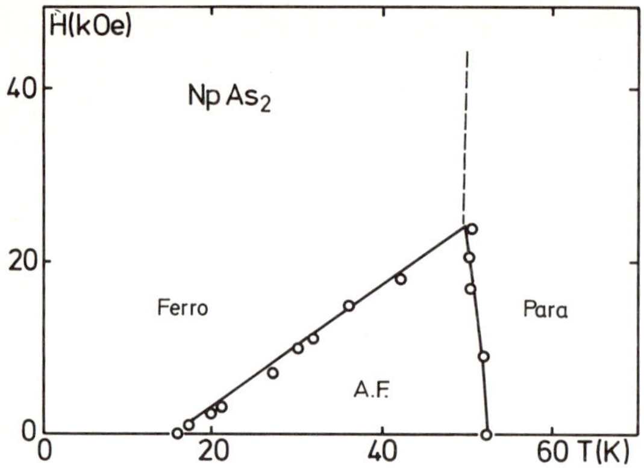

Fig. 8. Magnetic phase diagram of $NpAs_2$. (Blaise et al.[77])

3. Transport Properties

While magnetic measurements give direct macroscopic informations about the magnetic properties of a compound, indirect but very useful information may be obtained from other physical properties (transport properties, optical properties, elastic properties). In this paragraph we are going to restrict ourselves to transport properties and still more restrictively to electrical resistivity and specific heat; these measurements are more limited than magnetic ones in the case of the actinides: for resistivity, solid samples with well defined dimensions are needed and electrical contacts must be made to the samples, specific heat measurements require – up to now – large amounts of material (~ 1/g) and are complicated by self-heating.

a) Electrical Resistivity

The very first information obtained is whether a given compound is a conductor, semiconductor or insulator. Such a basic answer is of importance concerning the 5f localiza-

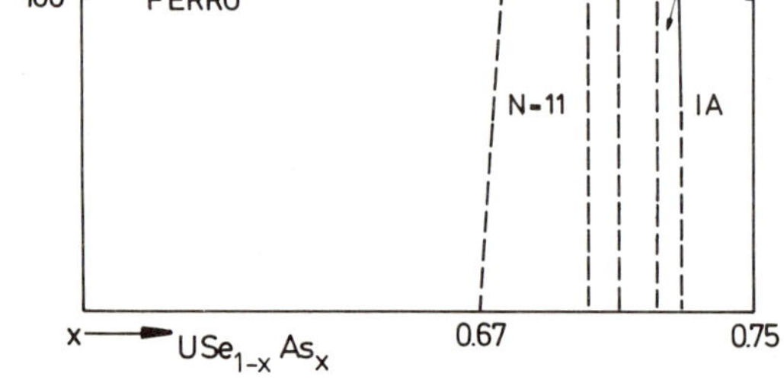

Fig. 9. Schematic magnetic phase diagram of the USe-UAs system. (Suski[79])

tion and the magnetic couplings. The electrical resistivity of a conducting compound may be written as a sum of separated contributions

$$\varrho_{exp} = \varrho_0 + \varrho_{el.-ph} + \varrho_{el-el} + \pi_{el-mag} \quad (19)$$

Apart from the temperature independent residual resistivity ϱ_0, mainly due to lattice defects, charge carriers are scattered either by phonons (ϱ_{el-ph}), or by interband electron-electron scattering or by magnons, spin fluctuations, Kondo scattering (ϱ_{el-mag}). Our interest is in ϱ_{el-mag} and the obvious problem is to extract it safely from the experimental resistivity. In principle it is possible to calculate ϱ_{el-ph} and ϱ_{el-el} but practically the easiest and probably most accurate way is to measure isostructural nonmagnetic compounds (by isostructural we mean not only the same crystal structure but also the same band structure near the Fermi level; this second condition cannot be satisfied in the case of itinerant 5f systems for which separation between ϱ_{el-el} and ϱ_{el-mag} is very puzzling). There is also a temperature dependent contribution due to the thermal broadening of the density of states, as for magnetic susceptibility, which gives a correcting factor to ϱ_{el-el}

$$\varrho_{el-el}(T) = \varrho_{el-el}(0)\left\{1 - \frac{\pi^2 k_B^2 T^2}{6}\left(3\left(\frac{1}{N(E)}\frac{dn(E)}{dE}\right)^2 - \frac{1}{N(E)}\frac{d^2N(E)}{dE^2}\right)\right\}_{E=E_F} \quad (20)$$

Resistivity has proved to be a very sensitive probe to the possibilities of magnetism: contrary to the susceptibility which samples the spin fluctuation propagator (see Sect. III) only for q = 0, the electrical resistivity integrates over all q values.

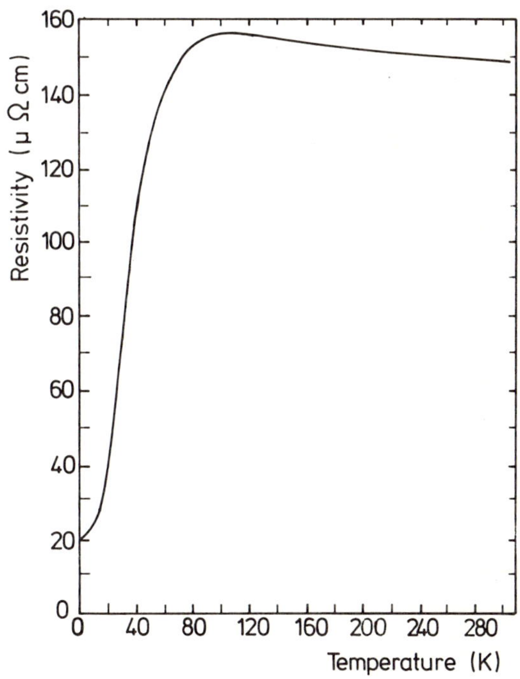

Fig. 10. Temperature dependence of the electrical resistivity of α-Pu (Meaden[80])

One of the most famous and most puzzling results has been the electrical resistivity of plutonium metal[80] (Fig. 10). It rises as T^2 at low temperature and saturates at high temperature, which is a typical magnetic behaviour. The absence of magnetic ordering having been established, physical interpretation came through spin fluctuation theory[81]: scattering by spin fluctuations gives a T^2 term for $T < T_{sf}$, the characteristic spin fluctuation temperature. As T increases, these fluctuations are destroyed and at high temperature one eventually recovers a spin disorder resistivity[82]. However, when applied quantitatively to Pu, $\varrho_{el\text{-}ph} + \varrho_{el.\text{-}el}$ was simply assumed to be equal to the thorium metal resistivity. This is surely much underestimated due to both strong interband scattering[83] and optical phonon scattering[84]. This overestimate of $\varrho_{el\text{-}mag}$ led to – in our opinion – unphysical values for parameters involved in the spin fluctuation formalism (very large $S = 10$ and very small 5f Fermi energy of $\sim 10^{-2}$ eV).

A better example of a magnetic resistivity term due to spin fluctuations in a non-magnetic compound is UAl_2[52]. $ThAl_2$ has not the same crystal structure so that we choose $LaAl_2$ as the reference non-magnetic compound. In Fig. 11 are plotted both the resistivity of UAl_2 and the difference $\varrho_{UAl_2} - \varrho_{LaAl_2}$. The resistivity difference, more representative of spin fluctuation scattering goes through a broad maximum around 100 K. Figure 12 shows that this resistivity difference has a logarithmic temperature dependence characteristic of a Kondo system above the maximum. Interesting enough this maximum temperature of 100 K – which is not at all apparent in the bare resistivity – corresponds to the change of behaviour in the susceptibility[56]. $T_K \approx 100$ K may be taken as a Kondo temperature above which the system behaves like a collection of non-interacting uranium impurities. Assuming localized spins in UAl_2 above 200 K, the spin resistivity is given by[85]

$$\varrho = \varrho_M\left(1 - \frac{3\,ZJ}{E_F}\ln T\right) \tag{21}$$

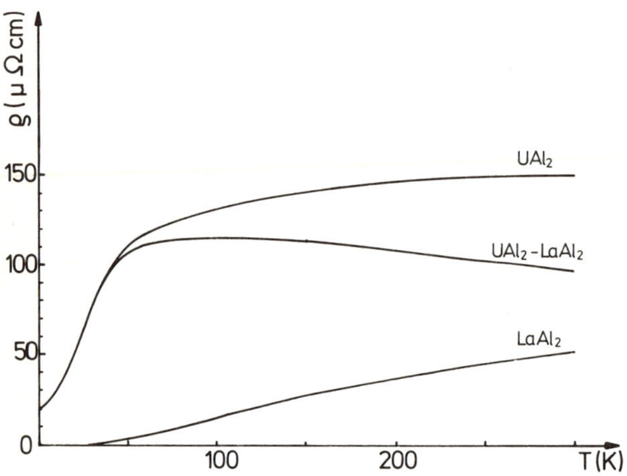

Fig. 11. Electrical resistivity of UAl_2 and its magnetic part taken as the difference with the electrical resistivity of $LaAl_2$

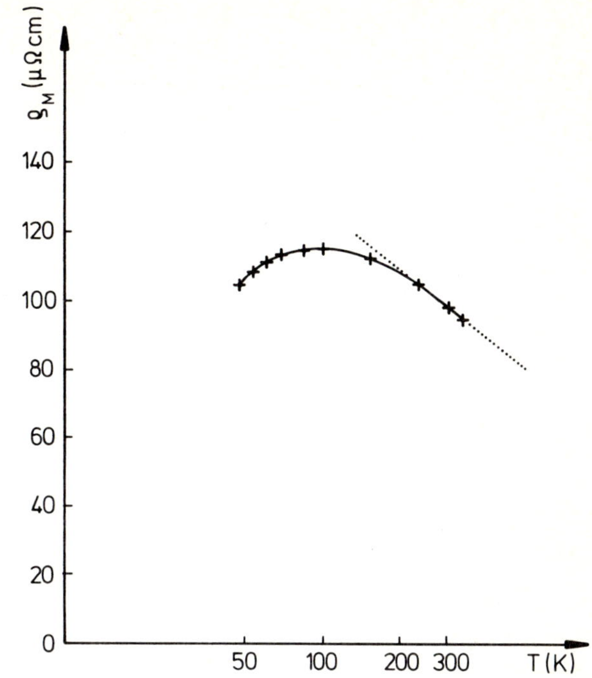

Fig. 12. Magnetic contribution to the electrical resistivity of UAl_2 as a function of the temperature (logarithmic scale)

where ϱ_M is the spin disorder resistivity, Z the number of conduction electrons per atom, J the exchange integral and E_F the conduction electrons Fermi energy. Fitting to Eq. (21) leads to $\varrho_M \sim 250\ \mu\Omega$ cm

$$\text{and} \quad \xi = \frac{3\,ZJ}{E_F} = 0.11 \ .$$

Curium metal is antiferromagnetic[10)] and its resistivity has been measured as well as that of isostructural non magnetic americium metal[86)]. The resistivity difference is taken as the magnetic contribution and is shown in Fig. 13. It passes through a maximum for

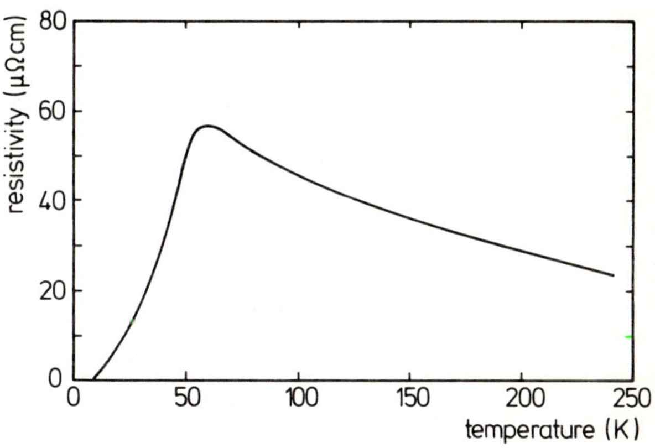

Fig. 13. Magnetic contribution to the electrical resistivity of Curium metal. (Schenkel[86)])

$T_N = 52$ K and above has a ln T dependence, in contrast to the classical spin disorder theory[87], but again in agreement with the Kondo formulation (Eq. 21).

Fitting the data for T greater than 80 K[90] leads to

$$\varrho_M = 156\ \mu\Omega\ \text{cm}$$

$$\xi = 0.15$$

Curium being mainly in a spin only $^8S_{7/2}$ state, a more detailed treatment in the frame of the RKKY model has been made[88] to fit the Curie-Weiss temperature $\theta_p \sim -350$ K. It follows that $Z \sim 1$, suggesting that f-d exchange interaction should be predominant over s-f interaction (contrary to Gd for example).

The exchange integral J is found to be about 0.13 eV. A value $J \sim 0.23$ eV was found[88] using de Gennes spin disorder theory[89].

The resistivity of antiferromagnetic UN ($T_N = 52$ K) was measured on a single crystal along the $\langle 100 \rangle$ easy axis[90] (Fig. 14). Below T_N, the spin disorder resistivity is proportional to $1 - m_n^2$, where m_n is the reduced sublattice magnetization, over nearly the whole ordered region. A small but distinct peak is observed at 0.25 K below T_N, being ascribed to new Brillouin zone boundaries introduced as a result of antiferromagnetic ordering[91]. The large non-linear increase above T_N confirms previous observation[92]. It was ascribed to crystal field effects[93]. On the other hand, assuming UN is an itinerant antiferromagnet, this increase may be ascribed to thermal broadening in interband scattering.

Ferromagnetic US and USe single crystals were also recently studied[94, 95]. While for US a T^2 dependence of the resistivity was observed up to 130 K ($T_c = 180$ K), in the case of USe, for $T < T_c = 160$ K, the resistivity may only be fitted by a $T^2 \exp(-\Delta/T)$ term. The quantity $\Delta = 180$ K is tentatively correlated with a $q = 0$ magnon energy.

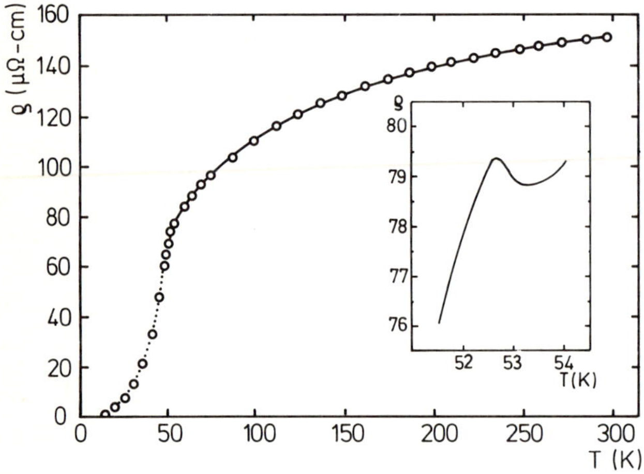

Fig. 14. Temperature dependence of the electrical resistivity of UN along the $\langle 100 \rangle$ direction. (Du Plessis and Van Doorn[90])

b) Specific Heat

As in the case of electrical resistivity, the experimental specific heat may be written as a sum of several separated contributions, leading to the same difficulties for interpreting data[96].

$$C_p = C_d + C_n + C_{latt} + C_{el} + C_{mag} + C_{Schottky} \quad (22)$$

The dilatation term C_d is due to anharmonic forces; at high temperature, it may reach several percent of C_p. The nuclear term C_n is usually negligible above 1 K. The lattice term C_{latt} is normally predominant above 10 K. At low enough temperature ($T < \theta_D/50$ when θ_D is the Debye temperature), the Debye model fits well and leads to $C_{latt} \alpha T^3$. The conduction electron contribution is $C_{el} = \gamma T$ the coefficient γ being proportional to the density of states at the Fermi level. Thus, plotting C/T versus T^2 at low temperature (typically below 10 K) will lead to the determination of γ. When the density of states is very high, $N(\mu_F)$ will be temperature dependent due to thermal broadening as for the susceptibility and the electrical resistivity and γ(300 K) may be as low as 50% $\gamma(0)$.

The magnetic term is associated with changes in internal energy related to magnetic transitions. Its thermal dependence leads to the magnetic entropy and enthalpy determination. The Schottky term comes from the excitation of higher lying crystal field levels in compounds with localized 5 f levels.

For non-magnetic actinide metals, specific heat data have been employed very usefully to corroborate 5 f localization starting with Am, as indicated by the sudden drop in γ values[97] ($\gamma_{Pu} \approx 12$ mJ/mol K^2, $\gamma_{Am} \approx 2$ mJ/mol K^2). It allowed also the discovery of superconductivity in Pa and Am metals[97, 98].

For spin fluctuation systems, the best example is again UAl_2[54]. In fact as shown in Fig. 3 it is the unique example where $T^3 \cdot \ln T$ dependence at low temperature, predicted by spin fluctuation theory, has been in practice observed (very recently, such behaviour

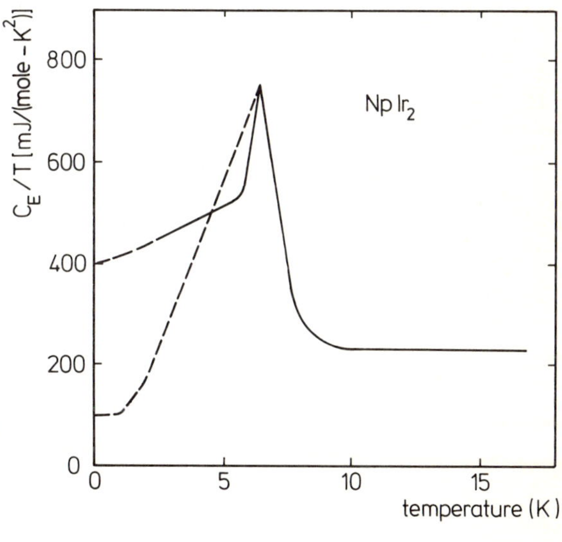

Fig. 15. Electronic specific heat C_E of $NpIr_2$. The *dashed line* for $T < T_N$ is $C_E/T = 100 + 1800 \exp(-6.6/T)$. (Brodsky and Trainor[101])

seems to have been observed in UCo_2 also[99]). The obtained γ value is huge ($\gamma(0)$ = 142 mJ/mol K^2) approaching values observed in some valence fluctuating compounds like $CeAl_3$[100]).

Figure 15 displays specific heat data for itinerant antiferromagnetic $NpIr_2$[101]) (T_N = 7 K) while results for $NpSn_3$[30]) are presented in Fig. 2, showing the BCS-like specific heat anomaly; the entropy change at the transition is almost 0. In the case of $NpIr_2$, the saturation is less clear: bad fitting below T_N is attributed to second phase contamination.

In the case of itinerant ferromagnets $NpOs_2$, results[38]) are shown in Fig. 16. $\gamma(0)$ is again extremely large (205 mJ/mol K^2) while above T_c = 7.5 K it is reduced to 92 mJ/mol K^2. The entropy change at the transition (ΔS = 0.2 Rln 2) was interpreted within the frame of weak itinerant magnetism[38]).

Going to more localized systems, the case of UAs is particularly interesting and has been studied in details. The surprisingly high γ value of 53 mJ/mol K^2 is not so sure since no data were taken below 5 K. In zero field the complete C_p versus T curve for both UAs and isostructured non-magnetic ThAs are reported in Fig. 17[102]) showing the two magnetic transitions. By contrast with reported first order transitions, rather long tails extend the sharp peaks, mainly around T_N.

Temperature dependence of the total magnetic entropy is shown in Fig. 18. Its maximum value 0.8 R ln 4 at 250 K is in between Γ_8 (Rln 4) and a Γ_6(Rln 2) crystal field levels values leading to a Γ_8 ground state with a Γ_6 first excited level in agreement with magnetic susceptibility analysis[67]). Measurements under magnetic field[103]) (up to 80 kOe) applied along the $\langle 001 \rangle$ axis of a single crystal are reported in Fig. 19. In higher field the peak splitting around T_N corroborates the occurrence of the ferrimagnetic state (see Fig. 6).

The last results we shall discuss concern antiferromagnetic UO_2 (T_N = 30.5 K) which is a semiconductor, and for which precise data exist as well as for ThO_2[103, 104]). The spin wave low temperature predictions fit the experiment data at low temperature with reasonable agreement and an excess found in the temperature range 30 K − 120 K is attributed to the Jahn-Teller effect. The entropy change at the transition is within 9% of the Rln 3 value expected for the Γ_5 triplet ground state.

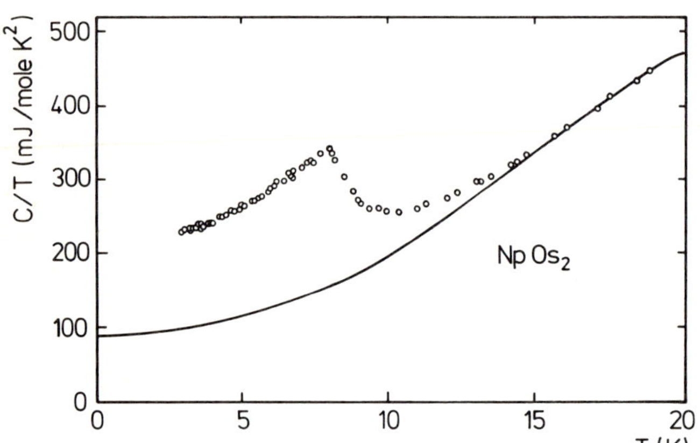

Fig. 16. Specific heat of $NpOs_2$ plotted as C/T versus T. The *solid line* is derived from the non magnetic isostructural $NpRu_2$. (Brodsky and Trainor[101]))

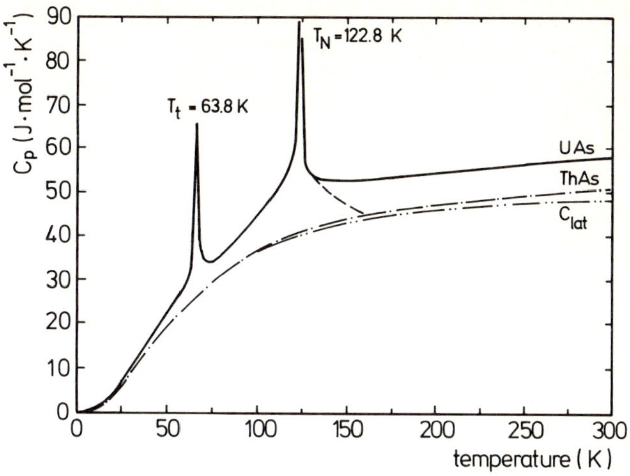

Fig. 17. Specific heat versus temperature for UAs (*solid curve*) and isostructural non-magnetic ThAs (*point-dashed curve*). The *dashed line* represents the estimate lattice specific heat of UAs and the *small dashed line* a T^{-2} law of short range magnetic ordering effect. (Blaise et al.[102])

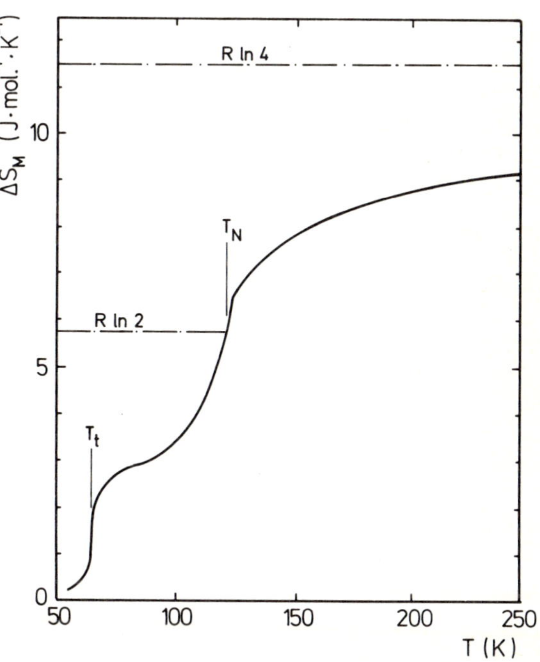

Fig. 18. Temperature dependence of the magnetic entropy ΔS_M of UAs (Blaise et al.[102])

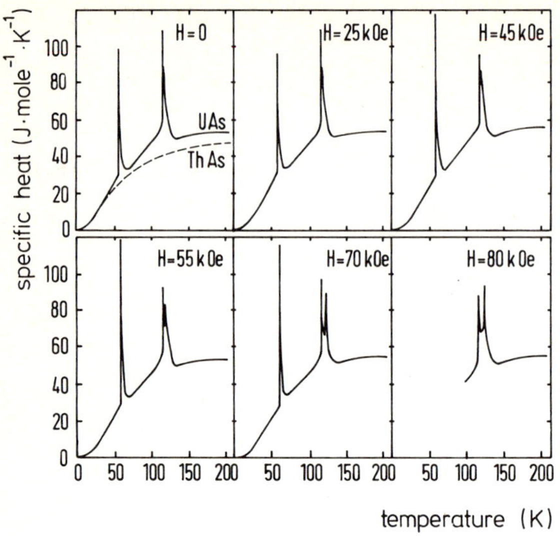

Fig. 19. Temperature dependence of the specific heat of UAs for constant magnetic field applied along the ⟨001⟩ axis. (Rossat-Mignod et al.[103])

V. Microscopic Properties

1. Neutron Scattering

Neutron scattering experiments play a crucial role in elucidating the magnetic properties of actinide compounds: elastic scattering allows to determine magnetic structures, to measure form factors and spatial correlations in the critical regime, while inelastic scattering is needed to study magnetic excitations.

a) Magnetic Structures

The elastic coherent cross section for nuclear and magnetic neutron scattering may be written

$$\frac{d\sigma}{d\Omega} = K(b^2 + 2\,bp(\vec{q}\cdot\vec{p}) + p^2q^2) \qquad (23)$$

where b is the coherent nuclear scattering amplitude, $\vec{p}$ is the neutron polarization, $\vec{q}$ the magnetic interaction vector and the magnetic scattering amplitude p is equal to $p = 0.27 \times 10^{-12}\, x\mu x\, f(\vec{k})$ where μ is the magnetic moment of the atom in Bohr magnetons and $f(\vec{k})$ the magnetic form factor.

Using unpolarized neutrons, $\vec{q}\cdot\vec{p}$ averages to zero and the study of the elastic magnetic cross section p^2q^2 allows, in principle, to determine the magnetic structure. However, this determination may not be complete (for example the phase between the different harmonics cannot be determined from neutron diffraction experiments). Different magnetic structures, with different domain population may also lead to the same

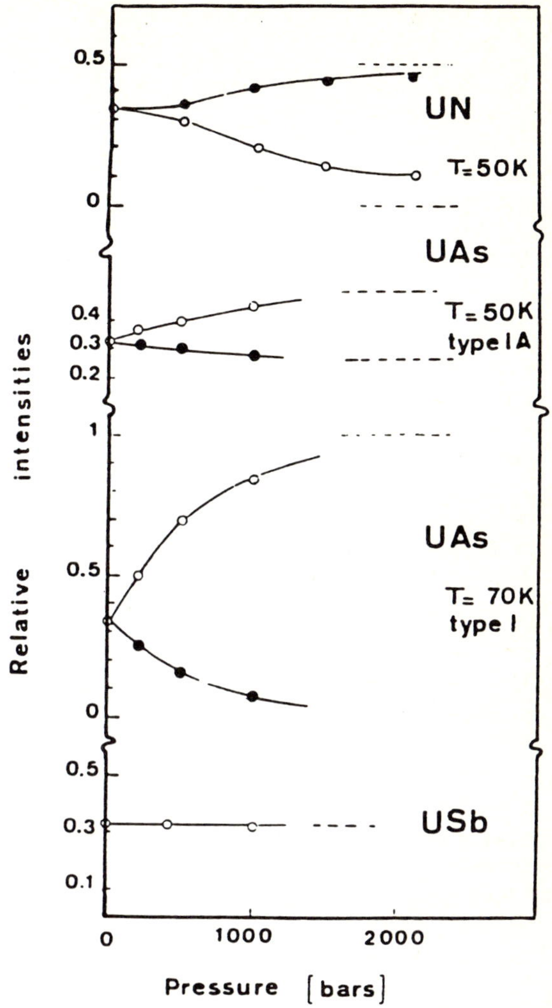

Fig. 20. Effect of an uniaxial pressure on the relative magnetic Bragg peak intensities associated with the three equivalent $\vec{k}$-vectors for UN, UAs and USb single crystals. The uniaxial pressure is applied along the ⟨001⟩ direction. *Full circles* correspond to magnetic peaks $I_\perp$, associated with a wave vector perpendicular to the stress and *open circles* to magnetic peaks $I_\parallel$, associated to a wave vector parallel to the stress. (Rossat-Mignod et al.[105])

diffraction pattern: the best example is given by the unusual "multi-$\vec{k}$" magnetic structures found in uranium monopnictides UP, UAs and USb[103, 105]. Experiments under unaxial stress were necessary, inducing domain motion, to establish the true magnetic structure without ambiguity (Fig. 20). Other types of complex magnetic structures are also found, such as a sinusoidal modulation of the magnetic moments along the ⟨001⟩ propagation axis in NpP[106] or incommensurate structures as in $NpAs_2$[70] (Fig. 21).

b) *Magnetic Form Factors*

Since the magnetic interaction vector $\vec{q}$ is known, it is possible to deduce the magnetic form factor $f(\vec{k})$. Although possible when using unpolarized neutrons, its measurement is much more precise using polarized neutron diffraction by a (single domain) ferromagne-

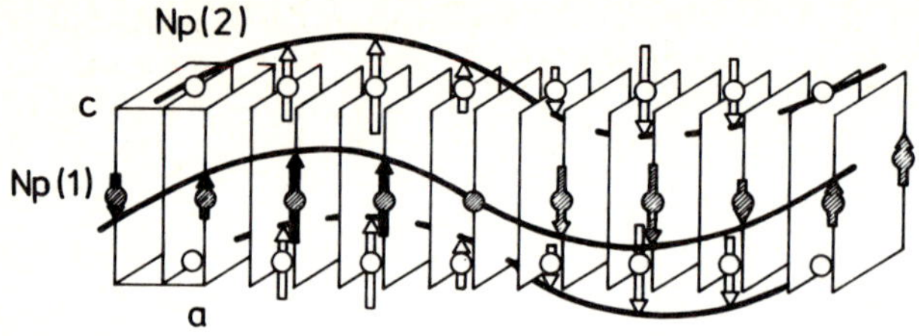

Fig. 21. Sine-wave modulation observed above T_{ic} = 18.5 K in $NpAs_2$. (Rossat-Mignod et al.[69])

tic crystal: in this case the cross term (2 bp($\vec{q} \cdot \vec{p}$) in Eq. (23) is non zero and the sensitivity comes from the fact that it is linear in p. Form factor measurements can give a detailed picture of the electronic structure since f($\vec{k}$) is directly involved. Thus an accurate low temperature measurement allows in principle to go back to the 5f electronic ground state in a compound under study. Moreover, the results may be Fourier transformed to obtain a spatial map of the spin density in the studied system. In the case of Pu compounds, even poorly resolved experiments are useful due to the very unusual shape of the $5f^5$ form factor which has its maximum away from sin θ/λ = 0. Such an experiment was carried out on PuP[68] and the result is displayed in Fig. 22, showing that Pu ions are effectively trivalent in this monopnictide. In the case of USb, a more precise experiment[107] showed through the anisotropy of the form factor that the ground state was the Γ_8^1 crystal field quartet of the $5f^3$ configuration, contrary to the analogous lanthanide compound, NdSb, where Γ_8^2 is the ground state[108].

Another feature of magnetic form factor measurements is that the extrapolated value at sin θ = 0 give the localized part of the magnetization, while bulk magnetization measurements include also diffuse contributions, especially that which is due to conduction electron polarization. For uranium monochalcogenides, which are all ferromagnets at low temperature, this polarization is negative in sign and increases in absolute value from US to UTe (Table 9).

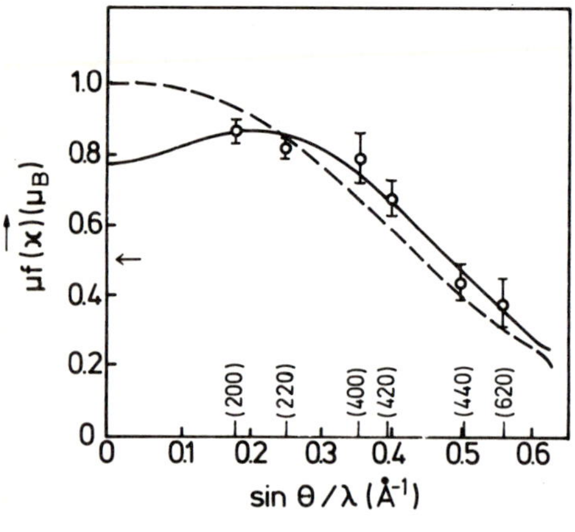

Fig. 22. Experimental points for the magnetic cross section from ferromagnetic ^{242}PuP at 4.2 K. The *solid* and *dashed curves* are the best fits to the data with $5f^5$ and $5f^4$ configurations, respectively. The *arrow* on the ordinate axis gives the total magnetic moment as determined by magnetization experiments. (Lander and Lam[25])

Table 9. Conduction electron polarization effect in uranium ferromagnetic monocompounds μ_{cep}: conduction electron polarization moment; μ_{sat}: saturation moment from magnetization studies; μ_{ord}: ordered magnetic moment as determined by neutron scattering

Compounds	Magnetic moment		Spin Polarization P % e	
	μ_{sat}	μ_{ord}	$\mu_{cep} = \mu_{ord} - \mu_{sat}$	
US[a,b]	1.55(2)	1.70(3)	−0.15(4)	−31
US[c,d]	1.81(2)	2.0(1)	−0.2(1)	−26
UTe[e]	1.91(5)	2.25(5)	−0.34(7)	−19
$USb_{0.8}Te_{0.2}$[e]	2.58(5)	2.64(5)	−0.06(6)	–

[a] Gardner and Smith[125)]; [b] Wedgwood[126)]; [c] Wedgwood and Kuznietz[127)]; [d] Busch und Vogt[128)]; [e] Busch et al.[75)]

c) *Critical Scattering*

Diffuse magnetic scattering just above the ordering temperature allows study of spatial correlations. Such measurements on UAs[109–111)] confirmed ideas about anisotropic exchange. In UAs, for example, it was found that correlations are much weaker along the ⟨001⟩ direction (more diffuse scattering), than perpendicular to it. Since the magnetic structure consists of ferromagnetic sheets stacked along the ⟨001⟩ axis in the + − + − sequence, this means that within each sheet the spins are strongly correlated, but that there is only weak correlation between the sheets. Another interesting result, in the case of UAs, is that the critical scattering, around the (1, 1, 0) reciprocal lattice point, is centered at (1, 1, 0.3) while the ordering is of type I, i.e. the elastic magnetic peak develops at (1, 1, 0) (Fig. 23).

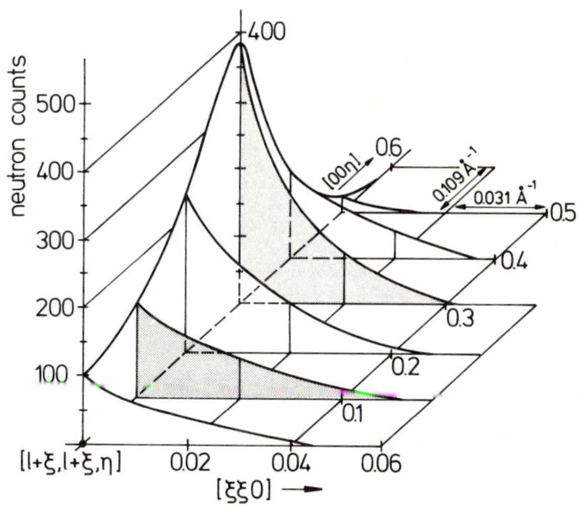

Fig. 23. Schematic representation of the diffuse scattering in UAs at T_N + 1.2 K. (Sinha et al.[109)])

d) Inelastic Scattering

Up to now, neutron inelastic scattering has not played a major role in the study of the actinides and there are good reasons for this: a large amount of material is needed (0.5 cm^3 crystals) so that studies started only recently and are confined to uranium systems. Moreover, the complexity of the results obtained has made interpretation of the experiments difficult.

While it was thought that transitions between crystal field levels could be seen easily in actinides, in the same way as in lanthanide systems, this has not been the case. In fact, the only system in which crystal field levels have been clearly found so far is UPd_3[112]. A rather detailed investigation of the collective excitations has been performed on USb[113]: phonons and magnons dispersion curves are shown in Fig. 24. This is one of the few materials that exhibit well-defined spin waves. However, these excitations vanish by about 100 K, which is less than 0.5 T_N, into a continuum of magnetic scattering. Such a diffuse magnetic scattering was, in fact, the only one observed in UN[112] and UAs[114] where no spin waves were found.

2. Mössbauer Spectroscopy

Mössbauer spectroscopy yields important information about the magnetism of actinide compounds and has been used with success concurrently with other techniques.

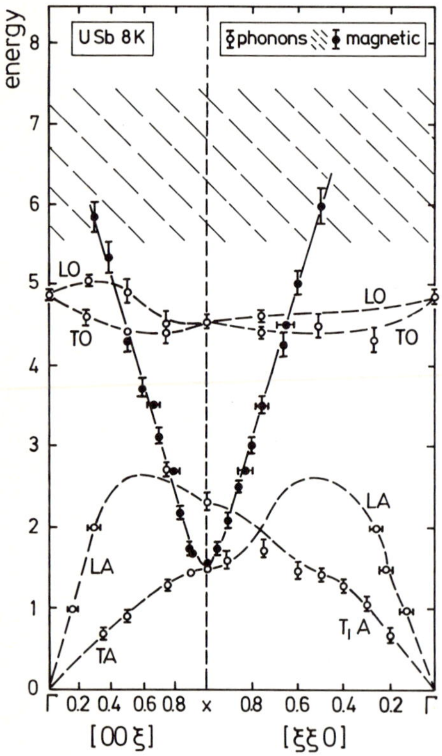

Fig. 24. The dispersion curves for USb; energy plotted against wave-vector transfer $\vec{Q}$ (in units of 2 π/a). The *dashed lines* represent the phonon dispersion and are based on the measured open points as well as on knowledge of phonons in NaCl structures. The magnetic modes are represented by *solid squares* (the collective excitation) and the *hatched area* (excitonic level). (Lander and Stirling[113])

While the bulk of work has been made on Np compounds, since ^{237}Np is a favorable isotope for Mössbauer effect measurements, such studies have also been made on ^{234}U and ^{236}U[115, 116], ^{238}U[117], ^{243}Am[118] and more recently on ^{231}Pa[119].

a) Isomer Shift

This shift in the Mössbauer line is related to the valence of the studied ion and is very large for ^{237}Np (40 mm/s). However, most compounds of interest for magnetism are not ionic, and covalency affects the isomer shift. Thus, although one observes some grouping into valence states, it cannot be used alone to assign uniquely a valence state – if any – for Np in non-ionic compounds[120].

b) Electric Quadrupole Interaction

This contribution, through the electric field gradient at the Np nucleus gives information about quadrupolar interaction and is useful in identifying non-equivalent sites of magnet-

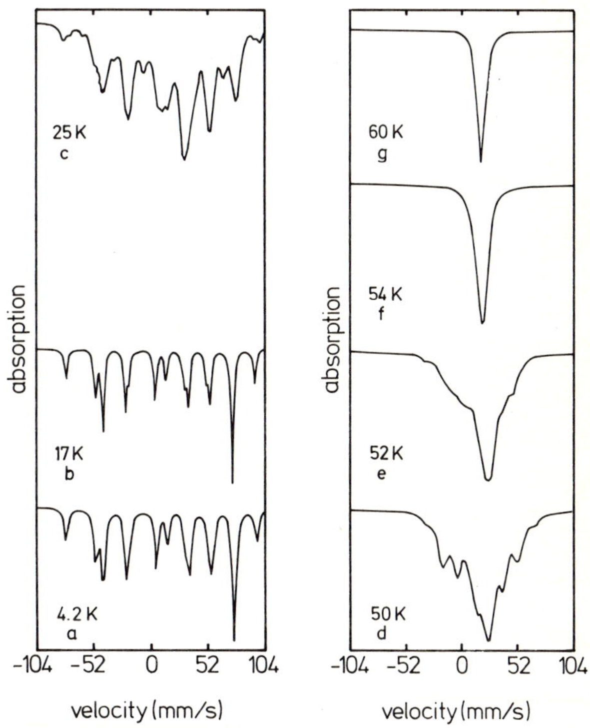

Fig. 25. Mössbauer spectra of $NpAs_2$ at various temperatures against a ^{241}Am metal source. The *solid lines* are fit to the data points (which have been omitted in this figure). (Bogé et al.[122])

ically ordered systems. In NpP, for example[121], the hyperfine spectrum at 4.2 K could only be fitted on the assumption of two non-equivalent magnetic sites; neutron diffraction measurements gave a magnetic structure with a varying longitudinal modulation. But it is only a combination of the two techniques which allowed the spatial arrangement of the two magnetic moments to be obtained. In the case of $NpAs_2$[122], a satisfactory fitting of the spectrum could be obtained using seven equivalent sites as deduced from the magnetic structure consisting of a pure sine wave modulation along the 100 axis of the tetragonal structure with a wave vector k = (0.14, 0, 0)[69] (Fig. 25).

c) *Magnetic Interaction*

This is the most directly useful hyperfine interaction for magnetic studies: the magnetic hyperfine field, $H_{h.f.}$, lifts the $2I + 1$ degeneracy of the I_Z component of the nuclear spin (in the case of ^{237}Np, $I = 5/2$, yielding 6 levels and sixteen allowed transitions). The value of $H_{h.f.}$ may be used to evaluate the ordered moment μ_0, since it has been found experimentally that μ_0 is proportional to $H_{h.f.}$ even for weak magnets like Np Mn_2[122]. This is particularly interesting for compounds where the ordered moments are too small to be observed by neutron diffraction as, for example, in $NpOs_2$[108] or $NpSn_3$[123]. For the same reason, the temperature dependence of sublattice magnetization in antiferromagnets may easily be measured up to very near T_N (if T_N is not too high) and the magnetization curve compared with Brillouin functions as was done, e.g., for $NpSn_3$ by Gal et al.[123] (Fig. 26).

Mössbauer effect measurements under high pressure are very interesting to study the pressure dependence of both T_N or T_c and μ_0. Results for $NpOs_2$ and $NpAl_2$[124] are very similar: both the hyperfine field and isomer shift decrease proportionately with decreasing lattice constant, thus with increasing 5f delocalization. By contrast, no change of hyperfine field was found in $NpCo_2Si_2$ where 5f electrons are expected to be well localized[124].

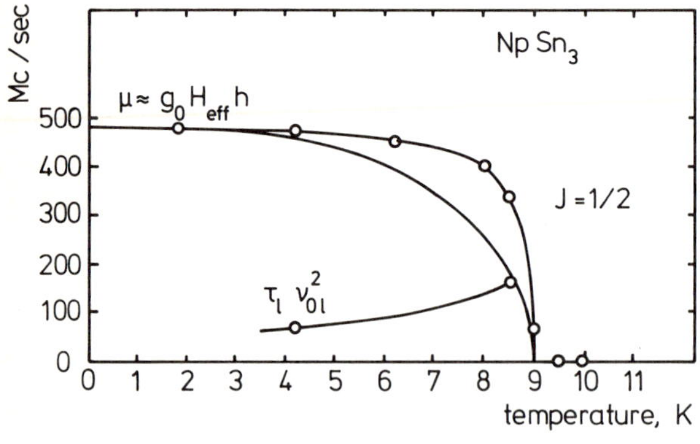

Fig. 26. Temperature dependence of the hyperfine magnetic field and of the longitudinal relaxation parameter in $NpSn_3$. The J = 1/2 Brillouin curve for the temperature dependence of H_{eff} is shown. (Gal et al.[123])

VI. Magnetic Data Tables

In this section, magnetic data for most actinide binary or ternary compounds are presented. The table presentation seemed to us the most convenient form for this reference section.

When an analysis of the paramagnetic susceptibility had been performed by the authors according to a M.C.W. (Modified Curie-Weiss) law, bare values for μ_{eff} and χ_0 are given with an asterisk and renormalized values are given below.

The symbols F, AF, mean that the compound under consideration becomes ferromagnetic (antiferromagnetic) at low temperature, the Curie (Neel) temperature being labeled $T_C(T_N)$.

The symbol P means that the compound follows a C.W. (Curie-Weiss) (or M.C.W.) law, but doesn't order magnetically.

The symbol TIP means that the compound has an almost temperature independent paramagnetism.

Finally, the symbol D means that the compound is diamagnetic.

When no symbol is indicated in the column "magnetism" either no analysis has been made or the behaviour of the susceptibility is more complicated.

μ_0 is the local ordered moment determined either by neutron diffraction or Mössbauer spectroscopy.

μ_s is the bulk saturation moment determined by magnetization measurements.

Table 10. Magnetic properties of cubic AnX_2 laves phases intermetallics with non-magnetic elements

Compound	Lattice constants (Å)	Magnetism	T_N(K)	T_C(K)	a) $\mu_0(\mu_B)$ b) $\mu_s(\mu_B)$	θ_p(K)	$\mu_{eff}(\mu_B)$	χ_0 (10^{-6} emu/mol)	Ref.
UAl_2	7.795					−250	3.0		54, 56
$NpAl_2$	7.785	F		56	a) 1.5 b) 1.21	+ 56	2.3		37,129
$PuAl_2$	7.840					−150	*1.0 2.58	*1320 3410	54,130
$AmAl_2$	7.861	TIP						~900	131
$NpRu_2$	7.446					~150	~2.8		106,132
$PuRu_2$	7.474	TIP						931	133, 134
$PuRh_2$	7.488	AF(?)	10(?)			− 49	*0.88 1.4	*370 439	133, 134
$AmRh_2$	7.51	TIP						~950	131
$NpOs_2$	7.528	F		7.5	a) 0.4 b) 0.44	+7	*2.37 2.36	*599 596	37, 132
UIr_2	7.509	TIP						1180	135
$NpIr_2$	7.509	AF	7.5		a) 0.6	− 28	*2.02 2.18	*1415 1520	106, 136
$PuIr_2$	7.531	TIP						575	133, 134
$PuPt_2$	7.652	F		6	b) >0.2	+ 6	*0.9 0.9	*240 237	133, 134
$PuZn_2$	7.760	AF(?)	15(?)			− 30	*0.71 0.85	*410 490	137

Table 11. Magnetic properties of cubic AnX_2 laves phases intermetallics with 3 d elements

Compound	Lattice constants (Å)	Magnetism	T_N(K)	T_C(K)	a) $\mu_0(\mu_B)$ b) $\mu_s(\mu_B)$	θ_p(K)	$\mu_{eff}(\mu_B)$	χ_0 (10^{-6} emu/mol)	Ref.
UMn_2	7.163							~2400 at 300 K	138, 139
$NpMn_2$	7.230	F		18	a) 0.3/Np; 0.2/Mn b) 0.4/f.u.	+ 21	*2.44 2.15	*4195 3699	37
$PuMn_2$	7.292	TIP						~3000	138
UFe_2	7.059	F		162	a) 0.03/U; 0.4/Fe b) 1.09/f.u.	+169	*3.03 2.49	*1200 988	140, 141, 142, 143
$NpFe_2$	7.144	F		492	a) 1.1/Np; 1.1/Fe b) 2.6/f.u.	+523	*4.22 4.09	* 130 126	141, 144, 145
$PuFe_2$	7.190	F		564	a) 0.45/Pu; 1.47/Fe b) 2.27/f.u.	+599	*3.65 3.30	* 270 244	141, 145
$AmFe_2$	7.30	F		613	a) −0.4/Am; 1.7/Fe b) 3.13/f.u.	630	~4		141, 145
UCo_2	7.005					−500	*2.6 4.1	*1000 1600	138, 143
$NpCo_2$	7.043	AF	15		a) 0.5/Np; 0.15/Co	− 46	*2.44 2.58	* 975 1034	144
$PuCo_2$	7.081							~2300 at 300 K	138
UNi_2	7.00							~2900 at 300 K	146

Table 11 (continued)

Compound	Lattice constants (Å)	Magnetism	T_N(K)	T_C(K)	a) $\mu_0(\mu_B)$ b) $\mu_s(\mu_B)$	θ_p(K)	$\mu_{eff}(\mu_B)$	χ_0 (10^{-6} emu/mol)	Ref.
$NpNi_2$	7.098	F		32	a) 1.2/Np; <0.3/Ni b) 1.0/f.u.	+ 40	*1.97 1.92	*320 311	144
$PuNi_2$	7.141					−1200	*4.1 7.4	~*1400 ~ 2500	138

Table 12. Magnetic properties of AnX_2 intermetallics with other structures

Compound	Crystal structure	Lattice constants (Å)	Magnetism	T_N(K)	T_C(K)	a) $\mu_0(\mu_B)$ b) $\mu_s(\mu_B)$	θ_p(K)	$\mu_{eff}(\mu_B)$	χ_0 (10^{-6} emu/mol)	Ref.
UB_2	AlB_2	a = 3.133 c = 3.987	TIP						500	147
NpB_2	AlB_2		F		100	b) ~0.3				148
UGa_2	AlB_2	a = 4.21 c = 4.01	F		126	a) 2.3 b) 2.71	126($\perp_c$) −148($//_c$)	3.0($\perp_c$) 3.5($//_c$)		149, 150
UGe_2	$ZrSi_2$	a = 4.12 b = 15.13 c = 3.98	F		52	b) 1.43	+53($//_c$)	2.5($//_c$)		151, 152, 153
$PuGe_2$	$ThSi_2$	a = 4.102 c = 12.81	F		35	b) >0.14				151

UNi_2	$MgZn_2$	a = 4.966 c = 8.252	F	21	b) 0.06		*2.6	*1000	43, 138, 154
$AmRu_2$	$MgZn_2$	a = 5.26 c = 8.73	TIP					~700	131
URe_2	URe_2	a = 5.600 b = 9.180 c = 8.460	TIP					~1000	155
$NpRe_2$	$MgZn_2$	a = 5.38 c = 8.77	F	47	b) 0.9	*47.4	*2.11 1.89	*1210 1085	155
$PuRe_2$	$MgZn_2$	a = 5.396 c = 8.729	TIP						155
UPt_2	$ZnNi_2$	a = 5.60 b = 9.68 c = 4.12					355		99, 156, 157
UHg_2	AlB_2	a = 4.98 b = 3.22	AF	~70		−200	3.2		158

Table 13. Magnetic properties of cubic $AuCu_3$-type AnX_3 intermetallics

Compound	Lattice constants (Å)	Magnetism	T_N(K)	T_C(K)	a) $\mu_0(\mu_B)$ b) $\mu_s(\mu_B)$	θ_p(K)	$\mu_{eff}(\mu_B)$	χ_0 (10^{-6} emu/mol)	Ref.
UAl_3	4.287							~1600 at 300 K	52, 159
$NpAl_3$	4.260	F		6.25	a) 1.2 b) >0.4	+62	*1.59 1.19	*1275 956	106
USi_3	4.035	TIP						~ 700 at 300 K	52, 159, 160, 161, 162
UGa_3	4.248	AF	70		a) 0.8				52, 159, 162, 163, 164
UGe_3	4.206	TIP						~1100 at 300 K	52, 152, 159, 161, 162
URh_3	3.988	TIP						973 at 300 K	165
$NpRh_3$	4.007	P				−483	3.6		165
$PuRh_3$	4.008	AF	6.2			−63	*1.0 1.31	*620 814	133
$NpPd_3$	4.095	AF	55		a) 2.0	−16	2.7		166, 167
$PuPd_3$	4.102	AF	24		a) 0.8	−34	*1.0 1.04	*140 141	133, 167, 168
UIn_3	4.601	AF	95		a) 1.0				52, 159, 162, 169

USn_3	4.626				~−50	2.5		52, 159, 161, 162, 170
$NpSn_3$	4.627	AF	9.5	a) 0.28				36
$PuPt_3$	4.103	AF	40		−36	*1.3 1.41	*490 531	133, 134
UTl_3	4.688	AF	90	a) 1.6				162, 169
UPb_3	4.793	AF	32	a) 1.7	−170	3.2		162, 171

Table 14. Magnetic properties of AnX_3 intermetallics with other structures

Compound	Crystal structure	Lattice constants (Å)	Magnetism	T_N(K)	T_C(K)	a) $\mu_0(\mu_B)$ b) $\mu_s(\mu_B)$	θ_p(K)	$\mu_{eff}(\mu_B)$	χ_0 (10^{-6} emu/mol)	Ref.
UPd_3	$TiNi_3$	a = 5.765 c = 9.618	AF(?)	7(?)				2.6		172, 173
$NpPd_3$	$TiNi_3$	a = 5.767 b = 9.544	AF(?)	32(?)			−40	2.8		166, 167
UPt_3	$MgCd_3$	a = 5.764 b = 4.898	AF(?)	15(?)			−50	2.6		99, 156, 157
$NpPt_3$	$TiNi_3$	a = 5.803 b = 9.566	AF(?)	22(?)			−47	2.95		174

Table 15. Magnetic properties of AnX intermetallics

Compound	Crystal structure	Lattice constants (Å)	Magnetism	T_N(K)	T_C(K)	a) $\mu_0(\mu_B)$ b) $\mu_s(\mu_B)$	θ_p(K)	$\mu_{eff}(\mu_B)$	χ_0 (10^{-6} emu/mol)	Ref.
UGa	UGa	a = 7.40 b = 7.60 c = 9.42	AF(?)	22(?)						175
UCo	UCo	6.355	TIP							176
PuNi	CrB	a = 3.59 b = 10.21 c = 4.22	TIP							177
UPt	CrB	a = 3.721 b = 10.772 c = 4.410	F		30	b) ~0.2		~3.4		74, 99, 157, 178, 179
NpPt	CrB	a = 3.782 b = 10.743 c = 4.378	AF	27			−20	~2.2		148
PuPt	CrB	a = 3.816 b = 10.692 c = 4.428	AF/F	44	19	b) ~0.2	~24	~0.84		148

Table 16. Magnetic properties of AnX_5 and miscellaneous intermetallics

Compound	Crystal structure	Lattice constants (Å)	Magnetism	$T_N(K)$	$T_C(K)$	a) $\mu_0(\mu_B)$ b) $\mu_s(\mu_B)$	$\theta_p(K)$	$\mu_{eff}(\mu_B)$	χ_0 (10^{-6} emu/mol)	Ref.
UNi_5	$AuBe_5$	6.783	TIP						~1300	180, 181
UCu_5	$AuBe_5$	7.021	AF	15		a) 0.9	−260	3.5		180, 181, 182, 183
UPt_5	$AuBe_5$	7.421					−310	2.7		156, 157
UBe_{13}	$NaZn_{13}$	10.257					−70	3 (T > 200 K)		184
$NpBe_{13}$	$NaZn_{13}$	10.254					−112	~2.7		185
$PuBe_{13}$	$NaZn_{13}$	10.282	AF	11			+11	*0.74 0.69	*430 400	185
UB_4	ThB_4	a = 7.075 b = 3.979					−680	3.4		184
NpB_4	ThB_4		AF	52			~−30	~1.4		148, 186
PuB_4	ThB_4	a = 7.10 c = 4.014	TIP						~ 700	148
PuB_6	ThB_6	4.113	TIP						~1600	148
UB_{12}	UB_{12}	7.473							~ 700	184
PuB_{12}	UB_{12}								~1000	148

Table 16 (continued)

Compound	Crystal structure	Lattice constants (Å)	Magnetism	T_N(K)	T_C(K)	a) $\mu_0(\mu_B)$ b) $\mu_s(\mu_B)$	θ_p(K)	$\mu_{eff}(\mu_B)$	χ_0 (10^{-6} emu/mol)	Ref.
UAl_4	UAl_4	a = 4.41 b = 6.27 c = 13.71	P				~−450	~3.5		52
U_3Si_5	AlB_2	a = 3.83 c = 4.07	P				−320	3.4/U		160, 158
U_2Ga_3	U_2Ga_3	a = 7.583 b = 9.398 c = 9.382	F		23	b) 1.1	+19	3.1/U		187
U_5Ge_3	Mn_5Si_3	a = 8.56 c = 5.80					~−520	~3.6/U		152
U_2Co_{11}	R 3 m	a = 4.76 c = 36.49	F		360	b) 2.4/f.u.				73
UPd_4	$AuCu_3$	4.067	AF	10		a) 0.8	−130	347		188
Np_3Sn	$AuCu_3$	3.28	AF(?)	140(?)			291	24/Np		189

Table 17. Magnetic properties of actinide monocarbides and moninitrides

Compound	Lattice constants (Å)	Magnetism	T_N(K)	T_C(K)	a) $\mu_0(\mu_B)$ b) $\mu_s(\mu_B)$	θ_p(K)	$\mu_{eff}(\mu_B)$	χ_0 (10^{-6} emu/mol)	Ref.
ThC	5.342	TIP						45.6 at 300 K	190, 191
PaC	5.057	D						−50 at 300 K	192
UC	4.960	TIP						800 at 300 K	193, 194
NpC_1	5.00	AF/F	320	220	a) 2.1 b) 1.4	+225	3.2		195, 196, 197, 198
PuC_1	4.98	AF	~96		a) 0.8	−70	* 1.24 ~1.6	~*880 ~1100	199, 200, 201
ThN	5.158	TIP						~50 at 300 K	190, 191
UN	4.889	AF	53		a) 0.75	−247	*2.66 2.76	*140 146	28, 202, 203, 204, 205
NpN	4.898	F		82	a) 1.4 b) >0.9	+82	*2.13 2.01	*400 377	78, 196, 206
PuN	4.905	AF(?)	13(?)		a) <0.3	−200	*1.08 1.05	*280 390	192, 207
AmN	4.995	TIP						780	208
CmN	5.041	F		109			7.02		208
BkN	4.951	F		88	b) >0.3	−50	7.85		209

Table 18. Magnetic properties of actinide monopnictides

Compound	Lattice constants (Å)	Magnetism	T_N(K)	T_C(K)	a) $\mu_0(\mu_B)$ b) $\mu_s(\mu_B)$	θ_p(K)	$\mu_{eff}(\mu_B)$	χ_0 (10^{-6} emu/mol)	Ref.
ThP	5.833	TIP						62 at 300 K	191
UP	5.589	CAF	125 22.5		a) 1.7 a) 1.9	+18	*3.34 3.34	*67 67	67, 210, 211, 212
NpP	5.610	CAF	130 74		a) 1.8 b) 2.3	125	2.8		78
PuP	5.651	F		126	a) 0.71 b) >0.42	+130	*1.06 0.87	*190 197	24, 25
CmP	5.743	F		73					213
PaAs	5.756	TIP						460	250
UAs	5.779	CAF	127 64		a) 1.92 b) 2.24	+50	*3.39 3.38	*86 85	67, 210, 214
NpAs	5.830	CAF F(H > 50 kOe)	172.5 155 142	175 (H > 50 kOe)	a) 2.5 b) 2.3	+185	3.0		78, 215
PuAs	5.86	F		125	a) 0.58 b) 0.67	*129	*0.98 0.63	*330 213	216, 217, 218
AmAs	5.876	AF(?)	13(?)				1.14		208
CmAs	5.905	F		88			6.58		208

USb	6.209	AF	213		a) 2.82	+140	3.64		78, 107
NpSb	6.254	AF	207		a) 2.5	~150	~2.3		78
PuSb	6.240	CAF/F	~90	~70	a) 0.63 b) 0.65	+90	*1.0 0.86	* 200 171	206, 217, 218
AmSb	6.299	TIP						1250	219
CmSb	6.243	F		162					213
UBi	6.364	AF	~285		a) 3.0	~+105	4.1		220, 221
PuBi		AF F(H > 70 kOe)	65	65 (H > 70 kOe)	b) 0.61		0.8		217

Table 19. Magnetic properties of actinide monochalcogenides

Compound	Lattice constants (Å)	Magnetism	T_N(K)	T_C(K)	a) $\mu_0(\mu_B)$ b) $\mu_s(\mu_B)$	θ_p(K)	$\mu_{eff}(\mu_B)$	χ_0 (10^{-6} emu/mol)	Ref.
ThS	5.683	TIP						38 at 300 K	191
US	5.489	F		177	a) 1.70 b) 1.55	+180	2.3		125, 126, 128, 222, 223
NpS	5.527	AF	23		a) 0.9	−35	*1.33 1.49	*739 825	224
PuS	5.536	AF(?)	4.5(?)					~400 at 300 K	186, 224, 225
ThSe	5.863	TIP						8 at 300 K	226
USe	5.740	F		160	a) 2.0 b) 1.81	+160	2.4		127, 128, 223
PuSe	5.773	TIP						~400 at 300 K	225
UTe	6.155	F		104	a) 2.25 b) 1.91	+120	2.7		128, 223, 227, 228, 229
PuTe	6.151	TIP						~400 at 300 K	225

Table 20. Magnetic properties of Th_3P_4 type compounds (° measured with a powder sample)

Compound	Lattice constants (Å)	Magnetism	T_N(K)	T_C(K)	a) $\mu_0(\mu_B)$ b) $\mu_s(\mu_B)$	θ_p(K)	$\mu_{eff}(\mu_B)$	χ_0 (10^{-6} emu/mol)	Ref.
Th_3P_4	8.617	TIP or D							191, 230
Th_3As_4	8.843	D							231
U_3P_4	8.214	F		138	a) 1.39/U b) 1.38⟨111⟩	+140	2.75/U		184, 232, 233
U_3As_4	8.521	F		196	a) 1.71/U b) 1.83⟨111⟩	+20	2.94/U		184, 232, 233
U_3Sb_4	9.112	F		146	b) 1.56/U	+155	3.04		184, 232
U_3Bi_4	9.368	F		108	b) 1.60/U	+110	3.14		184, 232
Np_3As_4	8.516	F		81	a) 1.74/Np b) 1.33	+83	1.64		77
U_3Se_4	8.76	AF(?)	35(?)			−80	2.5 3.1		234, 235
$U_{2.67}Te_4$	9.41	F		70	a) 1.75/U b) 1.4		3.14		184, 234, 235, 236
Pu_3S_4	8.416	AF	10				1.23		186
Am_3Se_4	8.782	TIP							219
Am_3Te_4	9.394	TIP							219

Table 21. Magnetic properties of $An_2(X, Y)_3$ and An_3Y_5 compounds

Compound	Crystal structure	Lattice constants (Å)	Magnetism	T_N(K)	T_C(K)	a) $\mu_0(\mu_B)$ b) $\mu_s(\mu_B)$	θ_p(K)	$\mu_{eff}(\mu_B)$	χ_0 (10^{-6} emu/mol)	Ref.
U_2C_3	Pu_2C_3	8.089	AF(?)	59			+138	1.8/U		202
Np_2C_3	Pu_2C_3	8.103	F		109	b) 0.8/Np	+114	1.7/Np		201
Pu_2C_3	Pu_2C_3	8.121	TIP						~450 at 300 K	202
αU_2N_{3+x}	Mn_2O_3	10.69	AF	94–13			+ 35(−15)	2.1(1.1)		204
βU_2N_{3-y}	La_2O_3	a = 3.700 c = 5.834	F		188		+192	2.4/U		237
U_2S_3	Sb_2S_3	a = 10.37 b = 10.58 c = 3.88	Fi(?)		82.8		~+100			238, 239
Pu_2S_3	Ce_2S_3	a = 3.92 b = 7.37 c = 15.65	AF	7				0.6/Pu		186
U_2Se_3	Sb_2S_3	a = 11.33 b = 10.94 c = 4.06	Fi(?)		800		~+100			238, 240
U_2Te_{3-x}	?	a = 11.79 b = 12.34 c = 4.43	Fi(?)		106		−3 to −60	~3.6/U		238

U_3S_5	?	a = 11.76 b = 8.09 c = 7.40	WF		29	−26	1.9/U	241, 242
ThU_2S_5	?	a = 11.86 b = 8.16 c = 7.47	AF	9		−27	2.4/U	242
U_3Se_5	?	a = 12.43 b = 8.45 c = 7.77	WF		~35			241
Np_3Se_5	?	a = 12.24 b = 8.43 c = 7.72	P			−28	3.2/Np	243

Table 22. Magnetic properties of An(X, Y)$_2$ binary compounds

Compound	Crystal structure	Lattice constants (Å)	Magnetism	T_N(K)	T_C(K)	a) $\mu_0(\mu_B)$ b) $\mu_s(\mu_B)$	θ_p(K)	$\mu_{eff}(\mu_B)$	χ_0 (10^{-6} emu/mol)	Ref.
UP_2	ZrSiS	a = 3.978 c = 8.154	AF	202		a) 2.0	+94 +86	2.0 2.5		244, 245
$PaAs_2$	ZrSiS	a = 3.978 c = 8.134	TIP						340	246
UAs_2	ZrSiS	a = 3.954 c = 8.120	AF	273		a) 1.6	+72 +34	2.4 2.9		247, 248
$NpAs_2$	ZrSiS	a = 3.962 c = 8.115	AF/F	52	18	a) 1.45	+53	1.88	900	68, 69
$PaSb_2$	ZrSiS	a = 4.277 c = 8.786	TIP							249
USb_2	ZrSiS	a = 4.272 c = 8.741	AF	205		a) 1.9	+72 +18	2.5 3.0		247, 253
$NpSb_2$	$LaSb_2$	a = 6.17 b = 6.04 c = 17.50	F		45	a) 1.9 b) 0.6	−19	2.87		251
$PuSb_2$	$LaSb_2$	a = 6.19 b = 6.05 c = 17.58	AF	21			−12	0.75	540	252
UBi_2	ZrSiS	a = 4.445 c = 8.908	AF	180		a) 2.1	−530	2.85 3.4		220, 253

αUS_2	?	a = 10.28 b = 6.32	P				−120	3.6	254, 255
βUS_2	$PbCl_2$	a = 4.13 b = 7.11 c = 8.48	P				− 30	3.1	255, 256
PuS_2	(Pnma)	a = 3.02 b = 7.37 c = 15.65	AF(?)	15(?)					186
αUSe_2	?	a = 10.73 b = 6.60	AF(?)	13(?)			−48	3.2	254, 255, 257
UTe_{2-x}	ZrSiS	a = 4.01 c = 8.95	F		73	b) 0.98	~0	2.3	258
UTe_2	(Immm)	a = 4.16 b = 6.13 c = 13.97	P				−70	3.1	259, 260
$NpTe_{2-x}$	ZrSiS	a = 4.355 c = 9.023	P				−56	3.04	251
$NpTe_2$	ZrSiS	a = 4.424 c = 9.004	P				−40	2.88	251

° above T = 300 K

Table 23. Magnetic properties of AnXY ternary compounds

Compound	Crystal structure	Lattice constants (Å)	Magnetism	T_N(K)	T_C(K)	a) $\mu_0(\mu_B)$ b) $\mu_s(\mu_B)$	θ_p(K)	$\mu_{eff}(\mu_B)$	χ_0 (10^{-6} emu/mol)	Ref.
UGeS	ZrSiS	a = 3.820 c = 8.323	AF	88		a) 1.3				261, 262
UGeSe	UGeTe	a = 3.932 c = 16.966	AF	40		a) 0.3				261, 262
UGeTe	UGeTe	a = 4.110 c = 17.599	AF	73		a) 1.5				261, 262
USnTe	ZrSiS	a = 4.259 c = 9.132	AF	68						263
UOS	PbFCl	a = 3.843 c = 6.694	AF	56		a) 2.0	0 −100	2.2 2.6		263, 264
UOSe	PbFCl	a = 3.904 c = 6.979	AF	75		a) 2.2	0 −120	2.5 3.3		263, 265
UOTe	PbFCl	a = 4.012 c = 7.501	AF	164		a) 2.0	0 −56	2.6		263, 265
UNSe	PbFCl	a = 3.837 c = 6.944	F		88		+43	2.03		263
UNTe	PbFCl	a = 3.929 c = 7.617	F		59	b) 1.12	+44	1.87		263
UPS	ZrSiS	a = 3.813 c = 7.981	F		118	a) 1.6				266

UPSe	ZrSiS	a = 3.951 c = 8.185	F	110	a) 1.4			266, 267
UPTe	UGeTe	a = 4.100 c = 17.026	F	85	a) 1.4			268
UAsS	ZrSiS	a = 3.884 c = 8.176	F	124	a) 1.2	−380	3.34	266, 267
UAsSe	ZrSiS	a = 3.962 c = 8.422	F	113	a) 1.5	−195	3.41	266, 269
UAsTe	UGeTe	a = 4.148 c = 17.256	F	66	a) 1.6	− 95	3.34	266, 268
USbS	ZrSiS	a = 3.937 c = 8.530	F	154				270
USbSe	ZrSiS	a = 4.173 c = 8.681	F	128	a) 1.5			266, 269
USbTe	ZrSiS	a = 4.321 c = 9.063	F	127				266, 270
UBiTe	ZrSiS	a = 4.434 c = 9.157	F	110				266
NpAsS	ZrSiS	a = 3.883 c = 8.344	F	30	b) >1.2			271
NpAsSe	ZrSiS	a = 3.972 c = 8.584	F	35	b) >0.2			271
NpAsTe	ZrSiS	a = 4.230 c = 8.982	F	40	b) >0.7			271

Table 23 (continued)

Compound	Crystal structure	Lattice constants (Å)	Magnetism	T_N(K)	T_C(K)	a) $\mu_0(\mu_B)$ b) $\mu_s(\mu_B)$	θ_p(K)	$\mu_{eff}(\mu_B)$	χ_0 (10^{-6} emu/mol)	Ref.
PuAsSe	ZrSiS	a = 4.002 c = 8.686	F		126	b) >0.1				271
PuAsTe	ZrSiS	a = 4.266 c = 9.067	F		125	b) >0.1				271
USSe	$PbCl_2$	a = 4.18 b = 7.33 c = 8.60	F		25	b) 1.0				261
USTe	$PbCl_2$	a = 4.29 b = 7.60 c = 8.85	F		87	b) 0.7				261
USeTe	$PbCl_2$	a = 4.41 b = 7.70 c = 9.08	F		83	b) 0.9				261

Table 24. Magnetic properties of $AnM_2(Si, Ge)_2$ compounds

Compound	Crystal structure	Lattice constants (Å)	Magnetism	T_N(K)	T_C(K)	a) $\mu_0(\mu_B)$ b) $\mu_s(\mu_B)$	θ_p(K)	$\mu_{eff}(\mu_B)$	χ_0 (10^{-6} emu/mol)	Ref.
UCu_2Ge_2	tetragonal $ThCr_2Si_2$	a = 4.058 c = 10.207	AF/F	25	100	a) 1.6				272
UPd_2Si_2	tetragonal $ThCr_2Si_2$	a = 4.097 c = 10.046	CAF	150		a) 2.0				273
UPd_2Ge_2	tetragonal $ThCr_2Si_2$	a = 4.200 c = 10.230	CAF	140		a) 2.0				273
URh_2Si_2	tetragonal $ThCr_2Si_2$	a = 4.009 c = 10.025	AF	137		a) 2.0				273
URh_2Ge_2	(P 4/mmm)	a = 4.155 c = 9.771	P							273
$NpCr_2Si_2$	$ThCr_2Si_2$	a = 3.863 c = 9.367	AF	73		a) 1.7				274
$NpMn_2Si_2$	$ThCr_2Si_2$	a = 3.895 c = 9.538	F		> 300	a) 1.1/Np				274
$NpFe_2Si_2$	$ThCr_2Si_2$	a = 3.870 c = 9.962	AF	87		a) 1.5/Np				274
$NpCo_2Si_2$	$ThCr_2Si_2$	a = 3.877 c = 9.732	AF	44		a) 1.5/Np				274, 275
$NpNi_2Si_2$	$ThCr_2Si_2$	a = 3.935 c = 9.466	AF	33		a) 1.0/Np				274

Table 24 (continued)

Compound	Crystal structure	Lattice constants (Å)	Magnetism	T_N(K)	T_C(K)	a) $\mu_0(\mu_B)$ b) $\mu_s(\mu_B)$	θ_p(K)	$\mu_{eff}(\mu_B)$	χ_0 (10^{-6} emu/mol)	Ref.
$NpCu_2Si_2$	$ThCr_2Si_2$	a = 3.922 c = 9.411	F		34	a) 2.0/Np				274, 275
$NpCr_2Ge_2$	$ThCr_2Si_2$	a = 3.909 c = 9.470	AF	62		a) 1.4/Np				276
$NpFe_2Ge_2$	$ThCr_2Si_2$	a = 3.918 c = 9.403	AF	28		a) 0.9/Np				276
$NpNi_2Ge_2$	$ThCr_2Si_2$	a = 3.903 c = 9.556	AF	27		a) 1.3/Np				276
$NpCu_2Ge_2$	$ThCr_2Si_2$	a = 3.888 c = 9.418	CAF	34		a) 1.2/Np				276
$NpCo_2Ge_2$	$ThCr_2Si_2$	a = 3.800 c = 9.556	CAF	36		a) 1.5/Np				276

Table 25. Magnetic properties of oxydes and hydrides

Compound	Crystal structure	Lattice constants (Å)	Magnetism	T_N(K)	T_C(K)	a) $\mu_0(\mu_B)$ b) $\mu_s(\mu_B)$	θ_p(K)	$\mu_{eff}(\mu_B)$	χ_0 (10^{-6} emu/mol)	Ref.
UO_2	CaF_2	5.470	AF	30.8		a) 1.8	−220	3.2		277, 278
NpO_2	CaF_2	5.434	AF(?)	25(?)		≤ 0.01	−70	*2.54 2.70	*730	198, 279, 280
PuO_2	CaF_2	5.396	TIP						480	281
AmO_2	CaF_2	5.377	AF(?)	8.5(?)		≤ 0.01	−41	1.5		118, 282
CmO_2	CaF_2	5.372					~+25	1.6 to 2.3	1900 to 4100	283
αU_3O_7	tetragonal	a = 5.472 c = 5.397	AF	30						284
U_3O_8	orthorhom.	a = 6.815 b = 11.825 c = 4.132	AF	25						285, 286
β-Pu_2O_3	hexagonal La_2O_3	a = 3.841 c = 5.958	AF	19		a) 0.85				287, 288
β-UH_3	β-W type	a = 6.631	F		180	a) 1.4 b) 1.2	+180	2.5		289, 290, 291
α-UH_3	BiF_3-type	a = 4.160	F		~180					292
NpH_{2+x}	CaF_2	a = 5.343 to a = 5.355	P							293

Table 25 (continued)

Compound	Crystal structure	Lattice constants (Å)	Magnetism	T_N(K)	T_C(K)	a) $\mu_0(\mu_B)$ b) $\mu_s(\mu_B)$	θ_p(K)	$\mu_{eff}(\mu_B)$	χ_0 (10^{-6} emu/mol)	Ref.
NpH_3	hexagonal	a = 3.771 c = 6.713	P							293
PuH_{2+x}	CaF_2		F		~60	a) 0.8 b) ~0.4	+45 to +68	0.8	390	293, 294, 295, 296
PuH_2	CaF_2	a = 5.359	AF	30			+ 5	0.96	270	293, 294 295
PuH_3	hexagonal	a = 3.78 c = 6.76	F		101		+98	0.63	440	293, 294

Table 26. Magnetic properties of actinide halides

Compound	Crystal structure	Lattice constants (Å)	Magnetism	T_N(K)	T_C(K)	a) $\mu_0(\mu_B)$ b) $\mu_s(\mu_B)$	θ_p(K)	$\mu_{eff}(\mu_B)$	χ_0 (10^{-6} emu/mol)	Ref.
AmI_2	monoclinic β = 98°46	a = 7.677 b = 8.311 c = 7.923					~+25	6.7		300
UF_3	LaF_3		P				−110	3.7		301
PuF_3	LaF_3		AF	9				1.0		302
UCl_3	$LaCl_3$		AF	6.5		a) 1.8	− 29	3.03		303, 304

$NpCl_3$	$LaCl_3$		TIP < 50 K				− 82	2.8	305
$PuCl_3$	$LaCl_3$		AF	4.5			− 7	1.1	305
UBr_3	$LaCl_3$		P				25	3.3	306
UI_3	$PuBr_3$	a = 13.98 b = 4.31 c = 9.99	AF	2.6		a) 1.98	5	3.3	306, 307, 308
UF_4	ZrF_4		P				−120	3.3	306, 309
PuF_4	ZrF_4		TIP < 50 K				−270	2.9	305, 310
$PaCl_4$	UCl_4		F		182		157	~1.0	311
UCl_4	UCl_4		TIP < 30 K				− 62	3.29	309, 312
$NpCl_4$	UCl_4		F		6.7	a) 1.28	+ 7	3.1	313
UBr_4	unknown		P				− 35	3.12	309
UCl_5	unknown		P						314

VII. Conclusions

At this point, the reader has been able to appreciate the richness and the variety of the magnetic properties of the actinide metals and compounds. The importance of magnetic measurements for the study of the actinides is thus clearly evidenced and will not decrease in the future.

On the contrary, new experimental processes are developping to obtain a deeper understanding of the particular properties exhibited by the actinides.

One example is given by the measurements under high pressure. This follows from the large magnetovolume effects which have been both calculated[35)] and already observed[30)]: in the case of UN, the pressure dependence of T_N by magnetic measurements under hydrostatic pressure and that of μ_{ord} by neutron diffraction measurements on a single crystal under quasihydrostatic pressure gave strong support for an itinerant antiferromagnetic state in UN.

Mössbauer spectroscopy under pressure has also been developped[126)]. Again results are decisive to test the 5f delocalization and the itinerant aspect of magnetic order.

As an example $NpOs_2$ and even $NpAl_2$ display a proportional decrease of the hyperfine field and isomer shift, indicative of increasing 5f delocalization while no change is detected in $NpCo_2Si_2$ where 5f electrons are expected to be well localized[126)].

Among promising future developments we would like to introduce two problems:

– In a band model, recent calculations[297)] have shown that a large orbital moment could be induced by the spin-orbit coupling thus reconciliating calculated and experimental form factors. Those calculations predict large changes in the form factor shape under pressure through different variations of the spin and orbital contributions: this has to be checked experimentally and represents a challenge for the experimentalist.

– On the other hand, recent measurements on UBe_{13}[298)] have shown this intermetallic to be a very special species of superconductor with a large density of electronic states at E_F (~ 1000 mJ/mol K^2). Such results will certainly motivate both theoreticians and experimentalists and the so called "heavy-fermion superconductors" family is already developing[299)].

VIII. References

1. Stoner, E. C.: Proc. Roy. Soc. *165,* 372 (1938)
2. Brooks, M. S. S., Glötzel, D. G.: J. Magn. and Magn. Mat. *15–18,* 873 (1980)
3. Hubbard, J.: Proc. Roy. Soc. *A 281,* 401 (1964)
4. Peierls, R.: Quantum theory of solids, Oxford 1955
5. Mott, N. F.: Phil. Mag. *6,* 287 (1961)
6. Johansson, B.: Phys. Rev. *B 11,* 2740 (1975)
7. Mott, N. F., Stevens, K. W. H.: Phil. Mag. *2,* 1364 (1957)
8. Freeman, A. J., Koelling, D. D.: in: The Actinides electronic structure and related properties (eds. Freeman, A. J., Darby, Jr., J. B.), Academic Press, New York, *Vol. I,* p. 5 (1974)
9. Jullien, R., Coqblin, B.: Phys. Rev. *B 8,* 5263 (1973)
10. Fournier, J. M. et al.: Physica *86–88 B,* 30 (1977)
11. Fujita, D. K.: Ph. D. Thesis, U. of California Report N., UCRL-19507
12. Fujita, D. K. et al.: in: Trans-Plutonium 1975 (eds. Müller, W., Lindner, R.), North-Holland, Amsterdam, p. 173 (1976)
13. Hill, H. H.: in: Plutonium 1970 and other actinides (ed. Miner, W. N.), Metal. Soc. AIME, New York, *2* (1970)

14. Brooks, M. S. S.: in: Rare-earths and actinides (eds. Tanner, B. K., Hoon, S. R.), J. Magn. and Mag. Mat. *29,* 257 (1982)
15. Arko, A. J. et al.: Phys. Rev. *B 12,* 4102 (1975)
16. Schadler, G., Hilscher, G., Weinberger, P.: Rare-earths and actinides (eds. Tanner, B. K., Hoon, S. R.), J. Magn. and Magn. Mat. *29,* 241 (1982)
17. Chan, S. K., Lam, D. J.: in: The actinides: electronic structure and related properties (eds. Freeman, A. J., Darby, J. B.), Academic Press, New York, *Vol. I,* p. 1 (1974)
18. Coqblin, B.: Proc. Conf. on Rare-Earths and Actinides, Durham, Inst. Phys., London, 1971, p. 92;
Galleani d'Agliani, E., Jullien, R., Coqblin, B.: ibid., p. 69
19. Carnall, W. T., Wybourne, B. G.: J. Chem. Phys. *40,* 3428 (1964)
20. Lea, K. R., Leask, M. J. M., Wolf, W. P.: J. Phys. Chem. Solids *23,* 1381 (1962)
21. Veal, B. W., Lam, D. J.: Phys. Rev. *B 10,* 4902 (1974)
22. Frazer, B. C. et al.: ibid. *A 140,* 1448 (1965)
23. Rahman, H. U., Runciman, W. A.: J. Phys. Chem. Solids *27,* 1833 (1966)
24. Lam, D. J., Fradin, F. Y., Kruger, O. L.: Phys. Rev. *187,* 606 (1969)
25. Lander, G. H., Lam, D. J.: ibid. *B 14,* 4064 (1976)
26. Hall, R. O. A., Mortimer, M. J., Lander, G. H.: in Proceedings of the 11ème Journée des Actinides, Jesolo Lido, Italy 1981
27. Fradin, F. Y.: in: Plutonium 1975 and other actinides (eds. Blank, H., Linder, R.), North-Holland, Amsterdam, 459 (1976)
28. Curry, N. A.: Proc. Phys. Soc., London *86,* 1193 (1965)
29. De Novion, C. H.: Thesis U. of Paris, Report CEA No. R 4113 (1970)
30. Fournier, J. M. et al.: in: The Physics of Actinides and Related 4 f Materials (ed. Wachter, P.), North-Holland, Amsterdam, 282 (1980)
31. Reihl, B., Hollinger, G., Himpsel, F. J.: in: Rare-Earths and Actinides, J. Magn. and Magn. Mat. *29,* 141 (1982)
32. Fedders, P. A., Martin, P. C.: Phys. Rev. *143,* 245 (1966)
33. Boeuf, A. et al.: in Proceedings of the 11éme Journée des Actinides, Jesolo Lido, Italy 1981
34. Martin, D. J. et al.: AERE Report HL 76/2599 (1976)
35. Brooks, M. S. S., Glötzel, D.: Physica *102 B,* 51 (1980)
36. Trainor, R. J. et al.: Phys. Rev. Lett. *37,* 1511 (1976)
37. Aldred, A. T., Dunlap, B. D., Lander, G. H.: Phys. Rev. *B 14,* 1276 (1976)
38. Brodsky, M. B., Trainor, R. J.: Physica *B 91,* 271 (1977)
39. Dunlap, B. D. et al.: AIP Conf. Proc. No. 24 (eds. Graham, Jr., C. D., Lander, G. H., Rhyne, J. J.), p. 351 (1975)
40. Murata, K., Doniach, S.: Phys. Rev. Lett. *29,* 285 (1972)
41. Edwards, D. M., Wohlfarth, E. P.: Proc. Royal Soc. London *A 303,* 127 (1968)
42. Brodsky, M. B.: Rep. Prog. Phys. *41,* 103 (1978)
43. Sechovsky, V. et al.: Physica *102 B,* 277 (1980)
44. Fournier, J. M.: one day meeting on Magnetism of Actinides, Bedford, Col. London 1980, unpublished
45. Arrott, A., Goldman, J. E.: Phys. Rev. *108,* 948 (1957)
46. Gignoux, D. et al.: J. of Magn. and Magn. Mat. *15,* 289 (1980)
47. Hilscher, G.: private communications
48. Lawrence, J. M., Riseborough, P. S., Perks, R. D.: Rev. Prog. Phys. *44,* 1 (1981) and refs. therein
49. Chazaviel, J. N., et al.: Phys. Rev. *B 14,* 4586 (1970)
50. Cohen, R. L., Eibsc Rutz, M., West, K. W.: Phys. Rev. Lett. *24,* 383 (1970)
51. Beal-Monod, M. T., Lawrence, J. M.: J. Phys. Rev. *B 21,* 5400 (1980)
52. Buschow, K. H. J., Van Daal, H. J.: in AIP Conf. Proc. No., *5* (eds. Graham, C. P., Rhyne, J. I.), *2,* 1464 (1971)
53. Fournier, J. M.: Solid State Commun. *29,* 111 (1979)
54. Trainor, R. J., Brodsky, M. B., Culbert, H. V.: Phys. Rev. Lett. *34,* 1019 (1975)
55. Naegele, J. R. et al.: Physica *102 B,* 122 (1980)
56. Fournier, J. M., Beille, J.: J. de Phys. *C 4,* 145 (1979)
57. Fournier, J. M.: Ph. D. Thesis, U. of Grenoble, No. A.O., 10568 (1975)

58. Van Vleck, J. H.: Electric and Magnetic Susceptibilities, Oxford Press 1952
59. Slater, J. C.: Phys. Rev. *36,* 57 (1930)
60. Desclaux, J. P.: At. Data and Nucl. Data Tables *12,* 312 (1973)
61. Kubo, R., Obata, Y.: J. Phys. Soc., Japan *11,* 547 (1956)
62. Place, C. M., Rhodes, P.: Phys. Stat. Sol. *47,* 475 (1971)
63. Mori, N.: J. Phys. Soc., Japan *31,* 359 (1971)
64. Jullien, R.: Ph. D. Thesis, Univ. of Orsay-Paris Sud 1974
65. Friedel, J.: J. Phys. Chem. Sol. *1,* 175 (1956)
66. Amoretti, G., Fournier, J. M.: to be published
67. Troc., R., Lam, D. J.: Phys. Stat. Sol. *65,* 317 (1974)
68. Blaise, A. et al.: J. of Magn. and Magn. Mat. *30,* 265 (1982)
69. Rossat-Mignod, J. et al.: ibid. *30,* 122 (1982)
70. Tillwick, D. K., Du Plessis, P.: ibid. *3,* 329 (1976)
71. Trzebiatowski, W.: in: Ferromagnetic Materials (ed. Wohlfarth, E. L.), North-Holland, Amsterdam, Vol. 1, 428 (1980)
72. Kaczev, J., Novotny, P.: Phy. Stat. Sol. *A 39,* 351 (1977)
73. Deryagin, A. V., Andrejev, A. V.: Sov. Phys. JETP *44,* 610 (1977)
74. Matthias, B. T. et al.: Proc. Nat. Ac. Sci. *64,* 459 (1969)
75. Busch, G., Vogt, O., Bartholin, H.: J. de Phys. *64* (1979)
76. Vogt, O., Wachter, P., Bartholin, H.: Physica *102 B,* 226 (1980)
77. Blaise, A. et al.: Proceedings of the 12éme Journée des Actinides, Orsay 1982
78. Aldred, A. T. et al.: Phys. Rev. *B 9,* 3766 (1974)
79. Suski, W.: in: Plutonium and other Actinides (eds. Blank, H., Linder, R.), North-Holland, Amsterdam, 621 (1976)
80. Meaden, G. T.: Proc. Roy. Soc. *A 276,* 553 (1963)
81. Arko, A. J., Brodsky, M. B., Nellis, W. J.: Phys. Rev. *B 5,* 4564 (1972)
82. Doniach, S.: in: Actinides, Electronic Structure and Related Properties (eds. Freeman, A. J., Darby, Jr., J. B.), Academic Press, New York, *Vol. 2,* p. 51 (1974)
83. Smoluchowski, R.: Phys. Rev. *125,* 1577 (1962)
84. Long, K. A.: Proceedings of the 5e Journée des Actinides, Fontenayaux-Roses 1975
85. Kondo, J.: Prog. Theor. Phys. *32,* 37 (1964)
86. Schenkel, R.: Sol. St. Comm. *23,* 389 (1977)
87. De Gennes, P. G., Friedel, J.: J. Phys. Chem. Sol. *4,* 71 (1958)
88. Amoretti, G., Dallacasa, V.: Proceedings of the 10éme Journée des Actinides, Stockholm 1980
89. De Gennes, P. G.: J. de Phys. Rad. *23,* 510 (1962)
90. Du Plessis, P. de V., Van Doorn, C. F.: Physica *86 B,* 993 (1977)
91. Elliott, R. J., Wedgwood, F. A.: Proc. Phys. Soc. *81,* 846 (1963)
92. Moore, J. P., Fulkerson, W., McElroy, D. L.: J. Amer. Ceram. Soc. *53,* 75 (1970)
93. De Novion, C. H.: C.R. Acad. Sci. Paris *273,* 26 (1971)
94. Schoenes, J. et al.: Physica *102 B,* 328 (1980)
95. Schoenes, J., Vogt, O.: Proceedings of the 12éme Journée des Actinides, Orsay 1982
96. Blaise, A.: J. de Phys. *C 4,* 49 (1979)
97. Smith, J. L. et al.: ibid. *C 4,* 138 (1979)
98. Smith, J. L., Spirlet, J. C., Müller, W.: Science *205,* 188 (1975)
99. Franse, J. J. M. et al.: in: Physics of Solids under High Pressure (eds. Schilling, J. S., Shelton, R. N.), North-Holland, Amsterdam 1981
100. Andres, K., Graebner, J. E., Ott, H. R.: Phys. Rev. Lett. *36,* 1779 (1975)
101. Brodsky, M. B., Trainor, R. J.: J. de Phys. *C 6,* 777 (1979)
102. Blaise, A. et al.: J. Low Temp. *38,* 79 (1980)
103. Rossat-Mignod, J. et al.: Physica *102 B,* 177 (1980)
104. Huntzicker, J. J., Westrum, E. F.: J. Chem. Therm. *3,* 61 (1981)
105. Rossat-Mignod, J. et al.: Physica *102 B,* 237 (1980)
106. Aldred, A. T. et al.: Phys. Rev. *B 10,* 1011 (1974)
107. Lander, G. H. et al.: ibid. *B 14,* 5035 (1976)
108. Furrer, A. et al.: ibid. *B 14,* 179 (1976)
109. Sinha, S. K. et al.: Physica *102 B,* 174 (1980)
110. Lander, G. H. et al.: Phys. Rev. Lett. *40,* 523 (1978)

111. Holden, T. M. et al.: Phys. Rev. *B 26,* 6227 (1982)
112. Buyers, W. J. L. et al.: J. Appl. Phys. *52,* 222 (1981)
113. Lander, G. H., Stirling, W. G.: Phys. Rev. *B 21,* 436 (1980)
114. Stirling, W. G., Lander, G. H., Vogt, O.: Physica *102 B,* 259 (1980)
115. Monard, J. A., Huray, P. G., Thomson, J. O.: Bull. Am. Phys. Soc. *17,* 86 (1972)
116. Meeker, R. et al.: Bull. Am. Phys. Soc. *17,* 85 (1972)
117. Ruby, S. L. et al.: Phys. Rev. *184,* 374 (1969)
118. Kalvius, G. M. et al.: Phys. Lett. *B 29,* 489 (1969)
119. Friedt, J. M. et al.: in Actinides 1981, Asilomar (USA) 1981
120. Dunlap, B. D., Kalvius, G. M.: in: The Actinides: electronic structure and related properties (eds. Freeman, A. J., Darby, J. B.), Academic Press, New York, *Vol. 1,* p. 237 (1974)
121. Lander, G. H. et al.: Int. J. Magn. *4,* 99 (1973)
122. Bogé, M. et al.: J. of Magn. and Magn. Mat. *30,* 127 (1982)
123. Gal, J. et al.: Solid State Commun. *13,* 647 (1973)
124. Moser, J. et al.: Physica *102 B,* 199 (1980)
125. Gardner, W. E., Smith, T. F.: in Proceedings 11th Int. Conf. Low Temperature Physics (eds. Allen, J. F., Finlayson, D. M., McCall, D. M. (St. Andrews)), p. 1377 (1968)
126. Wedgwood, F. A.: J. Phys. *C 5,* 2427 (1972)
127. Wedgwood, F. A., Kuznietz, M.: ibid. *C 5,* 3012 (1972)
128. Busch, G., Vogt, O.: J. Less Common Metals *62,* 335 (1978)
129. Dunlap, B. D. et al.: J. Appl. Phys. *40,* 1495 (1969)
130. Arko, A. J. et al.: Bull. Amer. Phys. Soc. *15,* 293 (1970)
131. Aldred, A. T. et al.: in Transplutonium 1975 (eds. Müller, W., Lindner, R.), North-Holland, Amsterdam, p. 191 (1976)
132. Brodsky, M. D., Trainor, R. J.: Physica *86–88 B,* 143 (1977)
133. Nellis, W. J., Brodsky, M. B.: AIP Conf. Proc. No. 5, Magn. and Magn. Mat. – 1971 (eds. Graham, Jr., C. D., Rhyne, J. J.), AIP – New York, p. 1483 (1972)
134. Harvey, A. R. et al.: Phys. Rev. *B 7,* 4137 (1973)
135. Brodsky, M. B. et al.: in AIP Conf. Proc. No. 29, p. 317 (1976)
136. Brodsky, M. B., Trainor, R. J.: J. de Phys. *39,* C 6, 777 (1978)
137. Brodsky, M. B.: Phys. Rev. *B 9,* 1381 (1974)
138. Lam, D. J., Aldred, A. T.: in: AIP Conf. Prov. No. 24, Magn. and Magn. Mat. – 1974 (eds. Graham, Jr., C. D., Lander, G. H., Rhyne, J. J.), AIP, New York, p. 349 (1975)
139. Fournier, J. M., Gordon, J.: unpublished (1982)
140. Yessik, M.: J. Appl. Phys. *40,* 1133 (1969)
141. Lander, G. H. et al.: Physica *86–88 B,* 152 (1977)
142. Aldred, A. T.: J. Magn. Magn. Mat. *10,* 42 (1979)
143. Franse, J. J. M.: ibid. *31–34,* 819 (1983)
144. Aldred, A. T. et al.: Phys. Rev. *B 11,* 530 (1975)
145. Aldred, A. T.: J. Magn. Magn. Mat. *10,* 53 (1979)
146. Fournier, J. M., Blaise, A.: unpublished (1982)
147. Troć, R. et al.: Bull. Acad. Pol. Sci. Ser. Sci. Chim. *19,* 427 (1971)
148. Smith, J. L., Hill, H. H.: AIP Conf. Proc. No. 24, Magn. and Magn. Mat. – 1974 (eds. Graham, Jr., C. D., Lander, G. H., Rhyne, J. J.), AIP, New York, p. 382 (1975)
149. Andreev, A. V. et al.: J. de Phys. *40,* C 4, 82 (1979)
150. Sechovsky, V. et al.: in Plutonium and other Actinides (eds. Blank, H., Lindner, R.), North-Holland, Amsterdam, p. 641 (1976)
151. Olsen, C. E.: J. Appl. Phys. *31,* 3405 (1960)
152. Trzebiatowski, W., Misink, A.: Rocz. Chem. *42,* 161 (1968)
153. Menovsky, A. et al.: to be published (1983)
154. Fournier, J. M., Beille, J.: unpublished (1980)
155. Brodsky, M. B. et al.: J. Appl. Phys. *49,* 1498 (1978)
156. Schneider, W. D., Laubschat, C.: Phys. Rev. *B 23,* 997 (1981)
157. Frings, P. H. et al.: J. Magn. Magn. Mat. *31–34,* 240 (1983)
158. Misink, A. et al.: Bull. Acad. Pol. Sci. Ser. Sci. Chim. *20,* 337 (1979)
159. Van Maaren, M. H. et al.: Solid State Commun. *14,* 145 (1974)
160. Misink, A., Trzebiatowski, W.: Bull. Acad. Pol. Sci., Ser. Sci. Chim. *18,* 633 (1970)

161. Mulak, S., Misink, A.: ibid. *19,* 207 (1971)
162. Misink, A. et al.: ibid. *20,* 459 (1972)
163. Ansorge, V., Menovsky, A.: Phys. Stat. Sol. *30,* K31 (1968)
164. Murasik, A. et al.: ibid. *a23,* K147 (1974)
165. Nellis, W. J. et al.: AIP Conf. Proc. No. 10, Magn. and Magn. Mat., 1972 (eds. Graham, Jr., C. D., Rhyne, J. J.), AIP, New York, p. 1076 (1973)
166. Lander, G. H. et al.: Bull. Amer. Phys. Soc. *17,* 338 (1972)
167. Nellis, W. J. et al.: Phys. Rev. *B9,* 1041 (1974)
168. Lander, G. H., Mueller, M. H.: unpublished (1972)
169. Murasik, A. et al.: Phys. Stat. Sol. *a20,* 395 (1973)
170. Rao, V. U. S., Vijayaraghavan, R.: J. Phys. Chem. Sol. *29,* 123 (1968)
171. Leciejewicz, J., Misink, A.: Phys. Stat. Sol. *a13,* K79 (1972)
172. Wernick, J. H. et al.: J. Appl. Phys. *36,* 982 (1965)
173. Andres, K. et al.: Sol. Stat. Com. *28,* 405 (1978)
174. Erdmann, B., Keller, C.: J. of Sol. Stat. Chem. *7,* 40 (1973)
175. Sternberk, J. et al.: Phys. Stat. Sol. *a28,* K45 (1975)
176. Turan, J. et al.: Acta Phys. Slov. *31,* 143 (1981)
177. Konobeesky, S. T. et al.: Proc. 2nd U.N. Int. Conf. on Peaceful uses of At. Energy (Geneva), Bd. 6, p. 194 (1958)
178. Luengo, C. A. et al.: J. Magn. Magn. Mat. *3,* 305 (1976)
179. Huber, J. G., Luengo, C. A.: J. de Phys. *C6,* 781 (1978)
180. Brodsky, M. B., Bridger, N. J.: in AIP Conf. Proc. No. 18, Magn. and Magn. Mat. – 1973 (eds. Graham, Jr., C. D., Rhyne, J. J.), AIP, New York, p. 357 (1974)
181. Van Daal, H. J. et al.: Phys. Rev. Lett. *34,* 1457 (1975)
182. Misiuk, A. et al.: Bull. Acad. Pol. Sci. Ser. Sci. Chim. *21,* 487 (1973)
183. Murasik, A. et al.: Phys. Stat. Sol. *a23,* K163 (1974)
184. Troć, R. et al.: ibid. *b43,* 147 (1971)
185. Brodsky, M. B., Friddle, R. J.: in AIP Conf. Proc. No. 24, Magn. and Magn. Mat. – 1974 (eds. Graham, Jr., C. D., Lander, G. H., Rhyne, J. J.), AIP, New York, p. 353 (1975)
186. Raphael, G., De Novion, C. H.: J. de Phys. *30,* 261 (1969)
187. Buschow, K. H. J.: J. Less Comm. Met. *31,* 165 (1973)
188. Murasik, A. et al.: Phys. Stat. Sol. *a28,* K107 (1975)
189. Brodsky, M. B., Aldred, A. T.: unpublished (1976)
190. Aronson, S., Auskern, A. B.: J. Chem. Phys. *48,* 1760 (1968)
191. Adachi, H., Imoto, S.: Techn. Rept. Osaka Univ. *23,* 425 (1973)
192. Hery, Y. et al.: in Proc. 2nd Int. Conf. on the Electr. Struct. of the Actinides (eds. Mulak, J., Suski, W., Troć, R.), Ossolineum, Wroclaw, p. 343 (1977)
193. Bates, L. F., Unstead, P. B.: Brit. J. Appl. Phys. *15,* 543 (1964)
194. Lam, D. J. et al.: in: Plutonium 1965 (eds. Kay, A. E., Waldron, M. B.), Chapman and Hall, London, p. 274 (1967)
195. Lam, D. J. et al.: J. de Phys. *32 c1,* 917 (1971)
196. De Novion, C. H., Lorenzelli, R.: J. Phys. Chem. Sol. *29,* 1901 (1968)
197. Lander, G. H. et al.: ibid. *30,* 733 (1969)
198. Ross, J. W., Lam, D. J.: J. Appl. Phys. *38,* 1451 (1967)
199. Lallement, R. et al.: J. Phys. Chem. Sol. *26,* 1255 (1965)
200. Green, J. L. et al.: J. Nucl. Mat. *34,* 281 (1970)
201. Lam, D. J., Aldred, A. T.: in: The Actinides: Electronic Structure and Related Properties, Vol. 1 (eds. Freeman, A. J., Darby, J. B.), Academic Press, New York, p. 109 (1974)
202. Raphael, G., De Novion, C. H.: Solid State Commun. *7,* 791 (1969)
203. Ohmichi, T. et al.: J. Nucl. Sci. and Techn. *9,* 11 (1972)
204. Troć, R.: J. Solid State Commun. *13,* 14 (1975)
205. Van Doorn, C. F., Du Plessis, P. de V.: J. Low Temp. Phys. *28,* 391 (1977)
206. Lam, D. J.: in AIP Conf. Proc. No. 5, Magn. and Magn. Mat. – 1971 (eds. Graham, Jr., C. D., Rhyne, J. J.), AIP, New York, p. 892 (1972)
207. Bœuf, A. et al.: in Proc. 11èmes Journées des Actinides (eds. Bombieri, G., De Paoli, G., Zanella, P.), Ist. di Chim, e Tecn. dei Radioelementi del CNR, Padova, p. 233 (1982)

208. Kanellakopoulos, B. et al.: in: Transplutonium 1975 (eds. Müller, W., Lindner, R.), North-Holland, Amsterdam, p. 181 (1976)
209. Nave, S. E. et al.: in: Actinides 1981, Asilomar, California, LBL-12441, p. 144 (1981)
210. Busch, G. et al.: J. de Phys. 40 *C4,* 62 (1979)
211. Curry, N. A.: Proc. Phys. Soc., London *89,* 427 (1966)
212. Heaton, L. et al.: J. Appl. Cryst. *3,* 289 (1970)
213. Nave, S. F. et al.: Physica *107B,* 253 (1981)
214. Leciejewicz, J. et al.: Phys. Stat. Sol. *38,* K89 (1970)
215. Blaise, A. et al.: J. Magn. and Magn. Mat. *31–34,* 233 (1983)
216. Blaise, A. et al.: J. Solid State Commun. *13,* 555 (1973)
217. Spirlet, J. C. et al.: in: 13èmes Journées des Actinides, Elat, Israel 1983
218. Burlet, P. et al.: in: 13èmes Journées des Actindes, Elat, Israel 1983
219. Dunlap, B. D. et al.: J. Appl. Phys. *42,* 1719 (1971)
220. Trzebiatowski, W., Zygmunt, A.: Bull. Acad. Pol. Sci. Ser. Sci. Chim. *14,* 495 (1966)
221. Kuznietz, M. et al.: J. Phys. Chem. Sol. *30,* 1642 (1969)
222. Tillwick, D. L., Du Plessis, P. de V.: J. Magn. Magn. Mat. *3,* 319 (1976)
223. Chechernikov, V. I. et al.: Sov. Phys. JETP *26,* 328 (1968)
224. Lam, D. J. et al.: in: Proc. Int. Conf. Magnetism – 1973, Vol. IV, Nauka, Moscow, p. 74 (1974)
225. Allbut, M. et al.: in: The Chemistry of Extended Defects in Non-metallic Solids, North-Holland, Amsterdam, p. 124 (1970)
226. Haessler, M., De Novion, C. H.: J. Phys. *C10,* 589 (1977)
227. Trzebiatowski, W., Sepichowska, A.: Bull. Acad. Pol. Sci. Ser. Sci. Chim. *8,* 457 (1960)
228. Longworth, G. et al.: J. Phys. *C6,* 1652 (1973)
229. Busch, G. et al.: ibid. *C12,* 1391 (1979)
230. Trzebiatowski, W. et al.: in: Proc. 11th Int. Conf. on Phys. Semicond., Polish Scient. Publishers, Warsaw, p. 1287 (1972)
231. Markowski, P. J. et al.: Rocz. Chem. *51,* 1027 (1977)
232. Sternberk, J. et al.: J. Phys. Sup. *32 C1,* 744 (1971)
233. Burlet, P. et al.: J. Solid State Commun. *39,* 745 (1981)
234. Suski, W. et al.: Phys. Stat. Sol. *14,* K157 (1972)
235. Suski, W., Janus, B.: Bull. Acad. Pol. Sci. Ser. Sci. Chim. *28,* 199 (1980)
236. Burlet, P. et al.: J. Solid State Commun. *43,* 587 (1982)
237. Matsui, H., Tamaki, M.: Phys. Stat. Sol. *a71,* K121 (1982)
238. Suski, W. et al.: J. Magn. Magn. Mat. *3,* 195 (1976)
239. Lagnier, R. et al.: Phys. Stat. Sol. *a57,* K127 (1980)
240. Lagnier, R. et al.: ibid. *a29,* K51 (1975)
241. Suski, W., Reizer-Netter, H.: Bull. Acad. Pol. Sci. Ser. Sci. Chim. *22,* 701 (1970)
242. Noel, H., Prigent, J.: Physica *102B,* 372 (1980)
243. Blaise, A. et al.: in: Plutonium 1975 and other Actinides (eds. Blank, H., Lindner, R.), North-Holland, Amsterdam, p. 635 (1976)
244. Trzebiatowski, W., Troć, R.: Bull. Acad. Pol. Sci. Ser. Sci. Chim. *11,* 661 (1963)
245. Troć, R. et al.: Phys. Stat. Sol. *15,* 515 (1966)
246. Hery, Y. et al.: Radiochem. Radioanal. Lett. *32,* 283 (1978)
247. Trzebiatowski, W. et al.: Bull. Acadm. Pol. Sci. Ser. Sci. Chim. *12,* 687 (1964)
248. Oles, A.: J. de Phys. *26,* 1561 (1965)
249. Hery, Y. et al.: Radiochem. Radioanal. Lett. *37,* 17 (1979)
250. Brooks, M. S. S. et al.: Physica *102B,* 84 (1980)
251. Blaise, A. et al.: J. Magn. Magn. Mat. *29,* 297 (1982)
252. Blaise, A. et al.: Phys. Stat. Sol. *a71,* K147 (1982)
253. Leciejewicz, J. et al.: ibid. *22,* 517 (1967)
254. Suski, W. et al.: J. Sol. St. Chem. *4,* 223 (1972)
255. Grønvold, F., Westrum, Jr., E. F.: J. Inorg. Nucl. Chem. *30,* 2127 (1968)
256. Suski, W. et al.: Phys. Stat. Sol. *a9,* 653 (1972)
257. Westrum, Jr., E. F., Grønvold, F.: J. Inorg. Nucl. Chem. *32,* 2169 (1970)
258. Klein-Haneveld, A. J., Jellinek, F.: J. Less Comm. Met. *8,* 123 (1969)
259. Suski, W.: J. Sol. St. Chem. *7,* 385 (1973)

260. Noel, H., Troć, R.: ibid. *27,* 123 (1979)
261. Zygmunt, A.: Phys. Stat. Sol. *43,* 573 (1977)
262. Ptasiewicz-Bak, H. et al.: ibid. *a 47,* 349 (1978)
263. Troć, R., Zolnierek, Z.: S. de Phys. *40 C 4,* 79 (1979)
264. Ballestraci, R. et al.: J. Phys. Chem. Sol. *24,* 487 (1963)
265. Murasik, A. et al.: Phys. Stat. Sol. *30,* 61 (1968)
266. Hulliger, F.: J. Less Comm. Met. *16,* 113 (1968)
267. Zygmunt, A. et al.: Phys. Stat. Sol. *a 25,* K 77 (1974)
268. Zygmunt, A. et al.: ibid. *a 22,* 75 (1974)
269. Leciejewicz, J., Zygmunt, A.: ibid. *a 13,* 657 (1972)
270. Zygmunt, A.: unpublished
271. Blaise, A. et al.: in: Proc. 2nd Int. Conf. on the Electronic Structure of the Actinide (eds. Mulak, J., Suski, W., Troć, R.), Ossolineium (Wroclaw), p. 475 (1977)
272. Leciejewicz, J. et al.: Sol. St. Comm. *41,* 167 (1982)
273. Ptasiewicz-Bak, H. et al.: J. Phys. *F 11,* 1225 (1981)
274. Gal, J. et al.: Solid State Commun. *20,* 421 (1976)
275. De Novion, C. H. et al.: J. Magn. Magn. Mat. *21,* 85 (1980)
276. Gal, J. et al.: Solid State Commun. *20,* 515 (1976)
277. Dawson, J. K., Lister, M. W.: J. Chem. Soc. 2181 (1950)
278. Faber, J., Lander, G. H.: Phys. Rev. *B 14,* 1151 (1976)
279. Dunlap, B. D. et al.: J. Phys. Chem. Sol. *29,* 1365 (1968)
280. Erdös, P. et al.: Physica *102 B,* 164 (1980)
281. Raphael, G., Lallement, R.: Sol. St. Comm. *6,* 383 (1968)
282. Karraker, D. G.: J. Chem. Phys. *63,* 3174 (1975)
283. Nave, S. E. et al.: Phys. Rev. *B 28,* 2317 (1983)
284. Westrum, Jr., E. F., Grønvold, F.: Thermodynamics of Nuclear Materials, p. 3 (1962)
285. Westrum, Jr., E. F., Grønvold, F.: J. Amer. Chem. Soc. *81,* 1777 (1959)
286. Leask, M. J. M. et al.: J. Chem. Soc. 4788 (1963)
287. Flotow, H. E., Tetenbaum, M.: J. Chem. Phys. *74,* 5269 (1981)
288. McCart, B. et al.: ibid. *74,* 5263 (1981)
289. Trzebiatowski, W. et al.: Rocz. Chem. *26,* 110 (1952)
290. Gruen, D. M.: J. Chem. Phys. *23,* 1708 (1955)
291. Bœuf, A. et al.: in 14ème Journées des Actinides, Davos (Switzerland) 1984
292. Sliwa, A., Trzebiatowski, W.: Bull. Acad. Pol. Sci. Ser. Sci. Chim. T 3, 675 (1962)
293. Aldred, A. T. et al.: Phys. Rev. *B 19,* 300 (1979)
294. Cinader, G. et al.: ibid. *B 14,* 912 (1976)
295. Ward, J. A. et al.: in Actinides 1981, Pacific Grove, California, LBL-12441, p. 86 (1981)
296. Bœuf, A. et al.: in 13ème Journées des Actinides, Elat (Israel) 1983
297. Brooks, M. S. S., Kelly, P. J.: Phys. Rev. Letters *51,* 1708 (1983)
298. Ott, H. R. et al.: ibid. *50,* 1595 (1983)
299. Stewart, G. et al.: ibid. *52,* 679 (1984)
300. Baybarz, R. D. et al.: J. Inorg. Nucl. Chem. *34,* 3427 (1972)
301. Berger, R., Sienko, M. J.: Inorg. Chem. *6,* 324 (1967)
302. Flotow, H. E. et al.: Thermodynamics of Nuclear Materials. p. 477 (1975)
303. Hinatsu, Y. et al.: J. Nucl. Sci. and Tech. *17,* 929 (1980)
304. Fischer, P., Murasik, A.: in 12éme Journées des Actinides, Orsay (France) (1982)
305. Hendricks, M. E.: Ph.D. Thesis, U. of South Carolina (USA) (1971)
306. Dawson, J. K.: J. Chem. Soc., 429 (1951)
307. Roberts, L. D., Murray, R. B.: Phys. Rev. *100,* 650 (1955)
308. Murasik, A. et al.: J. Phys. *C 14,* 1847 (1981)
309. Leask, M. J. M. et al.: J. Chem. Phys. *34,* 2090 (1961)
310. Lewis, W. B., Elliott, N.: ibid. *27,* 904 (1957)
311. Hendricks, M. E. et al.: ibid. *55,* 2993 (1971)
312. Vogt, O. et al.: Physica *102* B+C, 226 (1980)
313. Stone, J., Jones, E. R.: J. Chem. Phys. *54,* 1713 (1971)
314. Handler, P., Hutchison, Jr., C. A.: ibid. *25,* 1210 (1956)

Chapter E

Localization and Hybridization of 5 f States in the Metallic and Ionic Bond as Investigated by Photoelectron Spectroscopy

J. R. Naegele, J. Ghijsen* and L. Manes

Commission of the European Communities, Joint Research Centre, Karlsruhe Establishment, European Institute for Transuranium Elements, Postfach 2266, D-7500 Karlsruhe, Federal Republic of Germany

Localization versus itineracy and the degree of hybridization of 5 f states with orbitals of the actinide atom (especially 6 d) as well as with those of the ligand in compounds are central questions for the understanding of bonding in actinide solids. Photoelectron spectroscopy provides answers to these questions. In narrow band solids, like the actinides, the interpretation of results requires the use of band calculations in the itinerant picture, as well as models of final state emission in the atomic picture.

From the experimental viewpoint: 1. the analysis of the variation of photoionization cross sections (affecting the intensities of photoelectron spectroscopy), gives an insight into the orbital composition of the bands of the solid; 2. the combination of direct and inverse photoemission provides a powerful tool to assess the structure of occupied and of empty states, and, in the case of localized 5 f states, permits the determination of a fundamental quantity, the Coulomb correlation energy, governing the physical properties of narrow bands.

After a survey of the basic theory and some experimental aspects of photoelectron spectroscopy which are relevant to actinide solids, two systems are illustrated: elemental actinide metals, in which the Mott transition between plutonium and americium is evidenced in a "photographic" way by photoemission, and strongly ionic oxides, in which the 5 f localized behaviour is clearly seen, and indications of f-p or d-p covalent mixing are investigated.

* Present address: Max-Planck Institut für Festkörperforschung, Postfach 80 06 65, D-7000 Stuttgart 80, Federal Republic of Germany

I. Introduction

Recently a so called "Nearly Periodic Table of Transition Elements" has been published[1] that depicts the intermediate position of the actinides between the 4f and 3d series. Figure 1 shows this table. The crossover between electron bonding behaviour of the unsaturated shells and the formation of magnetic moments in atoms occurs at the beginning and near the end of the 4f and 3d series, respectively. In actinides, it occurs near the half filling of the actinide 5f shell, i.e. at the middle of the actinide series.

The exact position of the crossover depends however, on the type of compound that is formed from an actinide. It appears usually for the radioactive heavier actinides. In order to study it, therefore, compounds of Pu, Am, Cm have to be investigated. Compounds of elements preceding them in the actinide series present properties due to the itineracy of the 5f electrons, which are somewhat similar to the d-transition elements compounds (especially 3d). Heavier actinides are lanthanide-like although their properties may depart from a true lanthanide behaviour: unfortunately, their rarity and the difficulty of their handling is such that very few photoemission results are available for them.

This chapter is devoted to photoemission spectroscopy and the related inverse photoemission spectroscopy, which are well developed experimental tools to study occupied and empty electronic levels, respectively. Special emphasis is given to the 5f electrons and their localized or delocalized character.

The chapter deals with three main subjects: a survey of theoretical and practical aspects of photoelectron spectroscopy (Part II); an analysis of results on elemental metals (Part III); and an analysis of results on oxides (Part IV).

This chapter concentrates particularly on actinide compounds for which a large number of measurements have been performed in the vicinity of the crossover, i.e. the actinide metals and their oxides. They have in addition, the advantage that they display two different bonding types, opposite in nature, for what regards the localization vs. itineracy problem. The elemental actinide metals allow the study of f-f overlapping throughout the series without being disturbed by bonding effects from other atoms. The actinide oxides, which are known to have localized 5f electrons, are suited to the investigation of covalency effects (in contrast to metallic bonding) with the oxygen anions.

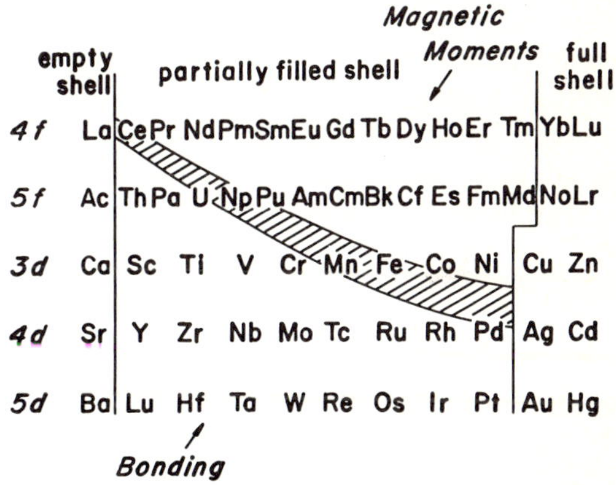

Fig. 1. Modified periodic table showing the itinerant vs. localized character of unsaturated transition shells. Bonding is contrasted to magnetic moment formation, to illustrate the band-like vs. atomic-like behaviour of wavefunctions. The *dashed region* represents the crossover between the two situations (from Ref. 1)

Within other classes of actinide compounds, as, e.g. intermetallics or NaCl-type compounds, no data on highly radioactive actinides have been reported so far (with the exception of PuSb measured by Baptist et al.[2,3]).

II. General Concepts of Photoelectron Spectroscopy

Photoelectron spectroscopy has been presented in many review articles and books. Therefore here only a brief description is given with special emphasis on actinides. For further reading, some selected literature is recommended[4-8].

1. *The Analysis of Binding Energies*

a. *Methods of Photoelectron Spectroscopy*

α. Photoemission: Spectroscopy of Occupied Electron States in a Solid

In photoemission spectroscopy monoenergetic (monochromatic) photons impinge on the sample to be investigated. The photon ejects core or valence electrons from the sample into the vacuum. This occurs when its energy hν is high enough to overcome the binding energy E_b of the electrons. In a simple picture, the kinetic energy E_{kin} of the photoemitted electrons in vacuum is given by

$$E_{kin} = h\nu - E_b - E_{WF} \quad (1)$$

wherein E_{WF} is the work function, i.e. the energy difference between the Fermi energy and the zero energy of the vacuum (vacuum level); (1) is the typical equation of the photoelectric effect. Thus, by analyzing the kinetic energy distribution of the photoemitted electrons, the binding energy of electrons can be determined directly.

Since the binding energy of core electrons is characteristic of each element, photoemission spectroscopy with X-rays (XPS: *X*-ray induced *P*hotoemission *S*pectroscopy) has long been employed for chemical analysis. Its original name ESCA reflects this: *E*lectron *S*pectroscopy for *C*hemical *A*nalysis. E_b is somewhat dependent on the specific chemical bond (ionic, covalent, metallic) of the solid, since even deeply bound energy states in an atom (or core) are affected by the nature of their environment. Therefore the energy shift ΔE_b of core lines gives additional information about the chemical bond (chemical shift)[9].

In contrast to core electrons which occupy discrete energy levels (large E_b), electrons taking part in the chemical bonding in a solid are distributed over energy bands (see Chap. A) (valence and conduction bands; E_b only a few eV). The electrons in these bands can be examined both by XPS and by photoemission spectroscopy with exciting photons in the UV region (UPS: *U*ltraviolet induced *P*hotoemission *S*pectroscopy).

In Fig. 2, the two methods (XPS and UPS) are illustrated. Very often, a combination of the two methods is employed (XPS/UPS): the advantages of this combination will be illustrated later when the properties of photoemission cross-sections will be discussed.

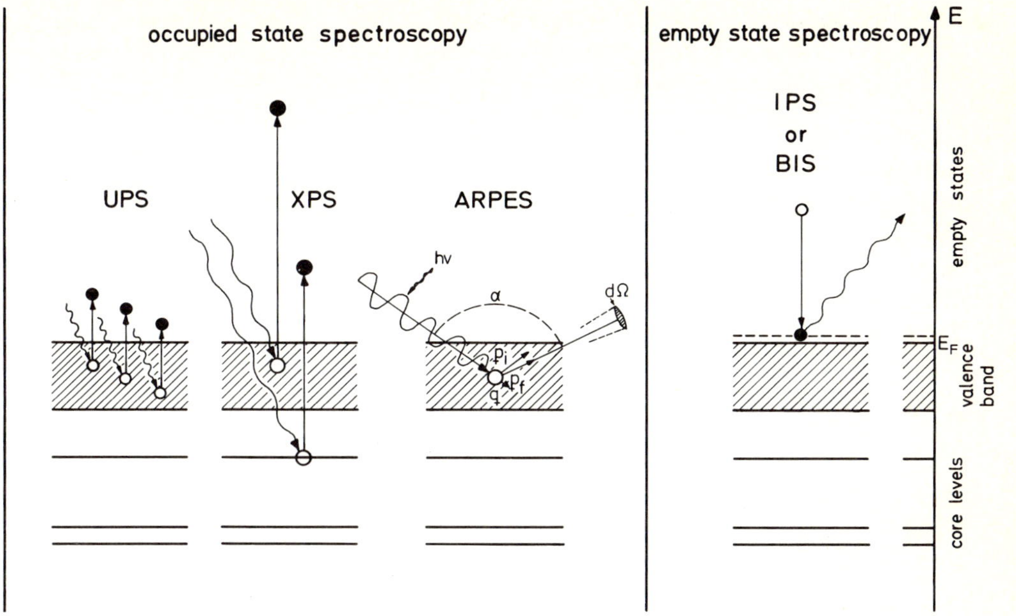

Fig. 2. Schematic representation of occupied state and empty state spectroscopies

β. *I*nverse *P*hotoemission *S*pectroscopy (IPS) or *B*remsstrahlung *I*sochromat *S*pectroscopy (BIS): a Spectroscopy of Empty Electron States in a Solid

In inverse photoemission spectroscopy or Bremsstrahlung Isochromat Spectroscopy, the sample is bombarded with monoenergetic electrons and the Bremsstrahlung radiation is recorded at constant photon energy for varying kinetic energy of the incident electrons. The electronic process involved is just the inversion of the photoemission process, and therefore, instead of investigating occupied states as in UPS/XPS, empty states are examined in BIS.

In Fig. 2, IPS or BIS is also described.

b. The Binding Energies and the Conservation Equations in Photoemission

In order to examine more deeply the information contained in the photoelectron processes described above, we begin by representing the photoemission as a process in which the initial state of a system of N electrons, having wavefunction Ψ_{in} is changed by a photon of energy hν to a final state, having wavefunction Ψ_{fin}

$$\Psi_{in}(N, E_{in}(N)) \xrightarrow{h\nu} \Psi_{fin}(N-1, h, E_{fin}(N-1, h)) \cdot \phi_\alpha^j(E_{kin}(j)) \qquad (2)$$

The process is accompanied by the emission of an electron j beyond the vacuum level of the system. The photoemitted electron has usually kinetic energy E_{kin} after emission, and its wavefunction ϕ^j_α in (2) is that of an unbound electron (α denotes the quantum number describing its motion in the unbound state). It leaves a hole (h), in the system, so that Ψ_{fin} describes a state in which N − 1 electrons and a hole are found. A large number of phenomena will depend on the rearrangement of the remaining electrons after the photoemission process. The rearrangement is usually referred to as the relaxation of their motion in the presence of the hole. E_{in} and E_{fin} in (2) are the total energies of the system at the initial and final states respectively. The *energy conservation* condition will therefore be:

$$E_{in}(N) + h\nu = E_{fin}(N-1, h) + E_{kin} \tag{3}$$

which may be rewritten as:

$$h\nu = E_{fin}(N-1, h) - E_{in}(N) + E_{kin} = E_b + E_{kin} \tag{4}$$

in which we have defined: E_b (“binding” energy) = $E_{fin} - E_{in}$ (We have put “binding” in inverted commas, because we shall discuss its meaning later). (This is of course just Eq. (1), where E_{WF} has been introduced in E_{fin} and E_{in}, which are now referred to the vacuum level.)

In BIS, the process (2) is reversed, and we analyze, as said before, the photon spectrum yielded by the system as “Bremsstrahlung” when electrons occupy empty states.

There is a condition of *momentum conservation* for photons and electrons which must also be satisfied in the photoemission process. For band electrons, for which the Bloch wavefunctions are characterized by the wavenumber k (proportional to the momentum p of the electron), the momentum conservation condition is important to determine the angular distribution of the photoemitted electrons. *A*ngular *R*esolved *P*hoto*E*mission *S*pectroscopy (ARPES), schematized in Fig. 2, is potentially able to provide, and has been used to obtain, the E(k) dispersion curves for solids.

c. *The Binding Energies Within Koopmans' Theorem Conditions*

As a first approximation, it may be assumed that the photoemission of the electron j does not perturb at all the wavefunctions of the other N − 1 electrons, which are therefore “passive” to the process. Hence Ψ_{in} in (2) can be factorized as:

$$\Psi_{in} = \phi^j_{in} \Psi_{fin}(N-1) \tag{5}$$

where $\Psi_{fin}(N-1)$ is the same as in (2).

In this case, the whole process is determined by the change of the wavefunction $\phi_{in} \to \phi_{fin}$ of the photoemitted j electron, and the energy conservation condition (4) can be written as:

$$h\nu = -E_{in}(j) + E_{kin}(j) \tag{6}$$

$$E_b = -E_{in}(j) \tag{6a}$$

This approximation assumes Koopmans' theorem (which has been discussed in Chap. A) to be valid. It is recalled that this theorem is valid for broad bands in solids, where electrons have a fully itinerant character. In this case, E_b as given by (6 a) is simply the one-electron energy $E(k)$ of an itinerant electron in the $E(k)$ band.

Within Koopmans' theorem approximation, therefore, photoemission probes the occupied bands of a solid, establishing their relative position in energy; BIS in the same approximation probes its empty bands. The combination of the two methods is therefore a very powerful tool for the elucidation of the band structure.

One has to be very cautious in interpreting, e.g., valence band spectra of actinides: generally a direct comparison of valence band spectra with the density of states as derived from band ground state calculations is not appropriate.

For the photoemission of an electron from an ion core M^{+v}, we have to consider a reaction of the type:

$$M^{+v} \xrightarrow{E_b} {}^*M^{+(v+1)} \tag{7}$$

We have marked with a star the final state ion $M^{+(v+1)}$ to indicate that the photoemission process leaves this ion in a state which is far from being its ground state.

Actually, if we could apply Koopmans' theorem conditions also to the ion core levels, E_b would be the negative of the eigenvalue of the electron $-\varepsilon_{nlm}$ (obtained, e.g., by a one-electron Hartree-Fock calculation – see Chap. A) in the initial state. In reality, the existence of a hole within the ion core implies that E_b is given rather by the expression[4)]:

$$E_b = -\varepsilon_{nlm} - \delta E_{relax.} + \delta E_{relat.} + \delta E_{corr.} \tag{8}$$

where the δE's are corrections to Koopmans' theorem conditions, which arise from the quantum description of the final state ${}^*M^{+(v+1)}$.

$\delta E_{relax.}$ describes the relaxation of the $(N-1)$ "passive" electrons when a hole appears in the ground state electronic configuration (a hole may be viewed as a positive charge being added to the ion). This term will tend to give lower binding energies than the Koopmans approximation and is often evaluated by calculating Hartree-Fock energy schemes for the final state.

$\delta E_{relat.}$ describes relativistic effects (such as variations in spin couplings – see Chap. A) and $\delta E_{corr.}$ other electron-electron (and also electron-vibrational) many-body correlation effects (which are not included in Hartree-Fock calculations).

d. Photoemission From an Atom or an Ion Core

It is known (Chap. A) that Koopmans' theorem is not valid for the wavefunctions and eigenvalues of strongly bound states in an atom or in the cores of a solid, i.e. for those states which are a solution of the Schrödinger (or Dirac) equation in a central potential. In them the ejection (or the emission) of one-electron in the electron system means a strong change in Coulomb and exchange interactions, with the consequent modification of the energy scheme as well as of the electronic wavefunction, in contradiction to Koopmans' theorem.

The photoexcitation process leading to photoemission is a very strong perturbation of the electron system. Therefore, after photoemission, one has to take into account the many-electron relaxation effects that follow. These effects do not permit the cancellation of Ψ_{fin} which we have performed previously.

In the photoemission spectroscopic method, it is customary to distinguish between a core level region, which is probed in XPS (see Fig. 2), and which contains the response coming from the bound or core levels, and a valence band region, which is explored by both XPS and UPS, and which contains the response coming from the outer electrons: in a solid, those of the ground state energy bands.

The two regions are usually distinguished by the binding energy ranges they cover: typically, some 50 eV (at most) from E_F for the valence band region, the rest being core level region.

In the valence band region the Koopmans' theorem conditions are usually satisfied, as long as broad band electrons are photoemitted. On the other hand, in open-shell solids such as the d-, 4f- and 5f-solids, the unsaturated d and f shells have, as we have discussed in other chapters, a very strong atomic-like (or localized) character, although their eigenvalues may lie not very far from those of the itinerant (s, d) band electrons. It is therefore evident that the Koopmans' theorem approximation may fail also for these cases, although their energies fall in the valence band region.

The δE's are *final state* many electron effects, depending on the electronic configuration of the $^*M^{+(v+1)}$ ion left by the photoemission process and which cannot be described in a one-electron picture.

Final state effects particularly relevant in photoemission from actinide solids, are now briefly reviewed.

e. *Multiplet Splitting*

Very important final state effects are those which are described in the final state multiplet theory. They occur in cores containing unsaturated, localized shells (open-shell cores). The theory has been successfully applied, e.g., in the interpretation of the spectral response from the 4f levels of lanthanides[10–13].

By making reference to this case, the photoexcitation is $f^N \rightarrow f^{N-1}$, where f^N (initial state) is an f-shell containing N electrons, and f^{N-1} (final state) is an f-shell containing $N - 1$ electrons. In a "sudden approximation", i.e. assuming the photoemission process to be so "sudden" not to leave time for the "passive" core electrons to relax to some lower energy (ground) state, the different coupling (see Chap. A and D) possibilities of orbital and spin moments of the remaining electrons in these cores may leave the core in a multiplicity of many-electron states (electronic configurations and levels). When these configurations and levels are non-degenerate, Eqs. (3) and (8) show that a peak is encountered in the photoemission curve for each of them. In order to evaluate the intensity of each peak, each emission has to be weighted according to its probability of occurring (see below the cross-section discussion). This is usually done in the so called fractional parentage scheme[11, 13, 14].

Multiplet splitting occurs in the core-level region as well as in the energy range which we called "valence region": in this case it can be due to photoemission from a *localized* state, although this state may be not very far from the Fermi edge. A typical example of

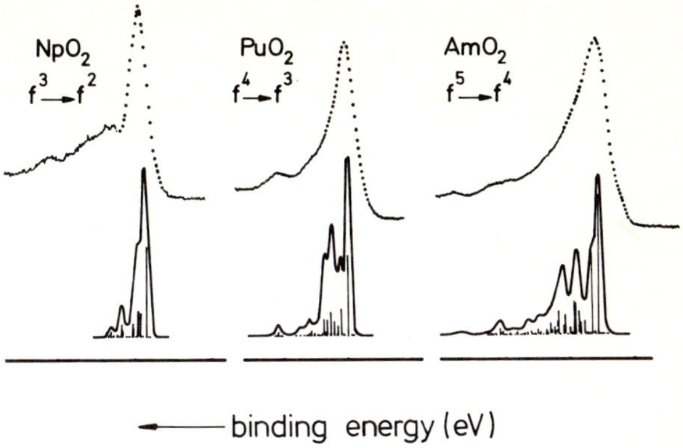

Fig. 3. An example of final state multiplet splitting interpretation of the XPS photoemission of actinide dioxides (from Ref. 15)

this case has been considered to be the 5 f occupied state that is found in the main energy gap of insulating solids such as actinide oxides (the electronic structure of which has been discussed in Chap. C and D). Veal et al.[15] have tried interpreting the measured emission from the 5 f state in these solids by fitting a multiplet scheme, as obtained by intermediate coupling calculations (see Chap. A and D), to the details of the valence band spectrum (Fig. 3). More recent multiplet splitting calculations for 5 f^n states have been performed by Gerken[16]. The valence band spectra of actinide oxides will be discussed later in this chapter in greater detail, and the two interpretations compared.

Final state multiplet splitting is not only observed for the 5 f spectra in the valence band range but also in core level spectra of actinides. The main interaction giving rise to the final state effects is assumed to be the exchange coupling between core states and the outer incomplete 5 f shell. The multiplet splitting is considerable for interacting states with the same principal quantum number. Therefore this effect will be observed mainly for 5 s, 5 p and 5 d core levels of the actinides. It will cause very complex spectra which are difficult to analyse because of the superposition with screening effects.

In inverse photoemission (and in case of high electron excitation energy) a final core state containing an additional electron (instead of a hole) may be created (for f-states: $f^N \rightarrow f^{N+1}$). Thus, and in the same lines as for photoemission, a multiplicity of non

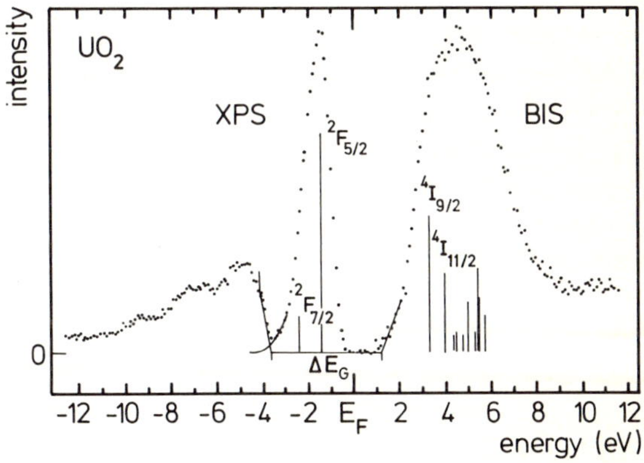

Fig. 4. The photoemission and inverse photoemission spectra of UO_2 (from Ref. 18)

degenerate configurations may arise for the final state and multiplet splitting may occur in the BIS response. This has been shown for lanthanide metals[17] and also assumed for UO_2[18] (Fig. 4).

f. Other Many-Body Effects in Photoemission

The apparent binding energies measured in photoelectron spectroscopy are affected by other many-body effects, which are not treated here[19, 20]. Only a few of them, relevant for open shell systems, hence for actinides, will be treated in Chapt. II. 3.

2. The Analysis of Intensities

a. Photoemission Cross-Sections

In order to analyze the intensity of the various photoelectron peaks, it is necessary to know their associated transition probabilities, or photoelectric cross-sections.

The photoelectric cross-section σ is defined as the one-electron transition probability per unit-time, with a unit incident photon flux per area and time unit from the state Ψ_{in} to the state Ψ_{fin} of Eq. (2). If the direction of electron emission relative to the direction of photon propagation and polarization are specified, then the differential cross-section $d\sigma/d\Omega$ can be defined, given the emission probability within a solid angle element $d\Omega$ into which the electron emission occurs. Emission is dependent on the angular properties of Ψ_{in} and Ψ_{fin}: therefore, in photoelectron spectrometers for which an experimental set-up exists by which the angular distribution of emission can be scanned (ARPES, see Fig. 2), important information may be collected on the angular properties of the two states. In this case, recorded emission spectra show intensities which are determined by the differential cross-section $d\sigma/d\Omega$. The total cross-section σ (which is important when most of the emission in all direction is collected), is

$$\sigma = \int \frac{d\sigma}{d\Omega} \cdot d\Omega \tag{10}$$

Photoemission cross-sections are different for different orbital states of an atom. Furthermore, for a given orbital state, the photoemission cross-section depends strongly on the energy of the exciting photon. This provides a very useful tool in photoemission as we shall discuss later.

As known from the basic quantum theory of transitions[21], cross-sections may be written as[4]:

$$\frac{d\sigma}{d\Omega}(h\nu) = C|M_{in,fin}|^2 \tag{11}$$

where $h\nu$ is the incident photon energy and $M_{in,fin}$ are matrix element of the transition operator

$$-\mathrm{i}\hbar e \mathrm{A}_0 \exp(\mathrm{i} q \cdot R)\nabla \qquad (12)$$

(where $A_0 e \exp(iq \cdot R)$ is the electromagnetic field, e its polarization vector, q the momentum of the photon) between the final Ψ_{fin} and the initial Ψ_{in} state.

The evaluation of elements such as the $M_{in,fin}$'s is a very difficult task, which is performed with different levels of accuracy. It is sufficient here to mention again the so called "sudden approximation" (to some extent similar to the Koopmans' theorem assumption we have discussed for binding energies). The basic idea of this approximation is that the photoemission of one-electron is so "sudden" with respect to relaxation times of the "passive" electron probability distribution as to be considered instantaneous. It is worth noting that this approximation stresses the one-electron character of the photoemission event (as in Koopmans' theorem assumption).

This approximation is valid when $\tau/\tau_{relax.} \ll 1$, where τ is a characteristic time required for the photoelectron to leave the perturbed system into the vacuum, and $\tau_{relax.}$ is a time characteristic for relaxation of the "passive" electrons into possible final states. If $E_{fin} < E'_{fin}$ are energies for two possible final states, $\tau_{relax.}$ may be thought to be proportional to $h/(E'_{fin} - E_{fin})$ by uncertainty arguments. A rough indication of $\tau_{relax.}$ is given by the broadening of the emission lines.

In the "sudden" approximation, cross-sections are proportional to[4]:

$$|\langle \phi_{fin}(j)|\mathfrak{t}|\Phi_{in}(j)\rangle|^2 \cdot |\langle \Psi_{fin}(N-1)|\Psi_{in}(N-1)\rangle|^2 \qquad (13)$$

where $\mathfrak{t}$ is the monoelectronic transition operator, Ψ_{fin} and Ψ_{in} are the one-electron wavefunctions of the photoemitted electron j, and $\Psi_{in}(N-1)$ is the wavefunction of the $N-1$ "passive" electrons in the initial state from which $\phi_{in}(j)$ has been factored out ($\Psi_{in}(N-1)$ is a non-unique representation of the "passive" electron remainder in the initial state, to be chosen appropriately). Notice that $\Psi_{fin}(N-1)$ and $\Psi_{in}(N-1)$ need not to be the same, as is supposed in the Koopmans' theorem assumption. Thus, the cross-sections are proportional to:

i. a one-electron transition probability (the first factor);
ii. a $(N-1)$ electron overlap integral (the second factor).

The first factor will dictate, from the properties of the one-electron wavefunctions, the characteristic dependence of the cross-sections on the initial orbital character of the electron and on the exciting photon energy.

The second factor is at the origin of the so called "monopole" selection rule. Symmetry requirements impose that the two Ψ's must correspond to the same irreducible representation in order for the overlap integral not to vanish.

b. Atomic Cross-Sections

In the Koopmans' theorem limit the photoemission of one-electron from an atom or a core in a solid is given by a single line, positioned at the eigenvalue of the electron in the initial state. The intensity of this line depends on the cross-section for the event, which is determined by the one-electron atomic wavefunctions $\Psi_{in(n,l,m)}(-E_b)$ and $\Psi_{fin(n',l',m')}(E_{kin})$ (where the atomic quantum numbers are indicated as well as the eigenvalues $E_{n,l,m} = -E_b$ and E_{kin} of the initial and final state) (the overlap integral of (13) being 1 in the Koopmans' theorem's conditions). The transition operator in (13) is the usual dipole operator (as in optical transitions[21)]). Hence, the usual atomic selection rules, as for optical transitions, hold, the most important being:

$$\Delta l = l' - l = \pm 1 \tag{14}$$

It is usually stated that the photoemission occurs via two "channels" in the energy continuum of the atom, meaning that the electron stays in two unbound orbitals having $l' = l \pm 1$[4)]. (The $(l + 1)$ "channel" is usually the most important in XPS.)

The total cross-section for a photoemission event occuring in an (nl) initial state is given by:

$$\sigma_{nl}(E_{kin}) = \sigma_{nl}(h\nu - E_{nl}) = A(h\nu) \cdot |lR^2_{l-1}(E_{kin}) + (l + 1)R^2_{l+1}(E_{kin})| \tag{15}$$

In (15), the dependence on the energy of the exciting photon hν (at a given eigenvalue E_{nl}) is stressed. A is a proportionality factor containing fundamental constants. R_{l-1} and R_{l+1} are radial *dipole* matrix elements:

$$R_{l\pm1} = \langle R_{fin(l\pm1)}|r|R_{in(n,l)}\rangle \tag{16}$$

in which $R_{in(n,l)}$ and $R_{fin(l\pm1)}$ are the radial parts of the wavefunctions of the initial state and of the "channel" $(l \pm 1)$ respectively.

Evaluation of the σ_{nl}'s requires detailed atomic calculations, with a good knowledge of the one-electron wavefunctions[22, 23)] (the best results, especially for high l orbitals, are obtained by relativistic calculations). They are usually expressed as a function of the exciting photon energy hν.

For energies typically used in XPS (1 – 1.5 keV) (see Table 1) the photoexcitation cross-sections have been calculated for the different orbitals and elements[22)] and are used for quantitative chemical analysis. For the actinides the 4f cross section is highest and consequently the entire photoemission spectrum is dominated by the spin-orbit split 4f core level doublet. In the valence band region, the high 5f cross section leads to a dominant 5f emission if occupied states with 5f character are present. For energies typically used in UPS (10 eV < hν < 50 eV) (see Table 1) the photoexcitation cross-section for valence band states with different quantum character is shown in Fig. 5 (see[23)] and references therein). Simplifying, it can be said that, for low excitation energies the s, p cross-section is large but decreases with excitation energy whereas the f cross-section is small but increases with excitation energy; the d cross-section has a maximum around 20 eV.

It must be emphasized that these cross sections are only valid for an electron excitation into free-electron like final states (conduction band states with parabolic band shape) and not for resonance transitions as f → d or p → d excitations[23)]. If too low excitation energies (< 10 eV, see Table 1) are used in UPS, the final states are not free-electron like. Thus the photoemission process is not simply determined by cross-sections as discussed above but by cross-sections for optical transitions as well as a joint density of states, i.e. a combination of occupied initial and empty final states.

Table 1. A display of common X-ray and UV sources for photoemission (ΔE is the linewidth)

XPS sources	E (eV)	ΔE (eV)[a]
Al K_α	1486.6	0.85
Mg K_α	1253.6	0.7
YM_ζ	132.3	0.5
ZrM_ζ	151.4	0.8
UPS sources		
He I	21.22	< 0.02
He II	40.82	
He II*	48.37	
Ne I	16.67/16.85	
Ar I	11.62/11.83	

[a] The X-ray linewidth can be reduced by the use of monochromating devices

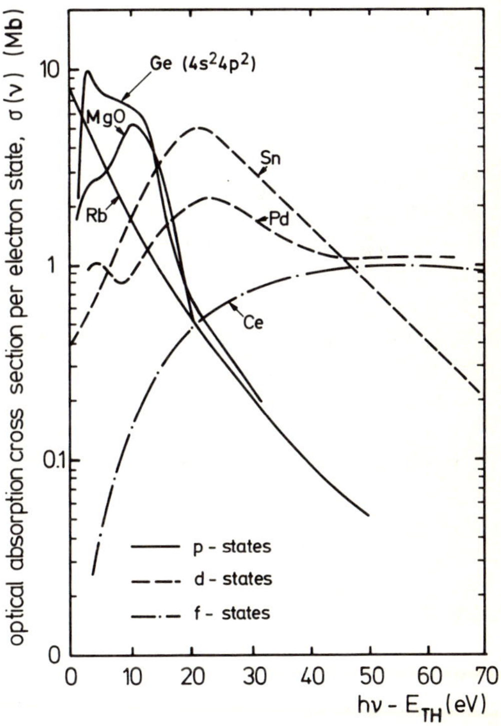

Fig. 5. Atomic cross-sections vs. photon energy in the ultraviolet range for different orbital states in various atoms (from Ref. 23); E_{TH}: threshold energy

c. *Cross-Sections of Itinerant Electrons*

The cross-sections for itinerant electrons, as, e.g., electrons in broad bands, are evaluated by taking into account that the electrons in the initial as well as in the final state may be represented by Bloch-wavefunctions $\Psi = u_k(R) \exp(ik \cdot R)$ (see Chap. A). In these wavefunctions "atomic information" is contained in the amplitude factor $u_k(R)$, whereas the wave part $\exp(ik \cdot R)$ is characterized by the wavenumber k of the propagating wave (proportional to the momentum of the electron).

The principal selection rule is provided as in optical transitions, by the momentum conservation condition[21)]:

$$k_{\text{fin}} = k_{\text{in}} + g + q \tag{18}$$

where g is a reciprocal lattice vector and q is the momentum of the photon (usually very small). This selection rule characterizes, as in the optical case, *direct* transitions (indirect transitions contribute also to the photoemission; they cause also phonon-assisted broadening of the emission bands). Equation (18) is important, together with Eq. (6a) of a preceding section, since it allows (experimental difficulties being for the time neglected) the determination of band dispersion E(k) for a solid in angular resolved photoemission.

A somewhat simplified form of the intensity of itinerant electron emission is:

$$I(h\nu) \propto \bar{\sigma}(h\nu) N(E) \tag{19}$$

where $\bar{\sigma}(h\nu)$ represents the mean photoelectron cross-section for the states composing the band, and N(E) is the density of states of the initial state. The analogy with optical transitions and optical spectroscopy in (19) is evident, the intensity for optical spectral phenomena being given by a somewhat similar function where N(E) is substituted by the "joint density of states", (taking care of the density of states in the final as well as in the initial state). In photoemission, however, the final density of states is the smooth density of states of the continuum, so that N(E) determines the emission spectrum. This is why photoemission is presented as being a photograph of the occupied bands density of states (conversely, BIS yields the "photograph" of the empty bands density of states).

d. *Excitation Energy Variation of Photoemission Intensities and the Analysis of the Orbital Character of the Bands*

The initial state amplitudes $u_k(R)$ of the Bloch-wavefunctions, carrying "atomic" information, introduce into the matrix elements (11) for the cross-sections of itinerant electron emission the orbital character still retained by the energy bands of the solid. Hence, the variation of emission with the exciting photon energy hν can be assumed, in first approximation, to be described by the cross-section atomic dependence given in the preceding subsection. In (19), this is taken care of by $\bar{\sigma}(h\nu)$. (To which extent this assumption is valid, is still a matter of debate. In bands having a pronounced atomic-like behaviour (such as 5f bands) the assumption is presumably fairly good.)

This is at the basis of much experimental work, in which the exciting hν is varied. The corresponding variation of intensities is correlated with the orbital character of the bands investigated. When comparing with theoretical band models, the *energy dependent photoemission method* is able to give therefore information about hybridization phenomena (giving rise to bands describable by means of partial degrees of mixing of atomic orbitals – see Chap. 4), provided the variation of cross-section with hν is known.

Synchrotron radiation sources are ideal for this purpose, since they allow a continuous variation of excitation energy in a wide range. Usually a combination of UPS and XPS is employed, which allows a series of excitation energies in UV in the 10–48 eV range. Typical excitation energies for the laboratory case are presented in Table 1.

The 5 f cross-section is particularly sensitive to the photon energy variation. This is a fortunate fact that has allowed a great deal of experimental and theoretical results to be obtained in actinides solids photoemission. In fact, due to it, 5 f states and 5 f character in bands are easily identified.

3. Localization vs. Itineracy in Photoelectron Spectroscopy

We now focus on relevant aspects of photoemission in actinides. For broad bands in solids (as, e.g., the (s, p, d) valence or conduction band in actinides), binding energies measured in photoemission (and in the strictly correlated inverse photoemission) are one-electron energies of the initial state, and the photoemission and inverse photoemission spectral responses are good "photographs" of the density of state function N(E). Band dispersion may be obtained by employing ARPES: this has not yet been done as sufficiently for actinides, since good single crystal samples are needed (see, e.g., the NaCl-structure compound of uranium USb[24)]). If band dispersion is measured, the bandwidth W is determined. In the discussion of localization vs. itineracy of Chap. A, W has been shown to be one of a couple of parameters, the relative values of which determine localized vs. itinerant behaviour.

By varying the excitation energy in direct and inverse photoemission, further important information is collected on hybridized states. Thus, it can be said that by photoelectron methods, the itineracy of open shells (in actinides 5 f) is well characterized.

Photoelectron spectroscopy is however able to give valuable information on the localized behaviour of open shell electrons. The second parameter, the Coulomb interaction energy U_H, to be balanced (Chap. A) with W to give a judgement on the localization vs. itineracy problem, has been shown to be experimentally accessible when XPS and BIS are applied on the same compound.

It should however be remarked that it is very difficult to measure both W and U_H with sufficient precision on the same electronic system. ARPES is very inprecise when dealing with very narrow bands (levels), typical of localization; the method for determining U_H, illustrated below, is best fitted when the photoemission response is treated within the atomic picture. This contradictory aspect is analogous with what is encountered in other physical measurements, and is particularly unsatisfactory when the state under observation is intermediate between localization and itineracy (see, e.g., discussion in Chaps. A and D about magnetism).

a. The Measure of the Coulomb Energy U_H

We have discussed, in Chap. A, the Hubbard model for localization vs. itineracy in narrow bands. In this model, it was shown, for a simple case ("hydrogen" case) involving one uncoupled electron in a shell, that a splitting of the narrow band in two sub-bands occurs when the Hubbard condition ($U_H \approx W$) is not satisfied. The two sub-bands describe two situations:

i. one in which electrons are found in the cores in a localized state, with their magnetic moment duly aligned according to Hund's rules ($\Psi(N)$);
ii. one in which "polar states", exist in the solid: a fraction of cores is depleted of one electron ($\Psi(N-1)$) and a fraction of cores enriched by one electron ($\Psi(N+1)$).

The difference between the two situations is U_H in the Hubbard model, and can be associated, in a thermochemical way, with the process (for an f level, e.g.)

$$f^N + f^N \xrightarrow{U_H} f^{N-1} + f^{N+1} \tag{20}$$

Equation (20) may be seen as the combination of the two processes of direct and inverse photoemission, when the 5f shell retains a strong character of localization (in case of "itinerant" 5f's, the Hubbard model predicts that the kinetic energy due to itineracy creates statistical fluctuations between the polar states, so that the itinerant character is lost).

Figure 6 shows the process in a schematic way. Baer[25)] describes the two electron excitations of direct and inverse photoemission as two-step processes in which first E_F is reached and then the emission of an electron (direct) or of a photon (inverse photoemission) to vacuum occurs from E_F. Δ_- and Δ_+ are the energies associated with the two first steps. They are counted from E_F, and they are to be considered as the *minimum energies* necessary to create the f^{N-1} and f^{N+1} final state. If the localized level response is a final state multiplet, therefore, $U_H = \Delta_+ + \Delta_-$ is given by the sum of the smallest measured energies of the multiplet.

The method has been used with lanthanide solids, and the U_H which is measured is in fair agreement with theoretical evaluations[26, 27)]. Figure 4 shows the method as applied for UO_2, a solid in which the 5f electrons have a good localized character (see Chaps. A, C and D).

b. Core Level Response in Open-Shell Metals: the (Z + 1) Concept, and the Thermochemical Method for the Evaluation of Their Binding Energies

We shall examine briefly the photoemission (XPS) binding energies (E_b, relative to E_F) of core levels for open shell metals. An ample discussion[6, 28, 29)] has been developed in this matter, leading to a satisfactory agreement between theoretical models and experimental spectroscopic results. For the purpose of this book, the theoretical model employed is relevant

i. because it prepares the ground for a more detailed analysis of core levels peaks and their satellites, which is important in the localization vs. itineracy problem;

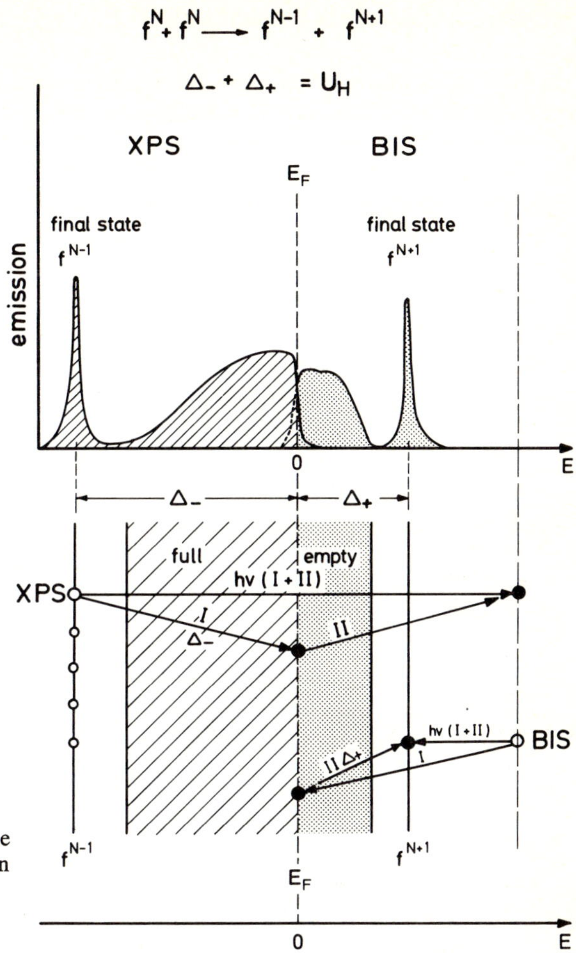

Fig. 6. Schematic representation of the final state model for the determination of the Coulomb interaction U_H from the complete spectrum of a system with localized f levels

ii. because it has provided a spectroscopically derived further piece of evidence for the occupation of the 5f shell in uranium metal, with the consequent rejection of the picture of uranium as a 6d-transition metal (see Chap. A, Part I)[28, 30].

The core level binding energies have been calculated for many metals by two different theoretical methods: the "renormalized atom" method (Herbst et al.[27]), and the "thermochemical method" (Johansson et al.[28, 30]). The agreement between the results of the two methods and with experimental results (when available) is fairly good (for a comprehensive review of the thermochemical method see Johansson[28]). We shall now concentrate on the last method, pointing out, however, that the basic assumptions of both are the same.

The model employed for the core-level excitation runs as follows[26]:

α) the process of photoemission is a two-step process: one electron is photoemitted from a conduction state at the Fermi level, and one electron is promoted from a core to a valence state of the solid;

β) the excitation takes place within a particular core of the metal;

γ) when the electron is ejected, that particular core site becomes *totally screened* by the surrounding conduction electrons of the metal.

The "efficient screening" approximation means essentially that the final state of the core, containing a hole, is a completely relaxed state relative to its immediate surrounding[28]. In the neighbourhood of the photoemission site, the conduction electron density of charge redistributes in such a way to suit the introduction of a core in which (differently from the normal ion cores of the metal) there is one hole in a deep bound state, and one valence electron more. The effect of a deep core hole (relative to the outer electrons), may be easily described as the addition of a positive nuclear charge (as, e.g. in β-radioactive decay). Therefore, the excited core can be described as an *impurity* in the metal. If the normal ion core has Z nuclear charges (Z: atomic number) and v outer electrons (v: metallic valence) the excited core is similar to an impurity having atomic number (Z + 1) and metallic valence (v + 1) (e.g., for La^{+3} ion core in lanthanum metal, the excited core is similar to a Ce^{+4} impurity).

With these assumptions, E_b can be written:

$$E_b = \Delta E_{v,v+1} + E^{imp}_{(Z+1),(v+1) \text{in } Z,v \text{ matrix}} \tag{21}$$

where the latter term, usually small, is due to the adjustment of the cell to the impurity. $\Delta E_{v,v+1}$ is a thermochemical quantity, that can be evaluated from the cohesive energies of the pure (Z, v) and ((Z + 1), (v + 1)) metals, and/or from spectroscopic atomic data (relative to the change in electronic configurations of the non-excited and excited cores), in much the same ways as similar problems which were discussed in Chaps. A and C.

The whole treatment turns around the efficiency of screening of valence electrons: we shall discuss this point more in detail. In light actinides, itinerant 5f electrons may provide efficient screening.

By this assumption and employing the above described (Z + 1) approximation, Johansson[28] was able to provide further evidence for the occupation of 5f states in uranium metal, via the analysis of its XPS-measured $6p_{3/2}$ peak.

For inverse photoemission, similar arguments lead to the interpretation of BIS structures by an analogous Z − 1, v − 1 approximation.

c. *The Final State Screening Model*

A number of recent XPS results for core-level peaks in d- and lanthanide metals and compounds has triggered much interpretational effort. The $2p_{3/2}$ core level peak in Ni, Pd and Ni/Pd alloys, e.g., has a satellite accompanying the main peak at higher binding energies (shake-up satellite)[31–33]. The $3d_{5/2}$ core level peak of La, Ce, Pr and Nd, and their intermetallic compounds has, on the contrary, a satellite accompanying the main peak at lower binding energies (shake-down satellite)[34–38] (see Fig. 7). The presence of this split core level response (main line + satellite) in solids with open electron shells has been interpreted by the so called "final state screening" model.

Another important feature which is met in d-systems is an asymmetric tailing of the main peak at the higher binding energy side.

The model employed for the interpretation of a split core level response is a two-electron process occuring within an ion core, which may be schematized[37–39]:

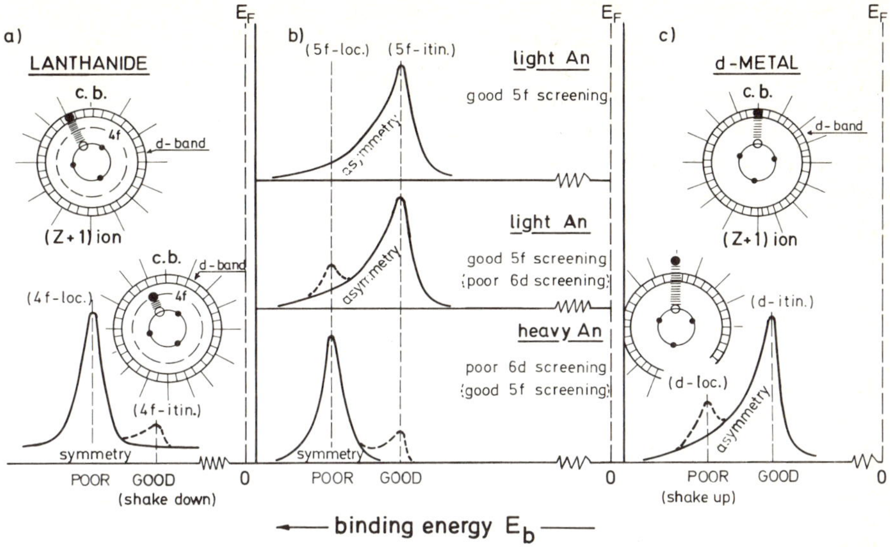

Fig. 7 a–c. Schematic representation of final state screening models for lanthanide and d-metal core level responses (**a**) and **c**)) (c.b. means conduction band). In part **b**), the possible situations for light and heavy actinides (before and after the Mott-Hubbard transition) are also represented

$$\sigma^q\beta^n\{B\}^m \rightarrow \sigma^{q-1}\beta^n\beta^{n+1^*}\{B\}^m\{B\}^{m+1^*} \begin{cases} \nearrow \sigma^{q-1}\beta^{n+1}\{B\}^m & (22\,a) \\ \searrow \sigma^{q-1}\beta^n\{B\}^{m+1} & (22\,b) \end{cases}$$

(Z) initial state (Z + 1) intermediate excited state (Z + 1) final state

where: σ^q, β^n and $\{B\}^m$ identify a core level, an outer level and a more band-like, hybridized state (e.g., Pu-metal: $4f^{14}5f^5\{6d\,7s\}^3$), respectively. The two-electron process passes through an intermediate state, in which empty states lying beyond E_F are pulled down below E_F (indicated by an asterisk; such initially empty states beyond E_F are seen in BIS measurements).

In open shell metals, these empty states can be d- or f-states somewhat hybridized with band states (see Chap. A). In a metal, these states may be pulled down into the conduction band (as a virtual state, see Chap. A); in a compound, presenting a ligand valence band (insulator or semiconductor), they may be pulled down to an energy position coinciding with or very near to this valence band (as a true impurity level). The two possible final states (Eqs. 22 a and 22 b) explain the occurrence of a split response: one of the crystal band electrons occupies either the outer hole level β^{n+1^*} (Eq. 22 a) or the more bandlike hole $\{B\}^{m+1^*}$ (Eq. 22 b).

The electron occupying the pulled down state screens the deep hole in the $(Z+1)$ core from its environment. The screening efficiency depends very strongly on the wavefunction describing the state: a charge in an outer atomic-like orbital screens better than one in a band state since the electron charge density is high at the atom (see Chap. A). In the core level split response, good screening is attributed to an observed lower binding energy, since a gain in energy U is obtained, e.g., through an alignment of spins in the pulled down state and in the core level containing a hole (Hund's rule type coupling).

If the atomic-like state hybridizes with band states, its eigenvalue broadens to a well defined bandwidth W. The electron becomes more itinerant (the U/W ratio being a good parameter to describe this trend – see Chap. A), and the screening becomes poorer. Moreover, the relative intensities of the two peaks in the split core level responses depend on the occupation probability of the pulled down states, which is enhanced when hybridization becomes larger. In fact, attempts have been made to establish correlations between the intensity ratio of the two peaks of the split response and localization or hybridization of the open shell states in a transition series.

In light lanthanides (La, Ce, Pr, Nd) the pulled down 4f state is nearly localized and hybridizes only weakly with conduction states. The bandwidth W_{4f} will be very narrow, U high and negative, and the occupation probability by conduction electrons rather low. This results in the occurrence of shake-down satellites at a lower binding energy for lanthanides, accompanying a poorly screened main peak (Fig. 7a). When proceeding to heavier lanthanides, the occupation probability and the intensity of the shake-down satellite are depressed: the symmetric, poorly screened core level is left, i.e. the 4f states are completely localized.

In d-metals, the opposite is true: the d-wavefunctions hybridize easily with conduction band states. The main peak can in this case be coordinated with the well screening outer d's, and the shake-up satellite, when observed, is due to the poorly screening process (Fig. 7c). For d-metals, furthermore, the very high density of d-states at E_F is the cause of many secondary electron excitation from just below E_F to empty states just beyond E_F which results in the asymetric high energy tailing of the main peak. Final state multiplet splitting, explained above, can in addition overlap the split response.

In actinides, a very efficient screening is provided by 5f electrons. As well known (see preceding chapters), they have a "d-like" character in the first half of the series and a 4f-like character in the second half: the crossover occurring at about the half-filling of the shell. Hence, Johansson[40)] could predict for actinide core level response in actinide metals at the half-filling a changeover when passing the Mott-Hubbard transition threshold at Pu (Fig. 7b). A good core level to be studied is, for actinides, the 4f spin-orbit split doublet ($4f_{5/2}$ and $4f_{7/2}$), due to its high cross section, and to the fact that its shape is only dependent on the screening process, because the multiplet coupling between 4f-electrons and the electrons in the incomplete 5f-shells is negligible.

d. The 6 eV Valence-Band Satellite in Ni

A satellite is found in the valence band spectrum of Ni metal (and compounds) at about 6 eV from E_F[32, 41)]. This satellite does not correspond to any feature in the calculated density of states. Therefore a brief discussion of this satellite is worthwhile, since its interpretation has inspired that of some structures found in light actinide metals photoemission spectra.

Despite the fact that there still remain unanswered questions, the 6 eV satellite can be explained in an atomic picture by two correlated d-holes on the same atom superimposing on the band-like spectrum from the itinerant 3d response at E_F.

Assuming a ground state composed of atomic-like d^9 and d^{10} configurations the satellite is thought to be due to the following process:

$$3\,d^9 4\,s \quad \text{—} \quad 3\,d^8 4\,s^2 \tag{23}$$

ground state Final state

The energy involved in this excitation is easily obtainable from atomic spectroscopy (see Chapt. A) and is found to be about 6 eV; i.e. coinciding with the energy position of the satellite.

The important point is that this interpretation introduces an atomic event in the interpretation of the photoemission from the metallic solid, which is in large part dominated by the band character of the 3 d electron emission. The underlying explanation is that 3 d-wavefunctions are, like 5 f, largely atomic-like in character (see Chaps. A, C, F), and that this partial localization makes the occurrence of the atomic event possible. In fact, similar satellites are encountered also in compounds where, the Ni atoms being far apart, no d-d overlapping, hence no band-like behaviour, is predicted: this proves the atomic character of the excitation process giving rise to the "6 eV" structures also in Ni metal.

Such details in photoemission spectroscopy may therefore give indirect, but very useful hints towards the solution of the localization vs. itineracy problem of narrow band solids.

4. *Experimental Aspects*

Several practical aspects of the photoelectron technique will be discussed here. First, we shall concentrate upon the surface specificity of XPS and UPS. Then the sample preparation procedures will be reviewed. Thereafter, the charging effect, the energy calibration and the problems of handling radioactive materials will be discussed. Lastly, a short review of similar topics applied to BIS will be given.

a. *Surface Sensitivity*

Figure 8 shows the attenuation length of electrons in solids as a function of their kinetic energy. The few theoretical calculations available[42)] are in good agreement with these empirical data[43)]. Only unscattered electrons convey useful information, while scattered electrons contribute to a structureless background[44)] (secondary electrons). From Fig. 8, it is clear that photoelectron spectroscopy probes at most a few tens of Ångströms.

Because of this surface specificity of photoelectron spectroscopy, two kinds of problems have to be taken into account: surface contamination by the residual atmosphere of the spectrometer, and the possibility that only surface properties are observed instead of the bulk ones.

In order to avoid surface contamination by the residual atmosphere of the spectrometer, it is necessary to work under UHV conditions (10^{-9} Pa). This is apparent from the Hertz-Knudsen relation:

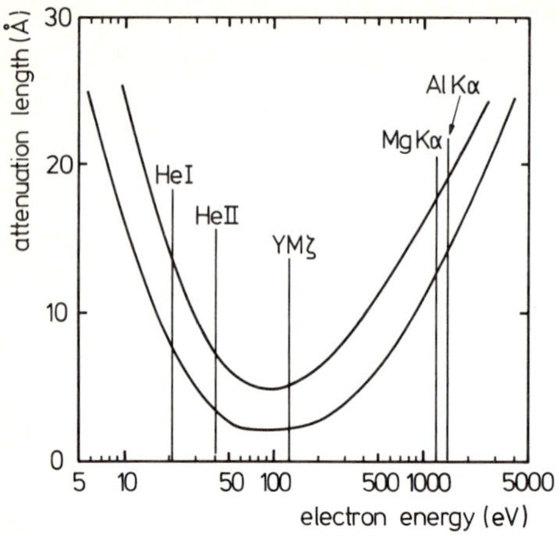

Fig. 8. A "universal" curve (or band) of electron attenuation length vs. electron kinetic energy, as resulting from an inspection of many experimental results on heavy metals. The energy of different laboratory sources, representing the maximal kinetic energy of the electron, is drawn for comparison in the figure (from Ref. 5)

$$\nu = p(2\pi mkT)^{-1/2}$$

where

ν is the number of molecules impinging on the sample surface per unit of time and area,
p is the residual gas pressure,
m is the mass of the residual gas molecule,
T is the temperature.

Assuming a sticking coefficient equal to unity, a residual atmosphere ($p = 10^{-9}$ Pa) composed of oxygen and nitrogen, and room temperature, about 10^3 sec are needed to build a hundredth of a monolayer, which corresponds to the detection threshold of photoelectron spectroscopy.

The modification of the electronic states energy due to the surface has readily been observed by photoelectron spectroscopy, over a wide range of excitation energies[45–47]. It manifests itself as a small shift of the electronic levels (a few tenths of an eV). It is at best detected by using a synchrotron radiation source, so tuned that the outgoing electrons have a kinetic energy corresponding to the minimum of the attenuation length curve[47]. Furthermore, a valence change of the surface atoms is possible, as has been observed for Sm[48, 49], and as will be discussed below for Am (Sect. III.3.a). A review of photoelectron spectroscopy as a tool for surface investigation will be found in Ref. 8.

In the case of the most common X-ray sources used for XPS (Mg and Al), and of the most widely observed core levels of the actinides (4f), the photoelectron kinetic energy is of several hundreds of eV. This ensures an analysis depth large enough to reveal the bulk properties.

For UPS, the access to bulk properties can be gained by using several radiation sources as e.g. He I and He II ($h\nu$ = 21.22 and 40.82 eV, respectively), and taking advantage of the electron mean free path variation. As can be seen by inspection of Fig. 8, the mean free paths of He I and AlK_α-excited photoelectrons are quite similar.

The possibility of a continuous variation of the radiation beam energy is the main advantage of a synchrotron radiation source with respect to this surface vs. bulk problem.

However, even a single line source, say AlK_α or He I, can be used successfully, if the analysis depth can be varied by performing angular-dependent measurements.

b. Sample Preparation

Before introduction into the spectrometer, a mechanical and/or electrochemical polishing of the sample surface has to be performed. While a good empirical knowledge has been gained for most current materials[50)], the right procedure to follow in the case of actinide compounds is still not well established.

After introduction into the spectrometer, several in situ methods are available to achieve a clean surface. First, the ion sputtering of samples is used in order to remove the surface contamination (adventitious carbon and, in the case of metals, an oxide layer). Here again, much experience has been gained on lighter materials[51)], but little is known of the right way to handle actinide compounds. Two difficulties are associated with this method: sputter-induced surface roughness, and preferential removal of lighter elements[52)]. A second method, scraping of the sample surface, does not induce a deviation from the bulk stoichiometry[52)], but results in a very rough surface. Both these problems are avoided if the examined material, in this case a single crystal, can be cleaved in situ: this results in a smooth crystalline surface and provides a composition similar to that of the bulk.

c. Charging Effect

In the case of conducting materials, each photoelectron leaving the sample is replaced, so that the neutrality of the sample is preserved. For insulators, on the contrary, there are no conduction electrons available to compensate for the lost photoelectrons. Therefore, a positive charge is built up on the sample surface. This charge alters the kinetic energies of the outgoing photoelectrons, which results in distorted spectra.

A widely used technique is the so called flood-gun. An auxiliary electron gun is tuned so that the ingoing (from the gun) and outgoing (photoemitted) electron fluxes compensate. In order to achieve this equilibrium, it is necessary to record spectra during a large interval of time and to search for steady state conditions. Therefore, in view of the short time needed to build one monolayer of contamination on the surface of the sample, the use of this technique is made difficult in the case of UPS. One should mention, that very recently a similar compensation electron gun has been successfully used for electronic vibrational spectroscopy, which is even more surface-sensitive than UPS[53)].

d. Energy Calibration

In XPS, the calibration of the spectrometer is achieved by measuring a series of core lines which are used as energy standards, viz. Au $4f_{5/2}$, Pd $3d_{5/2}$, Ag $3d_{5/2}$, Cu $2p_{3/2}$, Cu $3s_{1/2}$. These lines encompass the 100–900 eV binding energy range. Once the linearity of the analysing system has been checked, there is no problem of energy calibration, as long as conducting materials are measured. Indeed the Fermi edge is in this case clearly visible (see e.g. Fig. 9), so that the zero-point of binding energies is easily located.

For insulating materials, on the contrary, the Fermi level lies in the gap, so that it is not detected on the photoelectron spectra. A possible calibration method stems from the gold decoration technique: the binding energies can then be referred to the Au 4f doublet. This leads, however, to a poor reproducibility of the calibration[54]. Neither is the 1s line of adventitious carbon a better energy standard.

e. *Problems Specific to the Handling of Radioactive Materials*

When handling radioactive materials, the operator as well as the apparatus, especially the electron detector should be shielded from radioactive contamination.

On the basis of a commercial spectrometer (Leybold-Heraeus LHS-10) the system has been modified[55] for work with highly radioactive materials (the scheme is shown in Fig. 11 of Chap. B):

- A new preparation chamber (Pch) was designed for a base pressure in the low 10^{-8} Pa range and mounted into a closed glove box. It can be baked under glove box conditions without enhancing the temperature of the glove box atmosphere because of a water cooled shielding system. "In-situ" sample preparation is solely executed in the PCh.
- The μ-metal analysis chamber (ACh) is mounted in an open glove box frame making easy handling of the spectrometer feasible. The baking of the system is performed by commercial heating bands.
- The UHV pumping systems are mounted similarly in an open glove box frame and connected with the roughing pumps by vacuum lines containing absolute filters to keep back potential radioactive contaminants.

For maintenance or repair the open glove boxes are closed. After completing the work and potential decontamination, the glove box walls are removed and the system can be baked out easily.

In the case of BIS, monoenergetic electrons from an electron gun impinge on the sample; the reponse to be analysed is constituted of photons. No heavy and sophisticated electron energy analyser is necessary. Therefore, protection against radioactive hazards can be easily achieved by inserting the complete spectrometer into a glove box system.

III. The Photoelectron Spectra of Actinide Metals

1. *Introduction*

In this part, the photoelectron spectra of actinide elemental metals are reviewed. The organization of this part is the following:

- the valence band spectra of light actinide elemental metals are presented and interpreted through one-electron density of state calculated curves, focusing on the f-character of the bands, but also on their general orbital compositions and on their character (as shown by energy dependent photoemission) due possibly to hybridization;

- indications of the localized or itinerant character of the 5 f's are discussed in the frame of the (Z + 1) theory and of the discussion of the Ni 6 eV satellite presented in part II;
- a discussion of evidence from valence band spectra is presented for the Mott-like transition in the actinide series between Pu and Am (Chap. A), together with details of the spectrum of Am metal, which presents a well defined localized 5 f-character;
- the 4f core level lines of the metals are discussed, in the light of the final state screening model also presented in part II; the intention being to confirm from the 4f features the Mott-like transition between Pu and Am.

No photoemission spectra are unfortunately available for Np metal, in the light actinide 5 f itinerant side, and on Cm, Bk, Cf, etc. on the heavy actinide 5 f localized side. It is worthwhile to stress the need for good photoelectron evidence on these systems in order to shed more light on the elemental actinide metals series.

2. *The Valence Band Spectra of Light Actinides*

a. *Thorium Metal*

α. XPS and BIS Results

Figure 9 shows the photoemission valence band spectrum of Th metal. A comparison of the high resolution XPS valence band spectra of Th[41, 56] (resolving two distinct peaks at 1.8 and 0.6 eV below E_F) with a calculated total (s-d) density of states[57], (convoluted for broadening effects as lifetime and instrumental effects)[56] gives a nearly complete agreement. The two peaks are attributed to 6 d states.

The empty states of Th have been investigated by BIS[56] and are also shown in Fig. 9. Besides a small intensity at E_F, the spectrum is dominated by two broad peaks at 3.15 and 4.5 eV. The calculated partial density of 5 f states[57] which takes into account the high cross section for f states, and is broadened for lifetime and instrumental resolution[56], agrees qualitatively well with experiments, though the two peaks are slightly less sepa-

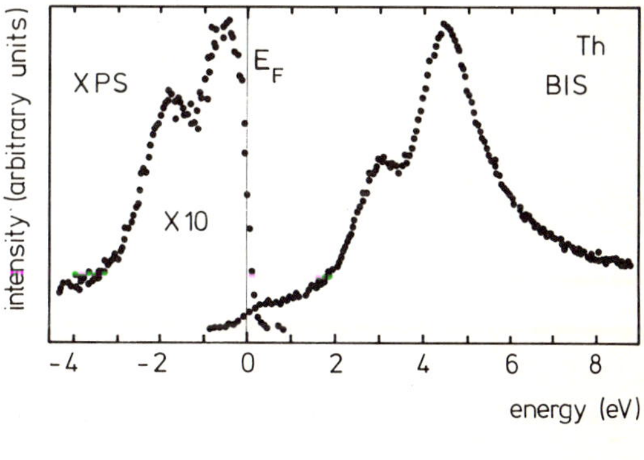

Fig. 9. The full photoelectron spectrum (direct and inverse photoemission) of the conduction band of Th metal (from Ref. 56)

rated in energy and about 0.7 eV closer to E_F. From a comparison with the BIS spectrum of La having 14 localized empty 4f states, it has been concluded[58] that the double peak structure is essentially 5f like. But in contrast to La, this structure has to be attributed to broad band-like states hybridized with d states and thus even contributing at E_F.

β. 5f-Mixing in the Valence Band

The possible contribution of 5f states to the BIS intensity at E_F suggests that some 5f character may exist (even in Th, with a $5f^0$ configuration) in the composition of the occupied valence band. In fact a small tail of occupied 5f states (0.5 states per atom, Ref. 57) is supposed to contribute to the intensity of the XPS experimental spectrum[56], the rather high intensity of which may be due to the high cross-section for 5f excitation. Energy dependent photoemission should then be able to identify this contribution.

The only UPS measurement on Th reported to date is for 40.8 eV excitation. This measurement surprisingly does not resolve the peaks at 1.8 and 0.6 eV from the emission at E_F[59], notwithstanding the higher resolution of 0.2 eV[59] than for XPS (0.4[56] and 0.6 eV[41], respectively). If there is a tail of 5f states reaching into the occupied part of the conduction band of Th, the UPS valence band spectrum should at least show at E_F some reduced emission, when compared with the XPS valence band spectrum. Therefore it is difficult to ascertain from the UPS spectrum for Th[59] whether a small contribution of 5f states at 0.3 below E_F is hidden in the slope of the main peak at 0.75 eV. The high intensity of this peak might in fact be attributed to some broadening of the Fermi edge by temperature effects and/or poor instrumental resolution.

However, in Ref. 59 also the first valence band spectrum of U has been measured by UPS, and shows a very sharp and, compared with Th, about 10 times more intense 5f peak at 0.3 eV below E_F. Thus the UPS peak at 0.75 eV and the XPS 0.6 eV peak for Th metal may be attributed to the same origin, namely 6d, as suggested by density of state calculations, the small shift between the two being induced by the different contribution of a possible 5f tail in the UPS and XPS spectra.

The situation for Th metal can be summarized as follows: the occupied part of the conduction band is dominated by band-like 6d states hybridized with 7s states contributing at the bottom of the conduction band; the empty part of the conduction band is formed by itinerant 5f states, hybridized in a broad fsd band. Possibly (but not conclusively), there is a small contribution of f character even in the occupied part of the valence band.

b. α-Uranium Metal

Figure 10 shows combined XPS-BIS results for α-U. All valence band spectra for U, (even if they suffer from poor resolution or if a slight oxygen contamination cannot be excluded) display a strong emission in a narrow band just below the Fermi edge E_F.

This emission is considerably enhanced as compared with the spectrum of Th metal (see Fig. 9). The comparison of XPS valence band for U and Th (with a $5f^0$ configuration) provides a straightforward and direct identification of the 5f emission. Consequently, the strong emission for U at E_F can be attributed to itinerant 5f states, hybrid-

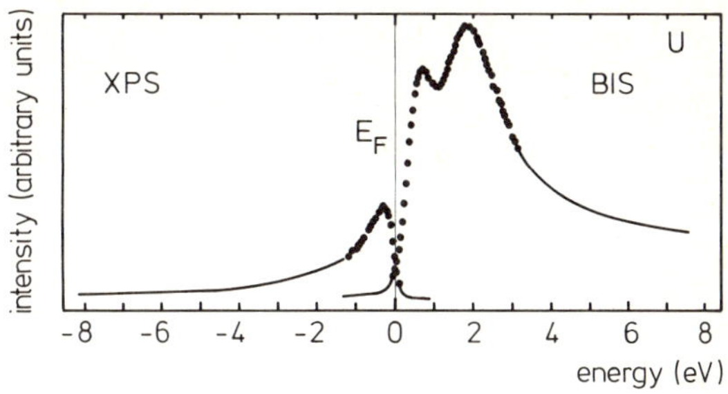

Fig. 10. The full photoelectron spectrum (direct and inverse photoemission) of the conduction band of U metal (from Ref. 56)

ized with 6d7s states, in contrast to Th, where mainly 6d states contribute to the spectrum. However, the more detailed interpretation of the valence band spectra of α-U is not so clearcut as for Th. Valence band spectra for U have been reported in a comparatively large number of papers in XPS[41, 56, 58, 60–65] and in UPS[59, 66]. Also, spectra have been measured using synchrotron radiation[67]. All the data have in common a narrow intense emission just at E_F, but differ for structures around 2.5 eV (Fig. 12). Table 2 summarizes the observations.

Spurious effects due to incompletely removed oxides layers are very likely to be recorded and misinterpreted in photoemission experiments from the very oxidizable U-metal surface. However, considering only high resolution XPS and UPS data for clean surfaces[56, 64, 66] as well as the measurements using synchrotron radiation[67], it can be

Table 2. Summary of valence band structures of α-U metal as reported by different authors (ΔE is the experimental resolution)

Technique	Resolution ΔE, eV FWHM	Type of structure and intensity	Energy below E_F, eV	Remarks	Ref.
XPS	1.0	broad shoulder, weak	ca. 3.0	clean	Fuggle et al.[60]
XPS	0.6	shoulder, pronounced	ca. 2.5	slight oxidation	Veal et al.[61]
XPS	0.6	small maximum	1.8	UO_2 subtracted	Verbist et al[62]
XPS	0.6	not detectable	–	poor stastistics	Nornes et al.[63]
XPS	0.4	shoulder weak	ca. 2.7	clean	Baer et al.[56]
UPS 40.8 eV	0.2	shoulder, very weak	ca. 2.4	weak O 2p contribution	Norton et al.[59]
XPS	0.6	shoulder, weak	ca. 2.8	clean	Grohs et al.[64]
UPS 21.2 eV	0.1	shoulder, pronounced	ca. 2.4	clean	Greuter et al.[66]
UPS 40.8 eV	0.2	shoulder	ca. 2.4	clean	Greuter et al.[66]
XPS	1.1	shoulder, weak	ca. 2.3	clean	Schneider et al.[58]
XPS	0.6	maximum	1.8	clean, poor statistics	Allen et al.[65]
variable synchroton source	0.3	maximum	2.3	weak O 2p contribution	Iwan et al.[67]

safely assumed that a broad structure at around 2.5 eV is genuine and not due to surface oxidation effects. In order to extract from the valence band spectrum all information about the electronic structure of α-U, three different lines of interpretation have been forwarded. Two of them are based on computed one-electron density of states functions[68, 69]. One considers in addition multielectron effects due to a beginning of a localization of the 5f electrons[30], in analogy with the discussion of the Ni 6 eV satellite, given in Part II. They are:

I. a comparison with the 5f partial density of states;

II. the study of the contribution of the 6d partial density of states, which is added to I;

III. the consideration of multielectron satellites due to localization of the photoexcited 5f hole, as an added effect to I and II.

Lines I and II interpret the photoemission spectrum of α-U in much the same way as we discussed the one of Th.

Line III has a great interest, because it takes into account in the case of U partial localization effects of 5f states. Such partial localization effects, if present in uranium metal, should be even more visible in the emission of plutonium metal. For this reason, line III will be discussed after the analysis of the valence band spectrum of plutonium metal.

The XPS valence band as shown in Fig. 11, and especially the narrow and intense peak just below E_F (observed in all experiments) have been discussed following mainly line I. Theoretical partial 5f density of states calculations[68, 69] agree in reproducing this feature, which can therefore be attributed to nearly pure 5f states. But these density of states curves predict additional structures which, although differing considerably in their position, are not observed experimentally. A maximum, observed only once at 1.8 eV[65] might be qualitatively described by one calculation[69]; however, relatively poor statistics (only 100 c/s) may have artificially introduced this structure since it is difficult to understand why other XPS valence band spectra (of comparable[64] or even higher[56] resolutions) do not show it.

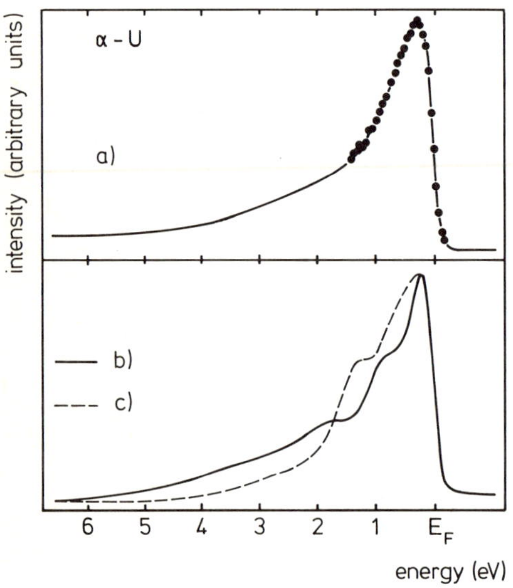

Fig. 11. The valence band spectrum of α-U metal, as measured in XPS (*a*), from Ref. 56 is compared with two one-electron density of states calculations (*b*) from Ref. 68, (*c*) from Ref. 69

Also the BIS results of Fig. 10 have been interpreted according to line I. However, the detailed agreement between experiment and theory becomes even worse when comparing the BIS data to calculated partial density curves of 5 f states[68, 69].

This spectrum has been compared with the BIS measurement on Nd metal[25], i.e. of the homologous lanthanide. Trivalent Nd has a localized $4f^3$ initial state configuration. For U, a $5f^2$ or a $5f^3$ initial state configuration are usually assumed, with a tetravalent or a trivalent core respectively. While in Nd the $4f^{N-1}$ and $4f^{N+1}$ multiplet states, as evaluated in an atomic-like Russell-Saunders scheme, can be well recognized in the XPS/BIS combined results, and are well separated from a (weak) d-emission at the Fermi edge, in U the occupied states and the empty states spectra join in a continuous band at E_F. Therefore, only the symmetry of 5 f states, given by the position of the main peaks in the joint spectrum, can be recognized with certainty.

By summing up all these observations, and considering that theoretical calculations[68, 69], firstly, provide different results and, secondly, describe less well the experiments on U metal than those on Th, the conclusion can be drawn that a fully 5 f band-like description may not be completely suitable[56]. An interpretation based on a localized behaviour (as for Nd) is, however, also unsuited (and not expected from other physical measurements and band-theory results). We conclude from this analysis, that the strong emission just below and beyond E_F originates manifestly from itinerant 5 f states.

Line II, based on the concept of 6 d-5 f hybridization, has been taken into consideration at the bottom of the valence band, where the 5 f density of states is low[56]. Thus, the broad structure around 2.5 eV (Fig. 12) may be tentatively attributed to a 6 d emission, similar to what is found in Th metal. A second 6 d-derived emission closer to E_F, as it is encountered in Th metal, may be hidden in the 5 f emission which is preferentially seen in XPS.

Energy dependent UPS data[66], together with XPS results, support this description: the satellite intensity becomes less pronounced compared with the 5 f emission at E_F

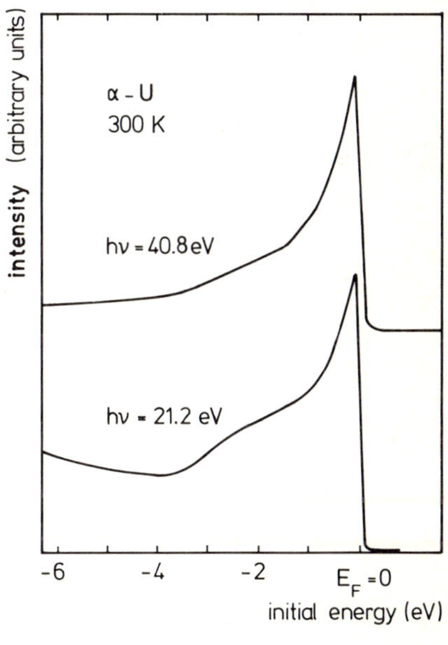

Fig. 12. UPS spectra of uranium metal (the surface of the metal is perfectly clean, as shown by the absence of an O2p signal in the spectrum; from Ref. 66)

when the excitation energy increases from 21.2 to 40.8 eV (Fig. 12) as expected for d vs. f emission, and the trend is respected when going to the 1486.6 eV of the X-ray excitation energy. The spectrum for 40.8 eV excitation (Fig. 12) is very similar to that for X-ray excitation (Fig. 11); but since the resolution in XPS is obviously less, the broad satellites are smaller in comparison with the sharp 5f emission when recorded with the same resolution. A more quantitative intensity analysis trying to normalize spectra for the same resolution and considering the energy variation of the photoionization cross-section for an adequate comparison with calculated 5f density of states would elucidate interpretation line II further.

From the comparison of the spectra with calculated one-electron density of states curves (lines I and II), therefore, the valence band spectra of U, from photoemission and BIS, are interpreted as due to a hybridized (d, f) continuous band, with a much large itinerant 5f contribution than in Th for the occupied part[57]. As for the broad structure at about 2.5 eV, qualitative agreements point to a 6d character.

c. *Plutonium Metal*

Within the actinide series Pu is the most intriguing element. On the basis of the Hubbard model, and taking into account an unhybridized bandwidth W_f (due only to f-f overlapping), the U_H/W_f ratio is 0.7 for U and 3 for Pu[27]: in fact, one would have expected already for Pu a 5f electron localization, since $U_H > W_f$. However, a hybridization of 5f with (6d7s) states broadens the 5f bandwidth and delays the Mott-like transition (see Chap. A) from Pu to Am[40]. This influences many properties of Pu metal[71, 72].

Only a few photoemission results (no BIS result) have been reported for this metal[73–76]. Figure 13 and Fig. 14 report the XPS and high resolution UPS spectra of α-Pu.

From a detailed analysis of the spectrum of Fig. 13, including a comparison with the calculated partial 5f density of states (broadened to take into account life-time effects and spectrometer resolution), it was concluded that the results could not be fully under-

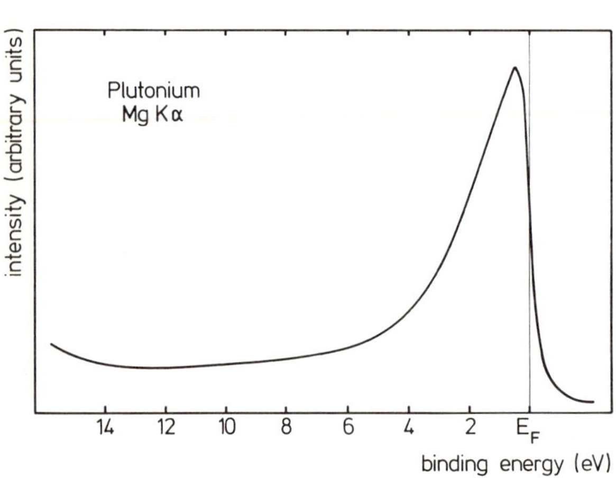

Fig. 13. The valence band spectrum of α-Pu metal (from Ref. 75)

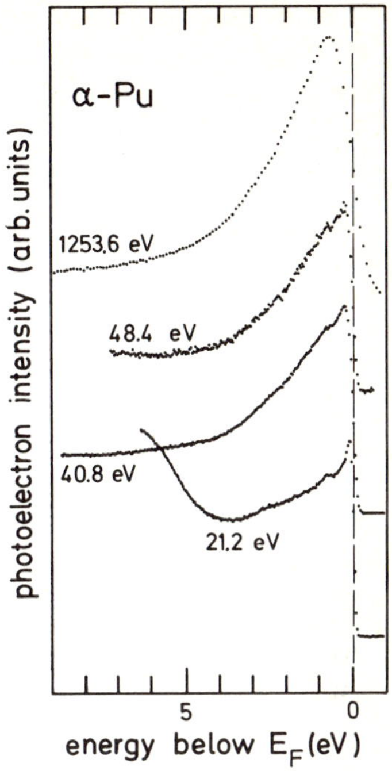

Fig. 14. Valence band spectra of α-Pu metal for different excitation energies (from Ref. 76)

stood without taking into account Coulomb correlation effects. Since the experimental spectra are much broader than the calculated ones, ground state calculations cannot describe them properly; excited states should be accounted for. This is not surprising, since already for U the ground state density of states calculation fails to describe satisfactorily the experimental spectra. Other valence band spectra of α-Pu[73, 74], even though of lower resolution, or presenting slight surface oxidation or bad statistics, all identify, as for U, itinerant 5 f states in a narrow band pinned at E_F; high resolution UPS valence band spectra[76] (Fig. 14) confirm the above description.

d. Localization Effects in the Valence Band Spectra of Uranium and Plutonium Metals

Partial localization of the 5 f states in the light actinides (line III of subsection b) might cause the appearance of satellite structures at energies not very far from E_F in their valence band photoemission spectra. If such structures could be convincingly demonstrated, important information would be added to the theoretical analysis of the localization vs. itineracy problem of the actinide metal series.

The interpretation of features comes back to a very wide discussion already taking place for 3 d metals, and, in particular for the case of Ni[31, 32]. Partial localization effects should be even more apparent in 5 f-metal spectra, since 5 f's are thought to be intermediate, in behaviour, between the (fairly itinerant) 3 d of the iron group and the (fully localized) 4 f of the lanthanides. (The ratios of the Coulomb energy U_H to the bandwidth

W_d for Ni and W_f for U gives values of ~ 0.5[26] and ~ 0.7[27]. This indicates that the U 5f electrons are even farther from an itinerant description than the Ni 3d electrons.)

The interpretation is based on the assumption that the U 5f states cannot be fully treated in an itinerant scheme, but many-electron effects have to be considered inducing satellites in the valence band[30]. The photoemission valence band spectrum is therefore considered as composed of three contributions: the itinerant 5f hole state with f electron screening (line I), the itinerant 6d7s hole state (line II) and the localized 5f hole state (line III) screened by 6d7s electrons. The first two contributions display the density of states as deduced from one-electron ground state calculations, whereas the third contribution results in final state multiplets typical for the specific $5f^N \rightarrow 5f^{N-1}$ transition. For U, selfconsistent band structure calculations[77, 78] and the measured position of the U $6p_{3/2}$ core level ([28]; see Part II) suggest an electronic configuration $5f^{2+\delta}(6d7s)^{4-\delta}$. Thus a satellite may be either due to $5f^3 \rightarrow 5f^2$ or $5f^2 \rightarrow 5f^1$ or both final states. Photoemission intensities for those states have been calculated in intermediate coupling[16, 79] showing that only one broad line is expected in both cases. The energy position of the $5f^1$ final state has been thermochemically evaluated[30] from the corresponding section: it is somewhat larger than 2 eV below E_F, i.e. exactly where the additional structure has been observed.

The crucial experiment to identify whether this satellite structure is due to a localized 5f hole, is claimed to be photoemission spectroscopy, in which the excitation (provided by synchrotron radiation) is tuned through the 5d-5f threshold energy[67]. At the threshold energy an empty 5f state just beyond E_F becomes occupied ($5d^{10}5f^N \xrightarrow{h\nu} 5d^9 5f^{N+1}$). Consequently, the emission at E_F could be suppressed considerably. On the other hand, since the 6d screening is still working[67], the satellite intensity is high enough to be detected. (This kind of experiment has been performed for the 6 eV Ni satellite[31, 32].

Resonance photoemission measurements have been recently made for U metal[67], and show indeed a resonant enhancement of the satellite at 2.3 eV only for the threshold energy ($5d_{3/2}$: $h\nu = 94$ eV) (Fig. 15). In addition the main peak at E_F shows the expected off-resonance behaviour. Further support for such an interpretation of the satellite is given by the analysis of the photon excited Auger emission. This is shown to be composed of two different bands also separated by 2.3 eV and due to the two screening channels by 5f or 6d states[67].

There are, however, arguments, which contradict the partial localization interpretation. This interpretation must assume that the 5f emission at E_F (itinerant state) and at the 2.5 eV satellite have different photon energy dependence of the cross section at the resonance. As recently discussed[80] this is difficult to explain since both structures are attributed to 5f states. Furthermore, the main asymmetric 4f core level should be accompanied by a shake-up satellite, induced by 6d screening of the localized hole, which has never been observed.

Other resonant photoemission studies for uranium compounds[24, 81] show, regardless of the degree of localization, that suppression (off-resonance) and enhancement (on-resonance) of 5f emission are always found for $h\nu = 92$ and 98 eV, respectively. On the other hand, localized 5f derived structures have been identified at 0.7, 0.9, 1.0 and 1.5 eV for USb, UTe, UPd_3 and UO_2 respectively[81], i.e. for compounds in which 5f's differ considerably with respect to localization.

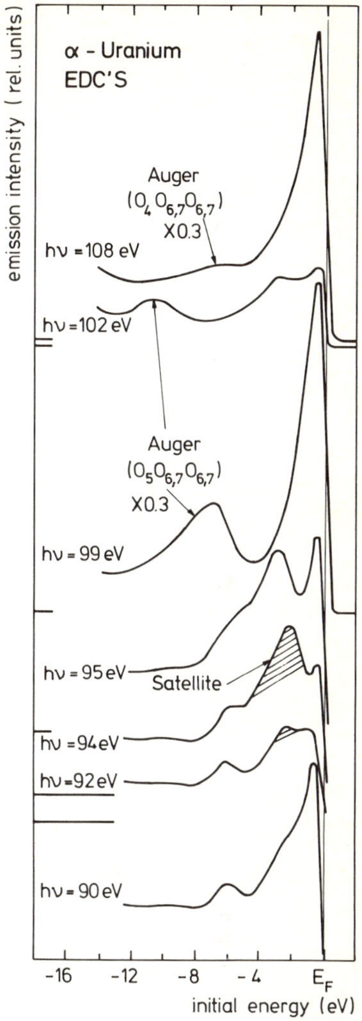

Fig. 15. Angle-integrated photoelectron energy distribution curves of uranium in the region of the giant 5 d → 5 f resonance (90 eV ≤ hν ≤ 108 eV). The 5 f intensity at E_F is suppressed by more than a factor of 30 at the 5 $d_{5/2}$ threshold (see the spectra for hν = 92 and 94 eV) and resonantly enhanced above threshold (see, e.g., the spectrum for hν = 99 eV). At an initial energy 2.3 eV below E_F a new satellite structure is observed which is resonantly enhanced at the 5 $d_{5/2}$ and 5 $d_{3/2}$ onsets. At threshold the satellite coincides with the Auger electron spectrum, which moves to apparently larger initial energies with increasing photon energy (from Ref. 67)

The other possible assignment of the 2.3 eV structure to 6d excitation (line II, Ref. 56, 66) cannot be ruled out. Resonant photoemission on Ce compounds[80, 82, 83] shows indeed that a resonant enhancement at the 4 d → 4 f threshold is not only present for 4 f but also for 5 d emission. Thus an easy identification of 4 f emission is not possible. Theoretical calculations of the cross section due to resonant enhancement result in a 2 eV shift to lower photon energy for the maximum 5 d enhancement compared with the 4 f enhancement[84].

Because of the much reduced itinerant 5 f character of Pu one would expect a similar, but even more pronounced multielectron satellite structure as observed at 2.3 eV in U. In contrast to U, this satellite due to a localized 5 f hole state screened by (6d7s) conduction electrons should not be a single line but show three or even four separate components as calculated for the 5 f^5 → 5 f^4 final state multiplet[16]. The fact that such a multiplet satellite is not observed in the XPS valence band spectrum is confusing. It could be that the overall multiplet satellite intensity is too small compared with the main 5f emission (itinerant 5f hole, screened by 5f conduction electrons) and that especially the high binding energy components of the $5f^4$ final state multiplet are not resolved because of low instrumental resolution and/or life-time broadening stronger than assumed in the calculations[16] (The life-time broadening will be very probably stronger in Pu than in the related rare-earth metal Sm because of the much higher density of states around E_F.). But even then one should observe the main ^5I multiplet line that has about the same intensity as the ^3H or ^2F line in U[79] (2.3, 2.2 and 1.9 units, respectively; Ref. 16). Since it is difficult to understand that the intensity of the main multiplet satellite is reduced for the more localized case of Pu in comparison with U, the interpretation of the 2.3 eV satellite of uranium as being due to 6d emission[56, 66] is favoured, the much increased 5f occupation (from about 2.5 in U to about 5 in Pu) with respect to the nearly unchanged 6d occupation explaining the reduced 6d emission relative to the 5f emission.

Recently low resolution XPS ($\Delta E \approx 1$ eV) and high resolution UPS ($\Delta E \approx 0.15$ eV) measurements have been performed on α-Pu[76] in the low 10^{-9} Pa range resulting in valence band spectra without any oxygen 2p signal even for the extremely surface sensitive UPS (Fig. 14). The XPS data corroborate previous results. The UPS data show the same strong emission of 5f states at E_F as found in XPS but a much sharper Fermi edge and additional weak structures at 0.8, 1.5 and 2.5 eV. Because of the difficulty of separating contributions of conduction band states from multiplet satellite no attribution of these structures has yet been attempted.

In conclusion, the partial localization effects in the valence band spectra of the light actinides, although extremely important if convincingly verified, still need much experimental and theoretical investigation. It is expected that the situation will improve considerably when resonant photoemission studies become available for actinide compounds in which the 5f and 6d emissions do not overlap. In addition, theoretical calculations of 5f and 6d cross sections near the 5d threshold will be very helpful.

3. Evidence from Photoemission Spectroscopy for the Mott-Hubbard Transition

a. The Valence Band Spectrum of Americium Metal

Am is known to be the first lanthanide like metal in the actinide series (References 27, 77, 85 and preceding chapters of this book).

The only UPS/XPS photoemission study of Am[55, 76, 86] shows a lanthanide like valence band feature as displayed in Fig. 16. The 5f emission is nearly completely withdrawn from E_F except possibly for some very weak 5f contribution seen only in high resolution He-I-spectra ($\Delta E \approx 0.12$ eV) as a very sharp peak just at E_F. The 5f intensity is concentrated in a structured peak around 2.8 eV binding energy (for MgK_α excitation, upper curve, the structures are not resolved) as deduced from the excitation energy dependence of the spectra. If one compares with Sm metal, the peaks at 1.8, 2.6 and 3.2 eV are attributed to the ^6H, ^6F, and ^6P states, respectively, of the $5f^5$ final state multiplet originating from the initial $5f^6$ ground state of "trivalent" Am.

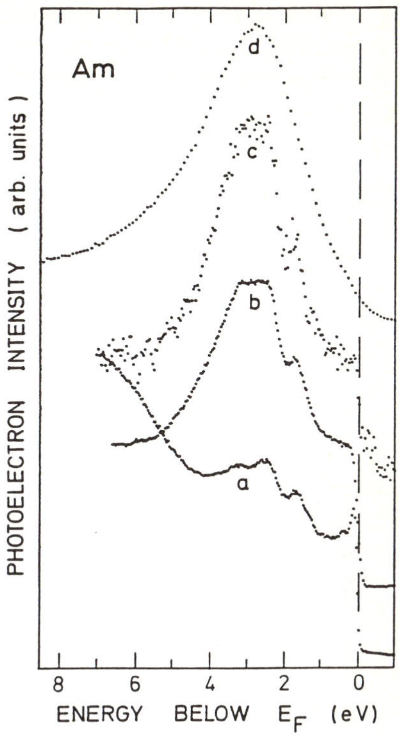

Fig. 16. UPS and XPS valence band of Am metal *a*) hν = 21.2 eV (He I); *b*) hν = 40.8 eV (He II); *c*) hν = 48.4 eV (He II*); *d*) hν = 1253.6 eV (MgK_{α}) (from Ref. 86)

There is a deviation of the multiplet shape from the one found for Sm. This is explained by the change of the main quantum number and the strong spin-orbit coupling; surface effects as broadening and shift of the multiplet as well as a contribution from a divalent surface can also be invoked.

The most recent calculations, however, of the photoemission final state multiplet intensity for the $5f^6$ initial state[16)] show also an intensity distribution different from the measured one. This may be partially corrected by accounting for the spectrometer transmission and the varying energy resolution of 0.12, 0.17, 0.17 and 1.3 eV for 21.2, 40.8, 48.4, and 1253.6 eV excitation. However, the UPS spectra are additionally distorted by a much stronger contribution of secondary electrons and the 5 f emission is superimposed upon the (6d7s) conduction electron density of states, background intensity of which was not considered in the calculated spectrum[16)]. In the calculations, furthermore, in order to account for the excitation of electron-hole pairs, and in order to simulate instrumental resolution, the multiplet lines were broadened by a convolution with Doniach-Šunjić line shapes (for the first effect) and Gaussian profiles (for the second effect). The same parameters as in the case of the calculations for lanthanide metals were used for the asymmetry and the halfwidths[13)].

Since the asymmetry parameter is dependent on the conduction band characteristics, i.e., specifically, on the density of states around E_F, it can be expected that the broadening effect is stronger for those actinides which still have some weak itinerant 5 f character, i.e., 5 f electrons at E_F even in the nearly localized situation.

Surface-induced effects can be expected to distort considerably more the spectra measured for excitation energies of 40.8 and 48.4 eV than for an excitation energy of 21.2 eV (and also 1253.6 eV) since surface sensitivity is higher for these exciting energies. Two different surface effects can contribute to the photoemission spectrum:

1. A $5f^6 \rightarrow 5f^5$ multiplet structure can be observed, produced by photoemission from a trivalent Am surface layer; this structure could be slightly shifted to higher binding energies and broadened (as observed in the rare-earth metals).
2. A valence change at the surface may occur as is observed for example for Sm metal[48, 49].

To decide whether a surface effect is present and, if so which, the experimental spectra shown in Fig. 16 have been corrected for the spectrometer transmission. The secondary electron contribution and the emission from conduction band states have also been subtracted. Comparing this spectrum with calculated multiplet intensities it seems that a contribution from a divalent Am surface resulting in a broad structureless $5f^7 \rightarrow 5f^6$ line at 1.8 eV is the most suitable explanation of the measured intensity distribution. Theory also supports this interpretation, since the empty $5f^7$ level of bulk Am lies only 0.7 eV above E_F within the unoccupied part of the 6d conduction band (as calculated from the difference of the Coulomb energy U_H and the $5f^6 \rightarrow 5f^5$ excitation energy Δ_-[27, 40, 87]). Any perturbation inducing an increase of E_F by that amount will cause an occupation of this $5f^7$ level, thus changing the occupation from $5f^6$ to $5f^7$, i.e., the valence of the surface layer. The reduced coordination number of surface atoms may be just such a perturbation, resulting finally in a narrowing of the 5f band which enables its electrons to flow into the $5f^7$ state[88].

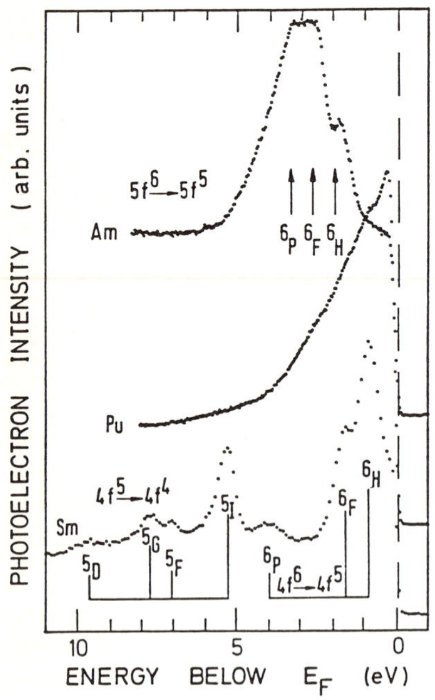

Fig. 17. Valence band spectra of Am, α-Pu, and Sm for He II excitation (40.8 eV) (from Ref. 86)

b. The Mott-Hubbard Transition in the Actinide Metal Series

Figure 17 is a clear illustration of the Mott-Hubbard transition in the actinide series: the 5 f emission occurs, for α-Pu, at E_F, indicating a high 5 f-density of states pinned at the Fermi-level, whereas the 5 f emission occurs at lower energy for americium metal. In this case, therefore, a theoretical concept deduced indirectly from the physical properties of the two metals, finds direct (one might even say: "photographic") confirmation in the photoemission spectra.

c. α-Pu ↔ δ-Pu: a Search for a Mott-Hubbard Transition

A sign of the onset of a localized behaviour of 5 f states is considered to be the appearance of a compact structure, such as hcp, fcc or dhcp. These structures are encountered in lanthanides and in heavy actinides (starting from Am). α-Pu is monoclinic, but δ-Pu has already the fcc structure. 5 f electron localization is then expected to occur for this phase. In this case, the UPS valence band (e.g., for 40.8 eV excitation) of the δ-phase should change considerably from that of the α-phase, and a final state multiplet structure (similar to the one for bulk Sm) might be expected[86)].

Figure 14, shows the valence band spectrum of Pu metal[76)], measured by UPS at different excitation energies. Similar UPS measurements at different temperatures, for which different allotropic forms of Pu are present, show two effects occuring in the spectrum when δ-Pu is reached: a breakdown of strong emission at E_F (due to 5 f band states) and an increase of the emission at higher binding energies. Neither the broadening of the Fermi edge at higher temperatures nor oxidation effects provide an explanation for the strong effect at E_F. Hence, it can be attributed to a decrease of 5 f character of the valence band, i.e. of a decrease of itineracy of 5 f electrons. This as well as the increase of the emission at higher binding energies can be explained in terms of a narrowing of the 5 f-band and a reduced hybridization with (s, d) states.

Since the interactinide distance in δ-Pu is larger than in α-Pu a narrowing of the 5 f-band is expected. The increased Pu-Pu distance leads naturally to a narrowing of the unhybridised 5 f-band (which can be interpreted as an increased 5 f-localization). The fcc structure (as opposed to the very directional monoclinic structure of α-Pu) is, on the other hand, an indication of a lower degree of hybridization. Recent band calculations, taking into account spin-orbit coupling[89)], give a much narrower 5 f-bandwidth (pinned at E_F) and therefore a very high 5 f-density of states (double than that of α-Pu).

Recent measurements of the electronic γ from low temperature specific heat in (Al-stabilized) δ-Pu phase show, in agreement with theory, a very high enhancement of the 5 f-density of state at E_F (for pure δ-Pu phase leading to an extrapolated $\gamma = 53 \pm 10$ mJ/g-atom · K^2 vs. 22–25 ± 1.0 mJ/g-atom · K^2 for α-Pu[90–92)]). This apparent contradiction with the decrease of 5 f emission at E_F seen in the spectra is, however explained, if one accounts for final state effects in δ-Pu due to the narrowing of the band, as is the increased emission at higher binding energies. One has to recall the discussion concerning the emission from band states, interpreted as a one-electron emission, evoked for the 2.3 eV satellite in U-metal as well as for the 6 eV satellite in Ni (line III).

From the photoemission viewpoint, the narrowing of the 5 f-band favours emission from a localized 5 f hole state (screened by (6 d 7 s) electrons) (as in line III) at the cost of the emission from an itinerant 5f hole state (screened by 5f electron) (as in line I). Emission following line I then would be dominant in α-Pu, emission following line III in δ-Pu.

However, no fingerprint such as a final state multiplet structure, as expected for fully localized 5f electron and found in Am metal, is observed. The localization of 5f-states is only weak (band-narrowing and widthdrawal from hybridization) in δ-Pu. This in fact is also consistent with the absence of a magnetic moment formation in δ-Pu (rather, a spin-fluctuation regime is observed[71, 72]).

Thus, Am and not δ-Pu remains the first actinide metal showing the characteristics of a fully localized, lanthanide-like, 5f behaviour.

4. Core Level Spectra

As outlined in Part II, core level spectroscopy is helpful in the determination of the bonding properties if the core levels are well chosen. Specifically, the actinide 4f levels contain important information on the character of the conduction electrons because the line shape is almost exclusively influenced by the screening mechanism. The superposition due to coupling of the 4f hole to the unfilled 5f shell and leading to multiplet splitting is negligible. On the contrary, the multiplet splitting is strong for other core levels: e.g., for the 5d core levels, producing complicated structures which are difficult to interpret.

a. The 4f Core Levels of Th and U Metals

4f core level spectra have been reported in a number of papers for Th[34, 58–64, 93] and U[58–66, 93]. Despite some minor scatter, the binding energies are 333.1 and 342.4 eV for Th and 377.3 and 388.1 eV for U for the $4f_{7/2}$ and $4f_{5/2}$ levels respectively. Figure 18 shows for Th as well as for U a strong asymmetry that is attributed to simultaneous excitations of electrons just below E_F to empty states just beyond E_F: this mechanism is strong for metals with a high density of itinerant states around E_F[94].

The main difference between the 4f core levels of Th and U is the presence of a satellite, appearing as a shoulder at about 2.2 eV higher binding energy from the main 4f lines of Th. On the basis of the conduction electron screening model, discussed in Part II, this satellite has been attributed to a screening process in which empty 5d states above E_F are pulled down into the occupied part of the conduction band and filled (poorly screened peak). The main asymmetric peak at lower binding energy is associated instead with a better screening (well screened peak) by an occupation of an empty $5f^1$ state, which is known, from BIS experiments, to lie around 4 eV above E_F[56] (Fig. 9) and is pulled down into the occupied part of the conduction band. Therefore, starting with an initial configuration $4f^{14}(6d7s)^4\,5f^0$ for Th, the final states are $4f^{13}(6d7s)^4\,5f^1$ (well screened) and $4f^{13}(6d7s)^5\,5f^0$ (poorly screened)[34] (the screening level is filled by one conduction electron).

Final state screening effects have been described generally, although qualitatively, in Part II. A more quantitative model interpreting the relative intensities of the different

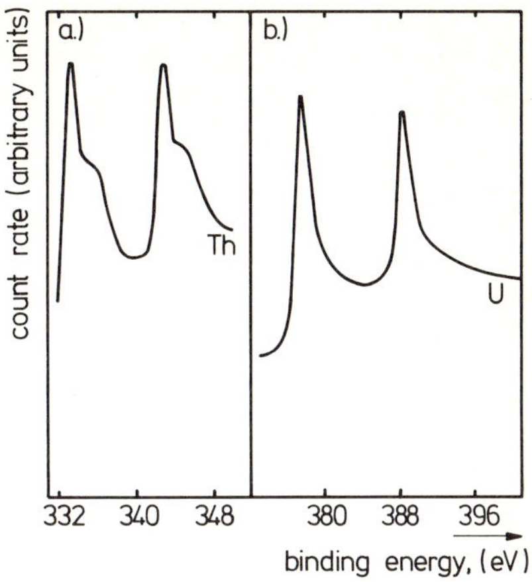

Fig. 18 a, b. 4 f core level photoemission spectra (hν = 1486.6 eV) of (**a**) Th metal; (**b**) U metal (from Ref. 59)

peaks caused by screening effects in a core-level excitation of an open-shell metal, has been developed by Schönhammer and Gunnarson for adsorbates[95]. Recently, Fuggle et al.[34] have applied to same ideas to compare the relative variation in intensity between the well screened and the poorly screened peak in lanthanide and d-transition bulk metallic systems.

The model predicts that the relative intensity varies in a continous way with the ratio W_{scr}/Δ_+ between two parameters. Δ_+ is the energy separation from E_F of the quasi-localized pure empty state, which can be measured by inverse photoemission (a $5\,f^1$ state in Th metal). This state is pulled down and its occupation provides the best screening. The Coulomb interaction U_H, which governs the localization property of this state is, as we know from Part II: $U_H = \Delta_+ + \Delta_-$. At least in principle[34], the greater is Δ_+, the greater is the localization character of the empty state.

W_{scr} is the width of the empty state, which is considered to be essentially determined by its degree of hybridization (or coupling) with other states of the metal. The two parameter picture, therefore, tries to separate the two main phenomena occuring in open-shell systems, and which have been discussed elsewhere in this book: the localization character of the state (as determined by Coulomb and exchange interatomic correlation) and its hybridization with other states.

When pulled down from beyond E_F, the quasi-localized state may be seen as being a "virtual state" in the conduction band (e.g., a 5 f state in the (s, p, d) conduction band). In this situation, the probability of occupation of the state by a conduction electron, is clearly related to the W_{scr}/Δ_+ ratio.

When $W_{scr} \ll \Delta_+$, this probability is small, and the relative intensity of the well screened peak is small; when $W_{scr} \gg \Delta_+$, the contrary occurs. The d-transition and lanthanide metal pictures represented in Fig. 7 are therefore generated.

Applying this model, the 4f doublet of Th metal would appear to be a typical 5f-transition metal spectrum (i.e., with a split doublet where the 5 f's play the screening role of the d electrons in a d-transition metal). This is consistent with strong hybridization of 5f and (d, s) states as predicted by theory[57, 68].

However, the $4f_{7/2}$ core level response of U metal does not show a poorly screened satellite, but only a pronounced asymmetry. This is somewhat in contrast with the greater localization predicted by theory[30, 68] when proceeding across the actinide series.

It is possible that the correlation between Δ_+ and the localization of the empty state is not as straightforward as indicated above. In fact, the BIS measurements of Th and U[56] display a shift of the empty 5f states of about 2.7 eV towards E_F for U, i.e., the Δ_+ is for U smaller than for Th. Due to this decrease for Δ_+, W_{scr}/Δ_+ has become probably larger for U even if W_{scr} has become slightly smaller due to a lesser degree of hybridization of slightly more localized 5f states.

b. The 4f Core Levels of α-Pu

4f core level spectra have been reported only in a few papers[73, 75, 76, 86, 96] and are displayed in Figs. 19 and 20. All published results agree very well and show the $4f_{5/2,7/2}$ doublet at binding energies of 435.1 and 422.2 eV; the strong asymmetry is again a clear indication of the very high density of band-like 5f states just around E_F; the dominant screening mechanism is the occupation of 5f screening levels (well screened)[75, 86]. Weak shoulders are observed on the high binding energy tail of the $4f_{5/2}$ and $4f_{7/2}$ lines (as for d-transition metals in Fig. 7). However, the doublet structured shoulder on the $4f_{7/2}$ tail is obviously induced by a satellite of the $4f_{5/2}$ line originating from the Mg $K_{\alpha 3,4}$ satellite doublet of the Mg X-ray excitation source. The other one on the $4f_{5/2}$ tail has been attributed to a 6d screening (poorly screened), although a very weak oxide contamination of the sample was not completely excluded[75].

Recent measurements in our laboratory under improved vacuum conditions[86] showed that this shoulder becomes more pronounced with increasing oxidation of the Pu surface, finally ending up with a Pu_2O_3 layer. Even for repeated extensive sputtering, the shoulder could not be removed completely. But one has to keep in mind that the recording of a 4f spectrum needs an extended time (for which UPS spectra show already a weak

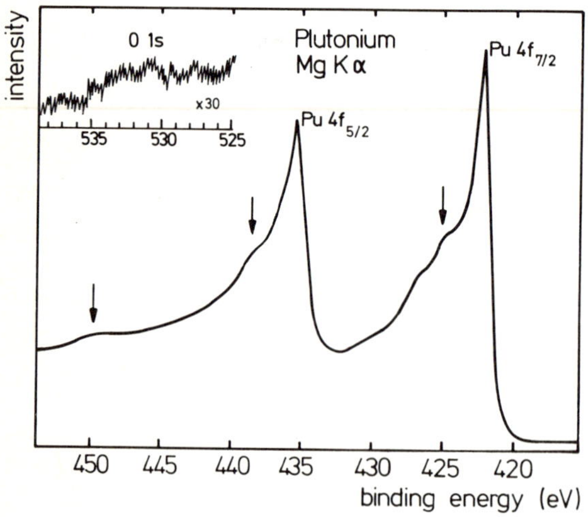

Fig. 19. The 4f spectrum of α-Pu metals (arrows indicate probable satellites). The structure appearing around 426 eV is also influenced by the Pu $4f_{5/2}$ level excited by the $MgK\alpha_{3,4}$ radiation (from Ref. 75)

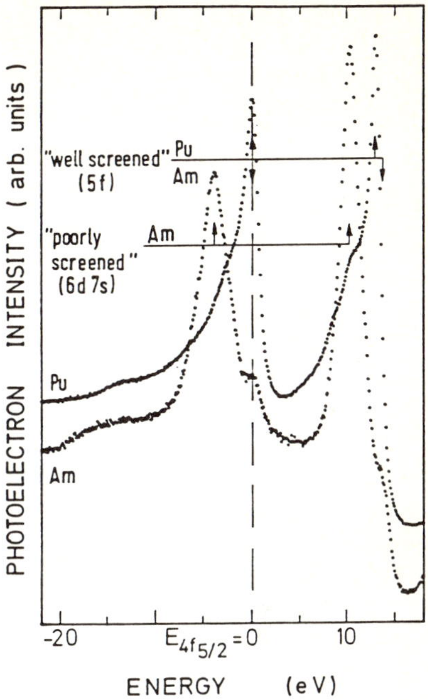

Fig. 20. XPS 4 f core level of Am and Pu metal (from Ref. 86)

oxygen 2 p signal, while in XPS no 1 s signal can be detected). Thus the origin of the shoulder on the 4 $f_{5/2}$ line at about 2.5 eV higher binding energy is not quite clear at present. Even if it is not induced by an oxygen contamination but due to (6 d 7 s) screening (poorly screened, 5 f electrons localized) the dominant screening occurs by 5 f states hybridized with conduction electrons, i.e., the 5 f electron localization is poor. This is consistent with the information from the valence band emission (Fig. 14).

c. *The 4 f Core Levels of Am*

As already described above and deduced from valence band spectra (Fig. 16) Am is the first rare-earth like metal in the actinide series. This is also well reflected in the 4 f core level spectrum reported recently[86]. The 4 f core level spectrum as shown in Fig. 20 is no longer dominated by asymmetric well screened peaks but by nearly symmetric lines which are attributed to (6 d 7 s) screening (poorly screened, 5 f electrons localized). The well screened peaks appear as satellites at 3.9 eV lower binding energy, their weak intensity being a clear indication for the poor 5 f hybridization. Therefore, the 5 f states are almost completely localized. The spectrum is very similar to the 3 d core level spectrum of light lanthanide metals[34, 37, 97, 98] for which 5 d screening is also dominant (poorly screened). The spectra may however be slightly distorted because the coupling of the 3 d hole to the incomplete 4 f shell produces a stronger multiplet contribution than in the case of the 4 f levels in Am. An energy separation of 3.8 eV between the poorly and well screened peaks has been recently calculated[29], in good agreement with the experimental result.

IV. The Photoelectron Spectra of Actinide Oxides

1. Introduction

a. The Actinide Oxides: General Description

Under the name "actinide oxides" a large number of compounds is included corresponding to the many possibilities of bonding and crystal structures offered by the large spread of oxidation states displayed by these elements (see Fig. 1 of Chap. A). In contrast to the lanthanide-oxygen system, where (with the exception of Ce, Pr and Tb), the Ln_2O_3 sesquioxides are the stable oxides (in correspondence with the stable trivalent state of these elements), all actinides form the fluorite-structure $An^{IV}O_2$ dioxides[1] (the only one for thorium). The light actinides form oxides with higher oxidation states: for instance $U^{IV}O_2$, U_4O_9, U_3O_8, $U^{VI}O_3$ are known for U, covering the range from tetravalency to hexavalency. In U_4O_9 and U_3O_8, U^{+5} and U^{+6} ions are copresent with U^{+4} (see the U-O phase diagramme in Chap. C). Sesquioxides become stable from Pu on in the series, following the general trend to a lanthanide-like behaviour of the heavier actinides.

The dioxides (and, in some measure, also the sesquioxides) are nonstoichiometric compounds: oxygen-rich AnO_{2+x} for light actinides; AnO_{2-x} from Pu (or Np) on; the nonstoichiometry being related to the relatively close oxygen potentials of An^{+4} and An^{+5} or An^{+6} in light actinides and of An^{+4} and An^{+3} in heavy actinides.

This is a source of difficulty in photoelectron experiments since the surface or even the bulk composition may be easily perturbed. In the high vacuum conditions (or under the perturbation of ion impact) typical of photoelectron experiments, they are very easy to form through reduction of the dioxides.

When more than one oxidation state of the actinide is present in the oxides, superimposition of the spectral response due to the different ions occurs.

This causes e.g. very complicated 4f spectra, which show for each specific ion the chemically shifted spin-orbit split doublet and its accompanying satellites. These in addition, depend on deviations from stoichiometry. Therefore an easy analysis is seriously prevented.

Other remarks concerning experimental difficulties in the photoemission of actinide oxides must be made. Actinide oxides are insulators or semiconductors. Charging effects therefore are usually compensated by flooding the sample with low energy electrons during the measurement. Since it is difficult to control the electron intensity necessary to compensate exactly the photoemission induced charge, reported binding energies have to be handled with caution; but relative binding energies, e.g. for satellites, are more reliable.

Another effect making the measurement of absolute binding energies difficult, stems from the easy variation of the stoichiometry of the sample, which changes the type of electronic conduction and the work function, i.e. the difference between the zero energy level of the vacuum and E_F. Such changes of E_F (E_F is usually taken as the reference zero for binding energies) can be caused also by sputter cleaning procedures and even by the

1 The Roman numeral superscript indicates the oxidation state

impinging X-ray beam. Small amounts of surface contamination also often change the work function.

All these effects are probably responsible for the discrepancies of reported photoelectron results in actinide oxides. Often, especially for the more radioactive and rare heavy actinides, dioxide samples are prepared for photoemission by growing oxide layers on top of the bulk actinide metal. These samples may then display features of trivalent sesquioxides since the underlying metal acts as a reducing medium.

b. *The Chemical Bond in Actinide Oxides, and the Research Directions of Photoelectron Spectroscopy*

In this chapter, the discussion will be limited to the actinide dioxides, and, in lesser extent, sesquioxides. Also, no special attention will be given to nonstoichiometry effects, except by indicating when they may be responsible for serious errors in the measurements or in their interpretation. In order to understand, however, the directions of research for the photoelectron spectroscopic technique, a short digression will be made on the nature of the chemical bond in these systems.

As already discussed in the preceding chapters of this book, actinide oxides are mostly ionic solids: this is shown by their physical properties, e.g. magnetism (Chap. D). In dioxides, due to the high electronegativity of oxygen, the An^{+4} ion is formed through the charge transfer to oxygen of 5 f electrons together with the s and d electrons of the outer configuration of the actinide atom (see Table 1 of Chap. A). This leaves in the actinide ion a $5f^n$ shell. Magnetic measurements are explained for all dioxides by assuming a magnetic moment of the ion as expected by a $5f^n$ ionic configuration: this is strong evidence of a full localization of the 5 f orbitals in these compounds. Trivalent sesquioxides also display an analogous charge transfer and are explained within an ionic, 5 f-localized picture.

Photoelectron spectroscopy verifies, as we shall see, this description. According to the fully localized picture of the $5f^n$ shell, therefore, great emphasis is laid, when analysing photoelectron spectra for the 5 f response, to all those phenomena, (shortly described in Part II, as e.g., final state multiplet splitting) which are typical of the response of localized states. Also, typical characteristics of $5f^n$ atomic states, already known from atomic considerations, such as their high energy position, and their spin-orbit (or j-j) splitting, are expected to be seen in the valence band region of photoemission spectra. Another important application of photoelectron spectroscopy, permitted by the localized behaviour, is the measurement of the Coulomb correlation energy U_H, by combined XPS-BIS techniques, as we have seen in Part II.

Metallic bonding, due to the f-f overlap (as found, e.g. in NaCl-structure actinide compound), is consequently to be excluded in oxide systems, as already shown by Hill plots (see Chap. A). Brooks and Kelly[99)] have recently calculated ground state properties for UO_2 in a LDA scheme by taking the hypothesis of an itinerant character of 5 f electrons (see Chap. C). This hypothesis leads to an excessive 5 f attractive contribution to cohesion, leading to an equilibrium volume for UO_2 35% lower than the experimental one.

The basic band scheme of oxides, in the most simplified picture, is the typical one of oxide insulators or semiconductors: an occupied oxygen 2 p-valence band and an empty actinide conduction band of, essentially (7 s 6 d) origin. In oxides of light actinides (as it is also the case for d shells in transition metal oxides), the $5f^n$ occupied state will lie somewhere in the band gap.

Thus, oxides are chosen in this chapter as an example of localized 5 f behaviour, and the confirmation of localization obtained from photoelectron spectroscopy is emphasized.

If metallic bonding does not contribute to the ground state properties of oxides, it is possible however, that the outer orbitals of the actinide atom might form hybridized bonds with the 2 p orbitals of oxygen atom. Brooks and Kelly have calculated the ground state properties of UO_2[100, 101], by including only d-p hybridization. They found a relevant contribution of the d-p covalent bond energy to the cohesive energy of this oxide. In the above cited paper[99], besides considering (and excluding) band-like 5 f states, they have included Mott-Hubbard localized (see Chap. A) 5 f states in the set of hybridized orbitals. They found a small amount of f-p mixing in the valence band, certainly less than the d-p mixing quoted above, and very little f contribution to the ground state properties of the oxide.

Cluster molecular calculations, a number of which has appeared on actinide oxides[102–105], are very sensitive to covalent mixing of the actinide-oxygen external orbitals which has a "local" bonding character. This is emphasized by these calculations which take into account a finite number of shells around the bonded species. It is just possible that the cluster method has a tendency to overestimate covalency.

This type of *covalency* (not to be confused with the metallic, itinerant 5 f-behaviour) is a departure from the simple ionic picture of oxides, and is important when considering some properties of the solids: e.g., mechanical properties such as shear properties which will depend strongly on the directionality introduced by the covalent bond[106], or defect formation, defect clustering and phase formation, a field particularly important for these oxides[107], of which we have previously recalled the nonstoichiometric behaviour and the richness of phases (see also Chap. C). Photoelectron spectroscopy can detect the orbital character of the occupied bands when energy dependent measurements are performed. Unfortunately, this method is less suited to identify d-p mixing, than f-p mixing, due to the large f cross section variation. Consequently, we will dedicate some space to the discussion of indications of f-p hybridization resulting from photoelectron experiments.

2. *The Valence Band Spectra of Actinide Oxides*

a. *Photoelectron Spectroscopy of ThO_2 and UO_2*

Here, the main features of the valence band results for ThO_2 and UO_2 will be illustrated. Since a large number of publications exists in this field (especially for uranium oxides), reference will be made only to a few selected investigations, chosen for the purpose of highlighting those aspects of the oxide bond discussed previously. A very comprehensive review of these results can be found[108] (and references therein; electronic and spectroscopic properties in Refs. 109–111). Figure 21 shows the photoemission spectrum of ThO_2 and UO_2 up to E_b = 45 eV[61]. The valence band region extends to about 10 eV. The marked difference is the appearance in UO_2 of a sharp and intense peak at E_b =

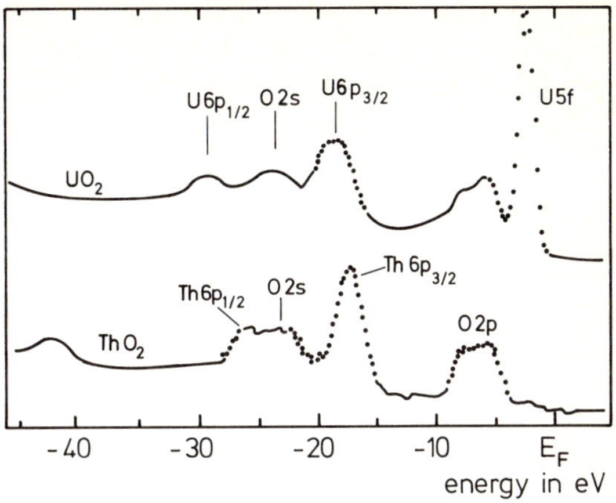

Fig. 21. XPS spectra of UO_2 and ThO_2 within 45 eV of the Fermi energy (from Ref. 108)

1.4 eV which is lacking in the ThO_2 spectrum. In an ionic picture, U^{+4} has a $5f^2$ state and Th^{+4} a non-occupied $5f^0$ state: hence the clear attribution of this peak to 5f emission. The insulating character of the two oxides is indicated by the lack of emission at E_F.

The broad band centered at about 6 eV and following this peak in the UO_2 spectrum, and appearing also in the ThO_2 spectrum, is much the same in the two oxides: hence, the attribution to emission from an essentially 2p occupied band, which, in an ionic picture, is the ligand valence band of the two oxides. There is some discrepancy between the results of different investigations about the detailed shape of this band. In Table 3 we have listed the main features of the valence band spectrum.

The 5f character of the peak at 1.4 eV in UO_2 is evidenced by varying the excitation energy in the UPS/XPS technique, as well as by other techniques such as ARPES and resonant photoemission (see Table 3).

Recent UPS/XPS measurements have been performed in our laboratory on $UO_{2.0}$ single crystals[112)], the surface of which had been prepared in different ways (Ar^+-sputtering; in situ scraping). These results are shown in Fig. 22 and illustrate the 5f character of 1.4 eV peak. When comparing the different methods of surface preparation, it is noted that very probably sputtering (a widely used method) reduces the UO_2 surface to a UO_{2-x} composition and modifies the relative intensity of the 2p-valence band to the 5f emission. A certain number of discrepancies between different investigations about this relative intensity might be attributed to stoichiometry deviations on the surface. Another important difference between in situ scraped and Ar^+-sputtered spectra is the disappearance in the former of the 10 eV weak satellite (see Table 3).

Figure 23 shows the different oxides of uranium[108)]. As expected in an ionic picture, the 5f emission decreases with decreasing occupation number of the 5f shell, to disappear completely in β-UO_3 ($5f^0$ configuration of the U^{+6} ion).

Figure 4 shows the combined XPS/BIS results for UO_2. The main peak in inverse photoemission, centered at approximately 5 eV above E_F is attributed to a 5f state, in part because of its dominating intensity (high cross section of states at 1500 eV electron excitation energy), in part by a comparison with the measured spectrum of Th (see Fig. 9), in which the 5f states are well separated from s and d states. Thus, the peak

Table 3. Summary of valence band structures of UO_2 (occupied and empty states) as observed and analysed by different methods

Structure	E_b(eV) Occupied states	XPS[18]	BIS	ARPES[113]		UPS/XPS this laboratory[112]		Resonant photoemission[81]	
p_1	1.37–1.5	$5f^1$, $^2F_{5/2}$	$U_H = 4.37$ eV	$5f^1$, $^2F_{5/2}$		$5f^1$, $^2F_{5/2}$		$5f^1$, $^2F_{5/2}$	strong resonance
p_2	~2.3	–	non-observed	$5f^1$, $^2F_{7/2}$	shoulder in second derivative of EDC; *no dispersion*	–	non-observed	–	non-observed
p_3	4.0–4.2	–	non-observed	2p?	minimum in second derivative of EDC; *large dispersion*	2p(6d) character of v.b.	v.b. maximum for hν = 21.2 eV; $I_f : I_d : I_p$ 20 : 20 : 120	–	non-observed
p_4	4.8–4.9	2p(6d) character of v.b.	–	2p(6d) character of v.b.	minimum in second derivative of EDC; *large dispersion*	2p + 5*f*	shoulder at hν = 21.2 eV; v.b. maximum for higher hν	2p(6d) character of v.b.	very weak resonance behaviour, i.e. no 5f character
p_5	7.0–7.5	2p(6d) character of v.b.	–	2p(6d) character of v.b.	minimum in second derivative of EDC; *large dispersion*	2p(6d) character of v.b.	–	2p(6d) character of v.b.	very weak resonance behaviour, i.e. no 5f character

p_6	9–10	2 p(Γ_{15} subband)	–	–	non-observed	surface composition effect?	observed for non cleaved or Ar-sputtered single crystals; non-observed for in situ scraped single crystals	2 p(6 d) character of v.b.	very weak resonance behaviour, i.e. no 5 f character
		Empty states							
q_1	~2	onset of 6 d empty band	shoulder						
q_2	~5	5 f^3, $^4I_{9/2}$	dominant peak; 5 f character proved by (low) energy dependent BIS[114]						

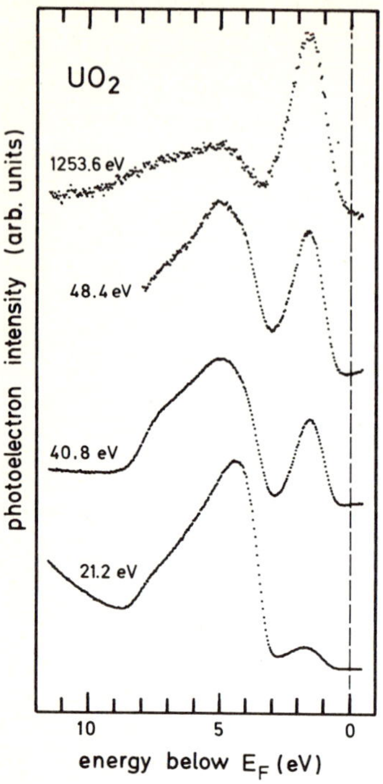

Fig. 22

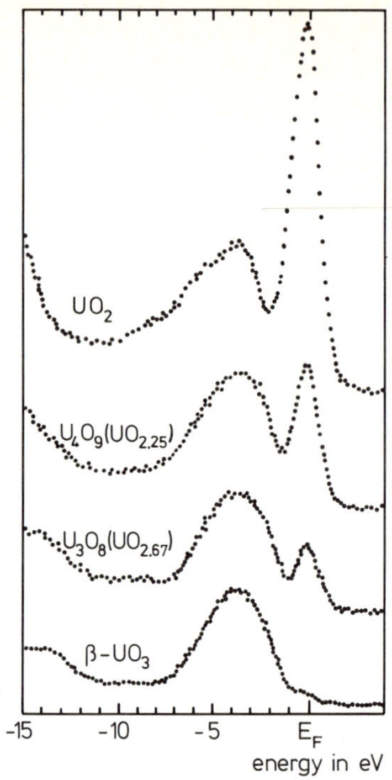

Fig. 23

Fig. 22. Valence band of UO_2 (single crystals) by UPS/XPS measurements (from Ref. 112)

Fig. 23. XPS valence spectra of different uranium oxides (from Ref. 108)

is attributed to excitation to a $5f^3$ final state (process: $5f^2 \rightarrow 5f^3$) and is analyzed in terms of multiplet lines of the $5f^3$ final state in the LS-coupling scheme, for which the $^4I_{9/2}$ state is the most intense one. (Considering the more appropriate intermediate coupling scheme for actinides, the $^4I_{9/2}$ state should become even more intense.)

BIS experiments may also be performed by varying the energy of the impinging monoergetic electrons (energy dependent BIS). On UO_2, these experiments have been made, with low energies between 20.6 and 49.6 eV by Chauvet and Baptist[114)]. The low energy dependent BIS experiments have generally the same complementary character to high energy BIS, as UPS energy dependent measurements have to XPS measurements, since they may be used to analyse the orbital character of the empty states. The spectra show a gradual increase with excitation energy of the main peak at 5 eV confirming its attribution to 5f states.

The shoulder at 2 eV in the BIS results of Fig. 4 has been attributed to the onset of the empty, predominantly d-like, conduction band. From the combined XPS and BIS spectra the p-d band gap, i.e., the energy difference between the upper edge of the mainly 2p valence and the lower edge of the mainly 6d conduction band, has been determined to be 5.0 ± 0.4 eV. This agrees with the value of 5.35 eV calculated in band calculations[101)]. It agrees also with the optically determined band gap by Schoenes[110)] for UO_2.

b. *Photoelectron Spectra of Other Actinide AnO_2 and An_2O_3 Oxides*

The XPS valence band spectra for the dioxides of the transuranium elements (from Np to Bk) have been presented in an extensive and pioneering work that also includes core level spectra and has been for a long time the only photoemission study on highly radioactive compounds. High resolution XPS spectra ($\Delta E = 0.55$ eV) were recorded on oxidized thin metal films (30 Å) deposited on platinum substrates with an isotope separator. (The oxide films for Pu and the heavier actinides may contain some oxides with lower stoichiometry, since starting with Pu, the sesquioxides of the heavier actinides begin to form in high vacuum conditions.)

Figure 24[15], shows the spectra together with previous results for ThO_2 and UO_2. The dominating peak next to E_F in all oxides but ThO_2 is attributed to the emission from localized 5f states. The mainly oxygen 2p derived valence band emission is observed around 5 eV and is clearly separated from the 5f emission only in the case of UO_2. In fact, for heavier actinide oxides, the 5f emission shifts towards higher binding energy as the atomic number increases, finally overlapping the valence band in energy.

This is to be expected for 5f orbitals having a localized, atomic-like behaviour: the increasing binding energy with increasing Z reflects the actinide contraction. The growing 5f emission intensity with increasing atomic number (compared with the $6p_{3/2}$ emission around 18 eV) reflects the increasing occupation number of the 5f states.

Due to the localization of the 5f electrons in the oxides a final state multiplet structure is expected. Therefore, these spectra have been compared with final state intensities calculated in an intermediate coupling scheme which accounts for the strong spin-orbit

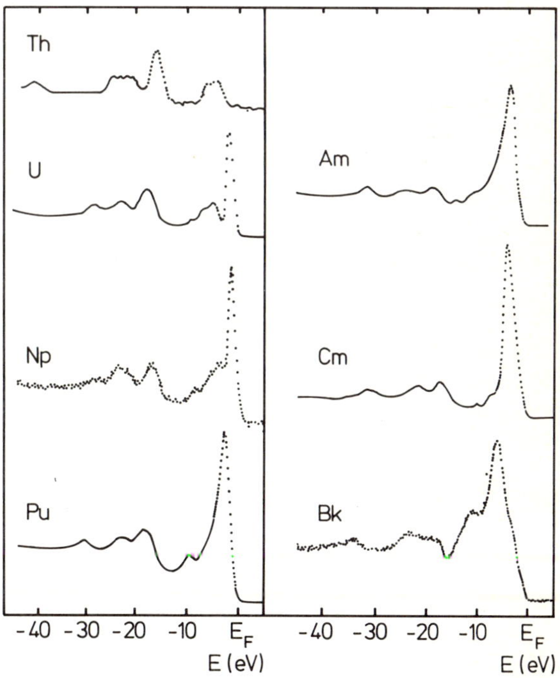

Fig. 24. XPS spectra for oxides of the actinides Th–Bk (from Ref. 15)

interaction. They were also broadened to simulate the experimental resolution; crystal field effects were excluded because of computational restrictions (Fig. 3).

This calculation leads to a much larger number of intense 5f photoemission multiplet lines than observed. (Also, previously calculated final multiplet intensities for the related 4f systems[11–13], have a lesser number of intense multiplet lines.) The multiplet manifold extends, e.g. to binding energies as high as 15 eV in the case of CmO_2 ($5f^6 \rightarrow 5f^5$). Even allowing for an additional phonon broadening, this multiplet structure should be expected to appear, at least in PuO_2 and AmO_2, with additional peaks on the high binding energy side of the main broad 5f emission. Recent intermediate coupling calculations[16, 79] disagree strongly with those used for the interpretation of the spectra of Fig. 3, mostly because they predict that the multiplets at high energy would have almost no intensity in photoemission. Hence, they describe much better the observed spectra.

Despite this controversial part concerning the final state multiplet description the importance of this pioneering work on highly radioactive actinide oxides must be emphasized: XPS valence band spectra recorded afterwards for NpO_2[116] and PuO_2[73, 116–118] confirm these early results.

Figure 25 shows one of these recent XPS results together with the valence band spectrum of Pu_2O_3. The two spectra are compared with results of a theoretical calculation employing a relativistic extended Hückel method by Lohr et al.[118], applied to a PuO_8^{12-}-cluster, considered to be representative of the dioxide lattice. The electronic diagramme is in good qualitative agreement with the experimental results. Only the satellite at about 9–10 eV is not explained. The slightly enhanced 5f emission in Pu_2O_3 reflects the 5f occupation which is higher in the sesquioxide ($5f^5$) than in the dioxide. (For this experiment, Pu_2O_3 and PuO_2 films were grown by oxidation of a Ga stabilized δ-Pu substrate.)

High resolution UPS valence band spectra for thin films of NpO_2 and PuO_2[116] are shown in Fig. 26; for comparison UO_2 is added. In contrast to XPS results, exhibiting only an intense broad asymmetric structure near to E_F, the UPS spectra separate the oxygen (mainly 2p) valence band from the actinide 5f level (closest to E_F). From the

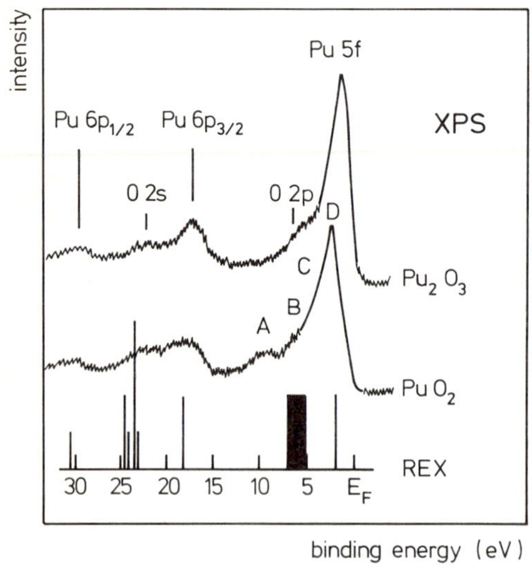

Fig. 25. Valence band of PuO_2 and Pu_2O_3 by XPS (from Ref. 117). (*REX*: Relativistic Extended Hückel method)

usual cross-section consideration, the f character of the latter is identified. In UPS, the predicted multiplet structures[108] should be observed, given the much higher resolution of the method, but are not. On the contrary, the results of the recent calculation of multiplet intensities[16], convoluted with a Lorentzian and Gaussian distribution to account for phonon broadening and the (known) instrumental resolution, reproduce very well the experimental results shown in Fig. 26. The width of the Lorentzian distribution has been chosen so as to reproduce the width of the 5 f main peak in UO_2 (practically, the only emitting $^5F_{5/2}$ component of the spin-orbit split initial state). The same width has been used for the other oxides, a justified procedure since the phonon frequencies are about the same throughout the actinide dioxides series[119].

Valence band spectra of Pu_2O_3 by UPS for different photon excitation energies, taken in our laboratory, show an enhancement of the 5 f peak with increasing excitation energy, identifying clearly the 5 f character of this peak.

Figure 27 shows the valence band emission for different oxidation stages up to PuO_2. The mainly O 2 p derived valence band shows a double structure with a maximum emission around 6.5 eV for Pu_2O_3 and 5.0 eV for PuO_2; the energy shift of the 5 f emission observed already in XPS (Fig. 25) is clearer in UPS (He II, $h\nu$ = 40.8 eV) and amounts to 0.9 eV. The change of the maximum 5 f emission intensity follows quantitatively the occupation of the 5 f level in its initial configuration, as required in an ionic picture.

However, this is not the only factor influencing the emission intensities. In fact, already PuO_2 shows a smaller 5 f emission intensity than NpO_2 (Fig. 26). This can only be explained by different final state multiplet intensity distributions for NpO_2 and PuO_2. This kind of considerations makes a more quantitative analysis of the experimental results difficult.

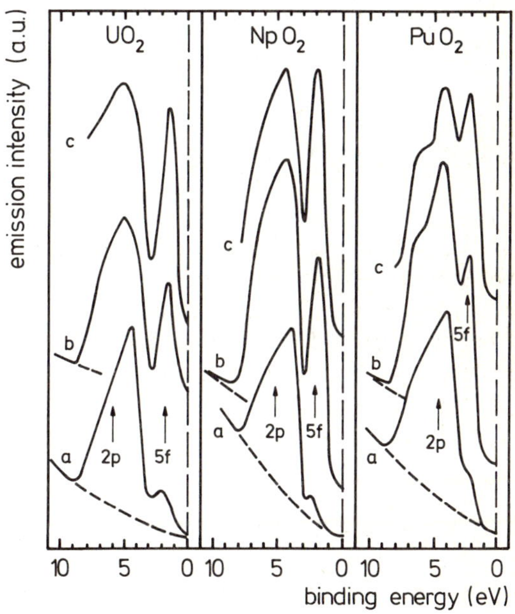

Fig. 26. UPS valence band spectra for UO_2, NpO_2 and PuO_2 *a*) 21.2 eV (He I); *b*) 40.8 eV (He II); *c*) 48.4 eV (He II*) (from Ref. 116)

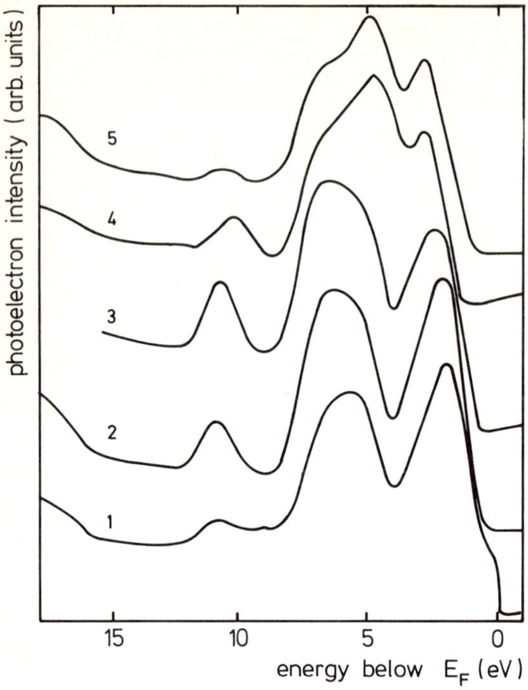

Fig. 27. Valence band spectra of oxidized Pu metal and sintered PuO_2 for He II excitation (40.8 eV), *curve 1*: Pu metal + Pu_2O_3; *curve 2*: Pu_2O_3; *curve 3*: Pu_2O_3 + PuO_2; *curve 4*: PuO_2 (all for oxidized Pu metal); *curve 5*: sintered PuO_2 (Argon sputtered)

The only UPS/XPS valence band spectra for Am_2O_3 were taken in our laboratory and are displayed in Fig. 28. The XPS valence band spectrum is very similar to the one previously reported for AmO_2[15, 120]. The spectra in Fig. 28 are very similar to those for Pu_2O_3 (Figs. 25 and 27). The oxygen 2p valence band and the Am 5f emission around 2.8 eV are again identified from their emission intensity variation with photon excitation energy and are found separated in UPS whereas in XPS the 5f emission dominates completely the spectrum. A comparison with the broadened multiplet intensities for the $5f^6$ initial state of Ref. 16 explains again why also for Am_2O_3 no multiplet components are resolved.

3. *Localization of the 5f States in Actinide Oxides*

The localization of 5f states in actinide oxides is well illustrated by the experimental evidence that has been reported above. From the photoemission results, this is sufficiently proved by

i. the intensity and shape of the main peak (p_1 in Table 3), its clear separation from E_F, its f-orbital character evidenced from the intensity variation when changing the excitation energy;
ii. the fact that the 5f emission follows the $5f^n$ occupation;
iii. the fact that the 5f emission wanders to higher binding energy in dioxides when the atomic number of the actinide is increased.

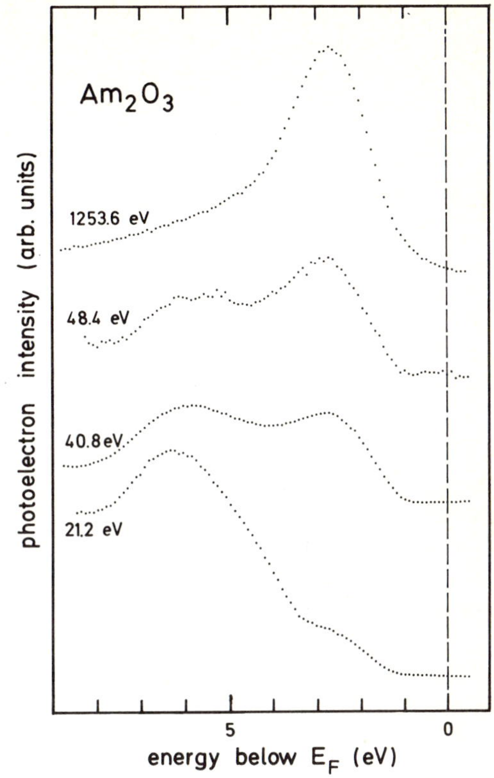

Fig. 28. UPS/XPS valence band spectra of Am_2O_3

A further argument was advanced by Veal et al.[121]. A linear relationship was found to exist between the XPS intensity of the main oxygen valence band and the oxygen-to-uranium ratio of the different uranium oxides investigated (Fig. 23). This was interpreted as indicating that this band consists entirely of 2 p states (perhaps 6 d hybridized) with no 5 f contribution, as expected in a localized 5 f picture. (The 5 f contribution, if present, would have caused deviations from this linear relationship, especially because of the very large photoionization f cross-section.)

Schneider and Laubschat[122] have introduced another argument in favour of the localization of the 5 f state in UO_2 based on a discussion of the final state screening mechanism already discussed in the previous sections. Itinerant 5 f states, due to f-f overlap, may hybridize with the 6 d-states of the actinide (as they do in U metal) and, in compounds, with outer electron states of the other component. The photoionization process, leaving a hole in the open shell core, pulls down quasi-localized empty states from above E_F to the valence band region. Their occupation depends on the degree of hybridization with other itinerant states. The pulled down state if occupied, screens the hole, the screening efficiency being a function of the detailed form of the screening charge density (see Fig. 7). If, however, the degree of hybridization is very small, the virtual level may also not be filled at all, so that no screening occurs. The final state of the photoionization process is in this case a two-hole state (one hole induced by the photo-ionization, one hole being constituted by the empty pulled down virtual state). To the d-

and f- ("poor" and "good") screening represented in Fig. 7, the possibility of a "no screening" (or two-hole) satellite should be added. Such a satellite should occur at higher binding energy than the emission due to screened states.

Schneider and Laubschat attribute to a two-hole satellite the ~10 eV structure which has been observed (see Table 3) by some authors. This structure is at 7 eV higher binding energy than the main 5f peak, in the spectrum of UO_2. It disappears, however, for in situ scraped single crystals. A similar satellite structure appears in PuO_2 at ~9 eV[117].

A process can be assumed:

$$5f^2 - 5f^1 5f^{3^*}$$

where by $5f^{3^*}$ is indicated a 5f empty state (hole) in the new energy position after the excitation process. Schneider and Laubschat correlate the intensity of the satellite as found in the spectra of a series of uranium compounds in Table 4 with U-U distances and the μ_{eff} from magnetic measurements. It is evident that the satellite appears at its maximum where the localization is the greatest and where the magnetic moment is a true atomic moment, as in ionic UO_2 and UF_4. If the interpretation of the satellite is correct, the emission should have an almost pure 5f character. Resonance experiments, however (see Table 3), fail to evidence this character. Therefore, a 2p attribution is preferred for the satellite, and the two-hole interpretation rejected by[81]. Since, in our experience, the satellite is present in UO_2 for Ar^+-sputtered samples, but not in scraped single crystals, and also weakens for bulk sintered PuO_2 after annealing or scraping, a possibility exists that it may simply be due to departures from stoichiometry (oxygen deficient surface oxide) induced by the Ar^+-sputtering.

We turn now to more direct evidence of 5f localization, as provided by more refined experiments, such as the combined XPS/BIS method and ARPES.

Table 4. Nearest interatomic U–U spacings, effective paramagnetic moments, and estimated 7 eV XPS satellite intensities for various binary uranium compounds (from Ref. 122). The references for the intensities and distances are given in[122]

Material	U–U distance (Å)	μ_{eff} (μ_B per formula unit)	Satellite intensity (%)
UF_4	4.52	3.3	10
UO_2	3.86	3.2	10
UGa_2	4.01	3.2	10
UAl_2	3.38	2.9	5
UAs	4.08	3.4	5
UPd_3	4.81	2.6	5
UCu_5	4.96	2.3	5
UAu_3			(5)
UPt	3.61	2.6	w
UPt_2	3.81	2.0	w
UPt_3	4.12	2.6	w
UPt_5	5.25	2.7	w
UNi_5	4.80	–	–
α-U	2.75	–	–

The most important information (by Baer and Schoenes[18)]) obtained when using the combined XPS/BIS method is the Coulomb interaction energy U_H that we have discussed in Part II. For UO_2, $U_H = 4.6 \pm 0.8$ eV has been obtained. This large separation between the two final states $(2(5f^2) \rightarrow 5f^1 + 5f^3)$ is in itself a hint to the localized character of the 5f states in UO_2. Baer and Schoenes compared the value for U_H with theoretical values: they found an agreement with $U_H = 4$ eV as calculated by Herbst et al.[27)] for a U^{+4} metal core. As discussed in Chap. A, intraatomic calculations of U_H in metals possibly underestimate screening by conduction electrons; nevertheless, they should be valid in the case of an insulating solid as UO_2.

A localized $5f^1$ final state, as observed in photoemission, should be spin-orbit split into two components: $^2F_{7/2}$ and $^2F_{5/2}$, the $^2F_{5/2}$ level being the more intense[2]. The $^2F_{7/2}$ weaker line is usually not observed in photoemission (see Table 3), possibly because of different broadening effects, englobing it in the main $^2F_{5/2}$ emission. In a UPS-ARPES study[113)], however, a shoulder was seen at 2.3 eV when taking the second derivative of a photoemission curve for excitation energy hν = 40.8 eV (Table 3). In ARPES, the intensities of all minima (corresponding to well defined peaks in the photoemission curve) and shoulders of the second derivative curve were monitored as a function of electron emission angles. The results (Fig. 29) show the dispersion of all these features along a ΓX direction in the Brillouin zone. It can be seen that not only p_1 (i.e. the $^2F_{5/2}$ emission) but also the shoulder at 2.3 eV (p_2) show practically no dispersion, as expected by localized states. Thus an attribution of p_2 as the second component $^2F_{7/2}$ of the spin-orbit split $5f^1$ state was suggested. If confirmed, this is clear evidence of 5f localization.

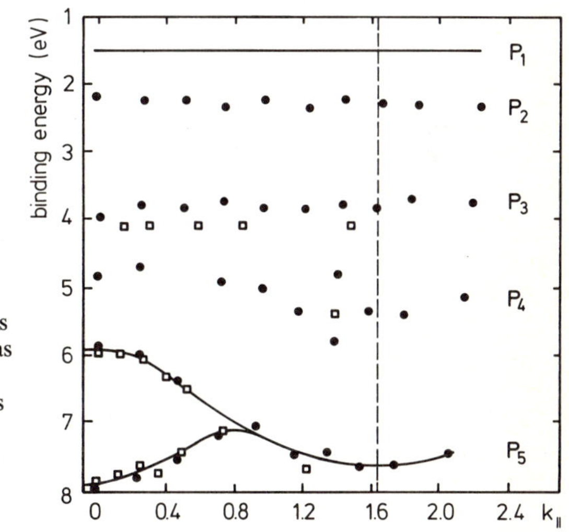

Fig. 29. Dispersion curves of some features of the valence band of UO_2 (see Table 3) as obtained by angular resolved photoemission. The Brillouin zone plane examined is the ΓXΓ′X′ plane. Data are for an excitation energy hν = 40.8 eV (He II); squares are for an excitation energy hν = 21.2 eV (He I) (from Ref. 15)

2 Intensity ratio for intermediate coupling, $I^2F_{7/2}/I^2F_{5/2} = 37/1$

4. *Indications of Hybridization From the Valence Band Spectra and Core Levels in Actinide Oxides*

The possibility of revealing f-p mixing by photoemission in valence band spectra (hence, the possible covalent bonding ensured by 5f electrons in oxides) has been briefly discussed in the introduction of this part.

Some evidence for this point is here briefly reviewed.

a. Analysis of the Valence Band

The best testing for f-p mixing is the analysis of the main valence band for f-character. In[123], Cox has compared the 2p/2s XPS valence band emission intensities in UO_2 and in a non-f oxide, Al_2O_3. Whereas in Al_2O_3 these intensities have roughly the ratio which is expected from cross section calculations ($I_{2p}/I_{2s} = 0.16$), the two intensities are roughly equal in UO_2, a fact that should be explainable only through the very high emission of f-character at X-ray excitation energies. Cox fortifies his suggestion of a relevant 5f character in bonding, by discussing the effect of 5f screening on the 4f core levels response.

A study of the 5f character in the (mainly) 2p valence band has been made by two of the authors[112] through the analysis of the feature of the UPS/XPS valence band results of UO_2 obtained on in situ scraped UO_2 single crystals (Fig. 22). They calculated the ratio of 5f, 6d, 2p intensities from known (atomic) cross sections at different excitation energies, by assuming the number of 5f, 6d, 2p states as given by non self-consistent and self-consistent band calculations[100, 101]. These values are reported in Table 3. The UPS experiment resolves (Fig. 22 and Table 3) two distinct structures at 4.2 and 4.9 eV. The first one is the peaking emission of the (mainly) 2p valence band at the lowest excitation energy (21.2 eV), and therefore can be assigned a predominant 2p character. The second one is the peaking and growing emission at higher UV excitation energy and also in XPS, therefore showing a 5f character (although not as marked as the main 5f peak, which is purely 5f).

Based on the results of Fig. 22, the admixture of f character in the valence band can also be indicated by the relative total emission intensities of the valence band and the 5f state at 1.5 eV in XPS. If one assumes the valence band to be purely 2p like and the 5f state purely 5f like, the emission ratio I_{5f}/I_{VB} as calculated from the XPS uranium 5f and oxygen 2p cross sections[22] for 12 valence band and 2 5f electrons yields $I_{5f}/I_{VB} = 5.5$. The experimental ratio as calculated from the areas of the 5f and valence band emission is on the contrary $I_{5f}/I_{VB} = 1.4 \pm 0.3$. Only in part, this considerable discrepancy between the theoretical and the measured intensity ratio may be induced by the fact that the atomic oxygen 2p cross section is not strictly applicable to band states. However, the assumption of a total number of 1.4 5f electrons contributing to the valence band, as determined in non self-consistent band structure calculations including 5f hybridization gives an intensity ratio of I_{5f}/I_{VB} (10.6p and 1.4f electrons) = 1.2 which is in surprisingly good agreement with the experiment.

The 5f contribution in the valence band is however, probably less than one state, which is equivalent to 8% f character in the valence band. This type of contribution is obtained by later, self-consistent calculations[101]. A contribution of a total number of 0.6

f states to the valence band has also been evaluated, from molecular cluster calculations[102–105]. However, in XPS a predominantly p emission is very sensitive to even a small f contribution: this is illustrated enough by the I_{5f}/I_{VB} ratio evaluated from the measured results.

An attempt of identification of 5f character has been recently reported for UO_2 by resonant photoemission[81] of the valence band region with excitation energies of 92 eV (off-resonance) and 98 eV (on-resonance). The core absorption edge used was the 5d ionization threshold, corresponding to a process:

$$5d^{10}5f^{N} \xrightarrow{h\nu} 5d^{9}5f^{N+1}$$

The main peak at 1.5 eV (p_1 in Table 3) disappears completely for off-resonance, thus confirming its 5f character. However, the structure at 4.8 eV, to which our UPS/XPS studies attribute also some 5f character, shows significant resonance character. Therefore, the attribution from the UPS/XPS study discussed above may be questioned.

In general, the UO_2 valence band spectrum shows very little resonance behaviour in comparison to that of other compounds also studied in the same experiment (UTh_{1-x}, Sb, UTe and even UPd_3, where 5f states are considered to be localized). It is concluded that a 5f contribution to the valence band, if any, must be uniformly distributed over the entire valence band.

In conclusion, photoelectron spectroscopy evidences only a weak 5f character in the valence band of UO_2, substantially in agreement with theory. The alleged much stronger 6d character of the valence band, especially at its bottom, has not yet been demonstrated by this method.

For heavier actinide oxides, the important XPS results of Veal et al.[15], confirmed later by other investigators for PuO_2 and AmO_2 (see Figs. 25, 26 and 27), have prompted a number of theoretical investigations (band as well as cluster molecular ones). As it has been already told, all calculations reproduce the shift to higher binding energy of the main 5f peak, due to localization and actinide contraction. The 5f peak overlaps the valence band in PuO_2. Courteix et al.[117] interpret the XPS spectrum of PuO_2, and especially the two shoulders B and C of Fig. 25 (resolved in UPS, Fig. 26) as indicating strong 5f hybridization with the two mainly 2p valence band Γ_{15} and Γ_{25}. The energy dependent UPS measurements of Fig. 26, which split the emission in three structures, do not seem to support this point. Only the 5f character of the peak closest to E_F is clearly indicated.

Increasing f-p hybridization when the 5f states overlap in energy the main valence band of the oxide cannot of course be excluded, even though there is no doubt (see above) about the increasing localized character of these states with increasing atomic number. In fact, the radial extension of the 5f wavefunctions as well as the matching of symmetry between 2p and 5f states, together with the overlapping in energy, are good conditions for the establishing of ligand-type hybridization, especially in PuO_2 and AmO_2. Cluster molecular calculations for dioxides[105, 124], particularly able to treat this kind of "local" covalent bonding (transferring 5f charge over some interatomic distance), have estimated that, when increasing atomic number, the cation ionic charge should go from + 2.5 in ThO_2 to only + 1.3 in AmO_2, to increase again to + 1.8 in BkO_2 (where the 5f states are supposed to be at higher binding energy than the 2p states). The departure of the cation ionic charge from the formal one is usually considered to be a good sign of covalency (already in UO_2, band calculations indicate a cation charge of +2 [99–101], mostly due to d-p mixing). Hence, in AmO_2 the covalency should be at its maximum.

The problem is that photoemission spectroscopy may meet some difficulty in unravelling the f-p mixing in cases when the broad valence band emission and the 5f emission overlap. The former, in fact, is described well by the ground state one electron density of states; the latter reflects, in general, not the initial state but rather the final state after the excitation. Band as well as cluster calculations yield at present, on the other hand, only ground state properties with sufficient precision.

Therefore, whereas the investigation of f-p mixing has given results in the case of UO_2, where the 5f emission is well separated, one must content oneself at present with qualitative statements (such as those presented for PuO_2) for the other oxides.

b. Core Level Spectra

Common features of the 4f core level spectra of actinide dioxides are the symmetry of the main lines and the appearance of a satellite at about 7 eV in their high binding energy side (Fig. 30). Similar satellites have also been found for UF_4, for which compound the intensity is even higher than for the dioxides. It is perhaps interesting to report some analysis of these features, on the basis of final state models.

The satellites of the 4f lines in dioxides have been first interpreted[18, 116, 117, 125] as due to a shake-up process, in which an electron is pushed into a non-occupied 5f level in the final state.

Weber and Gubanov[103] (see also Refs. 126, 127) have performed cluster molecular orbital calculations for the ground (initial) state and for the final state in a $(UO_8)^{12-}$ cluster, considered to be representative of the UO_2 lattice. They explain the structure as due to an excitation from an O 2p to an U 5f molecular orbital of the cluster in the new final configuration which includes a hole. The excitation should be accompanied by a charge transfer from the oxygen ligands to the uranium atom, i.e. be a change in the p-f mixing (which is found to be important in the bond by cluster calculations).

Cox[123] has commented instead on the $4f_{7/2}$ main line binding energies in UO_2, showing that the chemical shifts from their position in α-U in a series of uranium compounds of increasing ionic character indicate a charge transfer which is much less than that expected in an ionic picture. This implies of course a strong p-f covalency.

Contrary to these two interpretation, emphasizing p-f mixing, Schneider and Laubschat[122] have employed the two-hole picture for the 4f satellites in UO_2 as they did for the 10 eV satellite of the 5f main line in the valence band region, to emphasize localization. The binding energy difference between satellite and main line (about 7 eV) in both cases is an argument for the two-hole model. The interpretation (as we have seen) points to a 6d-5f dehybridization in UO_2, i.e., confirms the strong localization picture (without saying much, however, on f-p mixing).

If the 4f core level spectra of the sesquioxides are taken into account (Figs. 31, 32), an interpretation may be forwarded based on a comparison with the spectra of the metals. As we have seen in Part III, to the main 4f line of Am metal a 6d screening, and to its low binding energy satellite a 5f screening are attributed. For Pu metal, the main line is attributed on the contrary to 5f screening (in consistence with the good itinerant

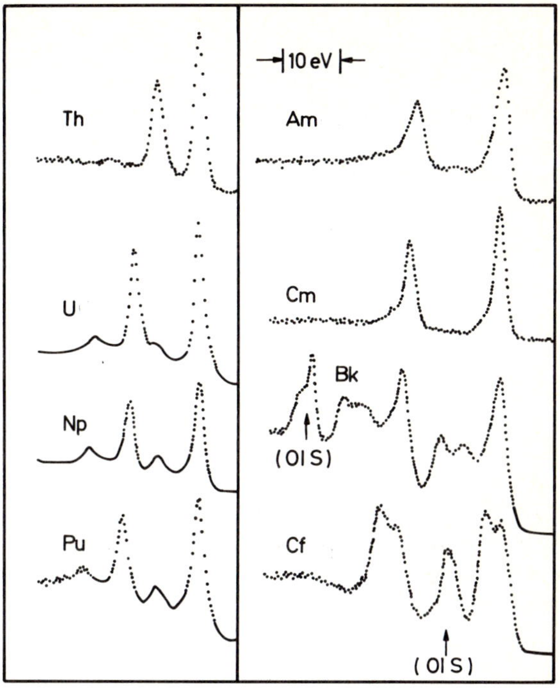

Fig. 30. 4 f core levels spectra of actinide dioxides measured by XPS, from ThO_2 to CfO_2 (from Ref. 15)

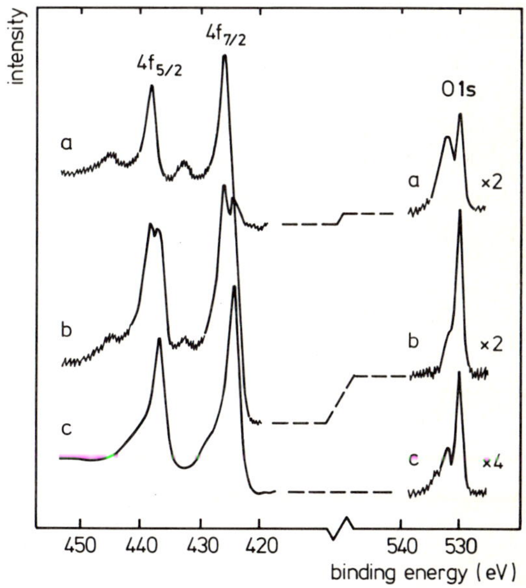

Fig. 31. Evolution of the Pu 4 f and O 1 s XPS-levels in PuO_2 thin films under Ar^+-bombardment; a) PuO_2; b) $PuO_2 + Pu_2O_3$; c) Pu_2O_3 (from Ref. 114)

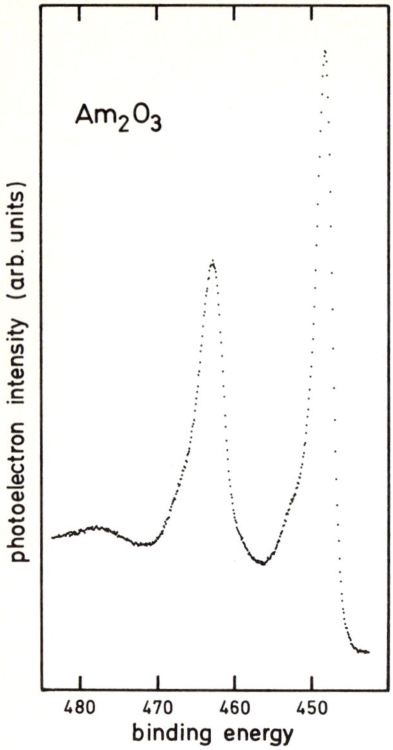

Fig. 32. XPS 4f core levels for Am_2O_3

character of 5f's in this system) and a 6d screening is assigned to the (somewhat dubious) satellite at 438 eV.

In the sesquioxide Am_2O_3 the 4f main lines occur almost at the same energy as in the metal: hence, it is reasonable to assign to them a screening similar to that of Am metal. The higher binding energy satellites may then be tentatively explained as being due to a two-hole state, on the lines of Schneider and Laubschat. However, because of the screening character assumed for the main peak and since the degree of localization of the 5f states when increasing their atomic number becomes higher, as showed by their moving to higher E_b, it is perhaps more likely for the screening level to be a pulled down 6d state rather than a 5f state. In this picture, the critical hybridization governing the occupation of the screening level is perhaps the 2p-6d (which band calculations have shown to be considerable in UO_2[99]). The change in intensity of the satellite can be taken as a measure of this hybridization.

If we apply the same model to the 4f main line and its 7 eV higher binding energy satellite in the dioxide spectra, the growing satellite to main line relative intensity from ThO_2 to PuO_2 would be in this picture a sign of decreasing 6d-2p hybridization. In Table 5 the attributions of the 4f response according to this screening model are listed.

Table 5. Binding energies of structures for XPS measured $4f_{5/2}$ levels (main line and satellite- the latter marked by (s)) for Pu and Am metals and for their oxides ($E_{relative}$ is calculated with respect to the lower binding energy $4f_{5/2}$ response in the metals). In the same table, a screening model is proposed for the different structures; energies are given in eV

Pu				Am			
$E_{rel.}$	Pu^{3+} Pu-met	Pu^{3+} Pu_2O_3	Pu^{4+} PuO_2	Am^{3+} Am-met	Am^{3+} Am_2O_3	$E_{rel.}$	screening mechanism
0	435.1			459.1(s)		0	5 f screening (metal)
(1.9)–3.5	(438.6)(s)	437.4		463.1	462.6	3.6–4.0	6 d screening (metal) sesquioxides
3.5			438.7			–	6 d(5 f?) screening (dioxides)
6.4		441.6(s)			466.1(s)	7.1	no screening or two-hole picture (sesquioxide)
10.4			445.5(s)			–	no screening or two-hole picture (dioxide)

V. Conclusions

In the chapter, we have illustrated some results of photoelectron spectroscopy on two classes of actinide materials, elemental metals and oxides, which we thought particularly relevant as they represent metallic and almost completely ionic bonding. Our interest having been focused on the localization vs. itineracy problem of the 5 f states, as well as on their hybridization with other electron states, we have particularly concentrated on those results which could throw light on these two aspects.

Photoelectron spectroscopy has long be considered as to be able to provide a "photographic" picture of the one-electron density of state of solids. In reality, the spectra of actinide solids (as of other narrow band solids) need very often more than this naive interpretation. In the case of 5 f response, final state effects are found to provide useful information even in the case of metals, as illustrated in this chapter. The general conclusion that the photoelectron spectroscopic response depends on many-electron excited final states as much as it depends on the initial states, when narrow bands are involved, must be emphasized. This points to the necessity both of better final state models and of band calculations giving reliable pictures of conduction bands.

On the basis of the known electronic properties of actinides (which have been discussed elsewhere in this book), theoreticians had distinguished the 5 f itinerant behaviour of light actinide metals from the 5 f localized behaviour of heavy actinide metals from Am on. The crossover, presented often as a Mott transition, had been predicted to occur between Pu and Am metal, due to the localized character of the 5 f state in the latter. Photoemission spectroscopy demonstrates this phenomenon directly with the observation of a 5 f multiplet away from the Fermi level. The detailed description of this peak is certainly complicated, as often happens for response of localized states in photoemission; on the other hand (Fig. 17) the contrast to the emission of Pu metal is convincing.

The 6 d-5 f hybridization of the conduction bands of light actinide metals, which is also predicted by theory, is also demonstrated in photoelectron spectra of Th, U and Pu. In Am metal, on the contrary, the emission at E_F is found to be non-f-like, as expected for a lanthanide-like actinide.

Turning to the mostly ionic oxides, the localized character of the 5 f emission in the valence band spectrum is illuminated by a series of experimental results. Direct and inverse photoemission for UO_2 gives the Coulomb energy U_H, which governs together with the bandwidth the localization vs. itineracy double aspects of the open shells.

It is worthwhile to mention the ample use of screening final states models in understanding core levels as well as valence band spectra of the oxides. The two-hole models, for instance, which have been described here, are certainly of relevance. Interpretational difference exists, for instance, on the attribution of the 10 eV valence band peak (encountered in other actinide dioxides as well), whether due to the non-screened 5 f final state, or to a 2 p-type characteristics of the ligand, or simply to surface stoichiometry effects. Although resonance experiments seem to exclude the first interpretation, it remains a question as to what extent a resonance behaviour other than expected within an atomic picture is exhibited by a 5 f contribution in the valence band region, and to what extent a possible d contribution may modify it. In fact, it has been shown that, for less localized states (as, e.g., the 3 d states in transition metals) the resonant enhancement of the response is less pronounced than expected.

As for the matter of f-p (or d-p) ligand hybridization in oxides, adding a covalent part to the bonding, a coherent and convincing picture has not yet been developed from experimental evidence.

Two final remarks which concern predictable future developments of photoelectron spectroscopy with regard to actinide solids, should be added.

- As we have seen, the most advanced photoelectron techniques, especially those which necessitate the use of synchrotron radiation sources, have been applied until now only to U and Th systems. Application on Pu and Am systems as well as to heavier actinides is to be expected in the future. The same development is likely to occur as for neutron experiments, where more and more these hazardous actinides are investigated at high levels of instrumental sophistication. Difficulties arising from handling and protection problems are, of course, much greater for photoelectron spectroscopy.

  For valence band studies, for instance, energy dependent UPS in the 30–300 eV excitation range is certainly of interest, and would be available in synchrotron radiation sources. This experiment, however, is difficult to realize, since it would involve the construction of safely enclosed systems, in which to handle dangerous α, β (and sometimes γ or neutron) emitters in considerable quantities. This systems must be well separated from the main beam lines. And all of this must be done in the laboratories of a storage ring, which are usually not adapted for radioactive hazards.
- Mixed valence phenomena, such as studied by photoelectron spectroscopy in lanthanide systems, are expected to become important especially (but not only) in the second half of the actinide series. It is to be expected that much of the photoelectron spectroscopic effort will be in the future devoted to the study of these phenomena in actinides, especially as soon as measurements on hazardous actinides will become more feasible.

Acknowledgments. One of us (J. R. N.) thanks Professor G. A. Somorjai (University of California, Berkeley) for his hospitality; a large part of this work was performed during the stay at Berkeley. We thank Drs. M. S. S. Brooks, C. Colmenares, and B. W. Veal for many helpful discussions.

VI. References

1. Smith, J. L., Kmetko, E. A.: J. Less-Common Metals *90,* 83 (1983)
2. Baptist, R.: Proceedings of the 12ème Journées des Actinides, Orsay (France) 1982
3. Baptist, R. et al.: J. Physique (Paris) *44,* 241 (1983)
4. Fadley, C. S.: Basic Concepts of X-Ray Photoelectron Spectroscopy, in "Electron Spectroscopy: Theory, Techniques and Applications" (Brundle, C. R., Baker, A. D., eds.), Academic Press, New York, Vol. II, p. 2 (1978)
5. Photoemission in Solids I, General Principles (Cardona, M., Ley, L., eds.), Topics in Applied Physics, Vol. 26, Springer Verlag, Berlin 1978
6. Photoemission in Solids II, Case studies (Ley, L., Cardona, M., eds.), Topics in Applied Physics, Vol. 27, Springer Verlag, Berlin 1979
7. Handbook of X-Ray and Ultraviolet Photoelectron Spectroscopy (Briggs, D., ed.), Hayden and Sons Ltd., London 1977
8. Photoemission and the Electronic Properties of Surfaces (Feuerbacher, B., Fitton, B., Willis, R. F., eds.), John Wiley and Sons, Chichester 1978
9. Jørgensen, C. K., Berthou, H.: Mat. Fys. Medd. Dansk. Viddensk. Selskab. *38,* 15 (1972)

10. Lang, J. K., Baer, Y.: Solid State Commun. *31,* 945 (1979)
11. Cox, P. A.: Structure and Bonding *24,* 59 (1975)
12. Cox, P. A., Lang, J. K., Baer, Y.: J. Phys. *F11,* 113 (1981)
13. Gerken, F.: J. Phys. *F13,* 703 (1983)
14. Jørgensen, C. K.: Structure and Bonding *22,* 49 (1975).
15. Veal, B. W. et al.: Phys. Rev. *B15,* 2929 (1977)
16. Gerken, F., Schmidt-May, J.: J. Phys. *F13,* 1571 (1983)
17. Lang, J. K., Baer, Y., Cox, P. A.: Phys. Rev. Letters *42,* 74 (1979)
18. Baer, Y., Schoenes, J.: Solid State Commun. *33,* 885 (1980)
19. Gadzuk, J. W.: Chapter V of Ref. 8
20. Kotani, A., Toyozawa, Y.: Japn. J. Phys. *37,* 912 (1974)
21. Cohen-Tannoudji, C., Diu, B., Laloë, F.: Mécanique Quantique, Vol. 2, Hermann, Paris 1977
22. Scofield, J. H.: J. Electron Spectrosc. *8,* 129 (1976)
23. Eastman, D. E., Kuznietz, M.: Phys. Rev. Letters *26,* 846 (1971)
24. Baptist, R. et al.: Physica *102 B,* 63 (1980)
25. Baer, Y.: in Actinides in Perspective (Edelstein, N. M., ed.), Pergamon Press, Oxford, p. 81 (1982)
26. Treglia, G., Ducastelle, F., Spanjaard, D.: J. Phys. (Paris) *41,* 281 (1980)
27. Herbst, J. F., Watson, R. E., Lindgren, I.: Phys. Rev. *B14,* 3265 (1976)
28. Johansson, B., Skriver, H. L.: J. Magn. Magn. Mat. *29,* 217 (1982)
29. Johansson, B.: to appear in Phys. Rev. B
30. Johansson, B. et al.: Physica *102 B,* 12 (1980)
31. Mårtensson, N., Johansson, B.: Phys. Rev. Letters *45,* 482 (1980)
32. Guillot, G., et al.: ibid. *39,* 1632 (1977)
33. Hillebrecht, F. U. et al.: Phys. Rev. *B27,* 2179 (1983)
34. Fuggle, J. C. et al.: Phys. Rev. Letters *45,* 1597 (1980)
35. Nagakura, J., Ishii, T., Sagawa, T.: J. Phys. Soc. Jpn. *33,* 754 (1972)
36. Fuggle, J. C. et al.: Phys. Rev. *B27,* 7330 (1983)
37. Crecelius, G., Wertheim, G. K., Buchanan, D. N. E.: ibid. *B18,* 6519 (1978)
38. Hillebrecht, F. U., Fuggle, J. C.: ibid. *B25,* 3550 (1982)
39. Kotani, A., Toyozawa, Y.: J. Phys. Soc. Jpn. *35,* 1073 (1973)
40. Johansson, B.: Phys. Rev. *B11,* 2740 (1975)
41. Fuggle, J. C. et al.: ibid. *B27,* 2145 (1983)
42. Lundqvist, B. I.: Phys. Stat. Solidi *32,* 273 (1969)
43. Seah, M. P., Dench, W. A.: Surf. Interf. Anal. *1,* 2 (1980)
44. Steiner, P., Höchst, H., Hüfner, S.: Z. Phys. *B30,* 129 (1978)
45. Citrin, P. H., Wertheim, G. K., Baer, Y.: Phys. Rev. Letters *41,* 1425 (1978)
46. Duc, T. M. et al.: ibid. *43,* 789 (1979)
47. van der Veen, J. F. et al.: Solid State Commun. *37,* 555 (1981)
48. Wertheim, G. K., and Campagna, M.: Chem. Phys. Letters *47,* 182 (1977)
49. Allen, J. W. et al.: Phys. Rev. *B21,* 1335 (1980)
50. Tegart, W. J. McG.: The Electrolytic and Chemical Polishing of Metals in Research and Industry, Pergamon Press, London 1956
51. Musket, R. G. et al.: Appl. Surf. Sci. *10,* 143 (1982)
52. Naegele, J. R. et al.: ibid. *4,* 510 (1980)
53. Thiry, P. A. et al.: Phys. Rev. *B29,* 4824 (1984)
54. Urch, D. S., Weber, M.: J. Electron Spectrosc. *5,* 791 (1974)
55. Naegele, J. R.: J. Phys. (Paris) *45,* C2–841 (1984)
56. Baer, Y., Lang, J. K.: Phys. Rev. *B21,* 2060 (1980)
57. Skriver, H. L., Jan, J. P.: ibid. *B21,* 1489 (1980)
58. Schneider, W. D., Laubschat, C.: ibid. *B23,* 997 (1981)
59. Norton, P. R. et al.: ibid. *B21,* 2572 (1980)
60. Fuggle, J. C. et al.: J. Phys. *F4,* 335 (1974)
61. Veal, B. W., Lam, D. J.: Phys. Rev. *B10,* 4902 (1974)
62. Verbist, J. et al.: J. Electron Spectrosc. *5,* 193 (1974)
63. Nornes, S. B., Meisenheimer, R. G.: Surf. Sci. *88,* 191 (1979)

64. Grohs, H. et al.: Solid State Commun. *33,* 573 (1980)
65. Allen, G. C., Trickle, J. R., Tucker, P. M.: Phil. Mag. *B43,* 689 (1981)
66. Greuter, F. et al.: Physica *102 B,* 117 (1980)
67. Iwan, M., Koch, E. E., Himpsel, F. J.: Phys. Rev. *B24,* 613 (1981)
68. Freeman, A. J., Koelling, D. D., Watson Yang, T. J.: J. Phys. (Paris) *40,* C4-134 (1979)
69. Skriver, H. L.: quoted in Ref. 56
70. The Actinides: Electronic Structure and Related Properties (Freeman, A. J., Darby Jr., I. B., eds.), Academic Press, New York 1974
71. Brodsky, M. B. et al.: in Ref. 70, vol. 2, p. 185
72. Dallacasa, V. J.: J. Phys. *F11,* 177 (1981)
73. Larson, D. T.: J. Vac., Sci. Technol. *17,* 55 (1980)
74. Cox, L. E., Ward, J. W.: Inorg. Nucl. Chem. Letters *17,* 265 (1981)
75. Baptist, R. et al.: J. Phys. *F12,* 2103 (1982)
76. Naegele, J. R. et al.: in Proceedings of the 12ème Journées des Actinides, Orsay (France) 1982
77. Skriver, H. L., Andersen, O. K., Johansson, B.: Phys. Rev. Letters *41,* 42 (1978)
78. Skriver, H. L., Andersen, O. K., Johansson, B.: ibid. *44,* 1230 (1980)
79. Beatham, N. et al.: Chem. Phys. Letters *63,* 69 (1979)
80. Lawrence, J. M. et al.: Phys. Rev. *B26,* 2362 (1982)
81. Reihl, B. et al.: ibid. *B26,* 1842 (1982)
82. Allen, J. W. et al.: Phys. Rev. Letters *46,* 1100 (1981)
83. Peterman, D. J. et al.: Phys. Rev. *B27,* 808 (1983)
84. Zangwill, A., Soven, P.: Phys. Rev. Letters *45,* 204 (1980)
85. Freeman, A. J., Koelling, D. D.: in Ref. 70, Vol. I, p. 51
86. Naegele, J. R. et al.: Phys. Rev. Letters *52,* 1834 (1984)
87. Herbst, J. F.: Physica Scripta *21,* 553 (1980)
88. Wertheim, G. K., Crecelius, G.: Phys. Rev. Letters *40,* 813 (1978)
89. Brooks, M. S. S.: private communication
90. Brodsky, M. B.: Rep. Progr. Phys. *41,* 1547 (1978)
91. Gordon, J. E. et al.: Proc. R. Soc., London *A351,* 179 (1976)
92. Stewart, G. R., Eliott, R. O.: Actinides 81, LBL report 12441, 206 (1981)
93. McLean, W. et al.: Phys. Rev. *B25,* 8 (1982)
94. Doniach, S., Šunjić, M.: J. Phys. *C3,* 284 (1970)
95. Schönhammer, K., Gunnarsson, O.: Z. Phys. *B30,* 297 (1978)
96. Larson, D. T., Haschke, J. M.: Inorg. Chem. *20,* 1945 (1981)
97. Wertheim, G. K., Campagna, M.: Solid State Commun. *26,* 553 (1978)
98. Johansson, B., Mårtensson, N.: Phys. Rev. *B24,* 4484 (1981)
99. Brooks, M. S. S., Kelly, P. J.: Solid State Comm. *45,* 689 (1983)
100. Kelly, P. J.: PhD-Thesis, Portsmouth Polytechnic 1980
101. Kelly, P. J., Brooks, M. S. S.: Physica *102 B,* 81 (1980)
102. Gubanov, V. A., Rosen, A., Ellis, D. E.: Solid State Commun. *22,* 219 (1977)
103. Weber, J., Gubanov, V. A.: J. Inorg. Nucl. Chem. *41,* 693 (1979)
104. Gubanov, V. A., Rosen, A., Ellis, D. E.: ibid. *41,* 975 (1979)
105. Ellis, D. E., Gubanov, V. A., Rosen, D. E.: J. Phys. (Paris) *40,* C4-187 (1979)
106. Blank, H.: in: Thermodynamics of Nuclear Materials, 1974, *Vol. II,* IAEA-Vienna, 45 (1975)
107. Manes, L.: A new method of statistical thermodynamics and its application to oxides of the lanthanide and actinide series, in "Nonstoichiometric oxides" (Sørensen, O. T. ed.), Academic Press, New York, Chapt. 3, 99 (1981)
108. Veal, B. W., Lam, D. J.: in Gmelin Handbook of Inorganic Chemistry, System Nr. 55, Uranium, Suppl., Vol. *A5,* Springer Verlag, Berlin, p. 176 (1982)
109. Manes, L., Naegele, J. R.: in: Plutonium 1975 and Other Actinides (Blank, H., Lindner, R., eds.), North-Holland, Amsterdam, p. 361 (1976)
110. Schoenes, J.: Phys. Rep. *63,* 301 (1980)
111. Teterin, Yu. A. et al.: Phys. Chem. Minerals *7,* 151 (1981)
112. Naegele, J. R., Manes, L.: private communication
113. Veal, B. W., Arko, A. J., Weber, L. W.: Actinides 1981, LBL-report 12441, p. 66 (1981)
114. Chauvet, G., Baptist, R.: Solid State Commun. *43,* 793 (1982)
115. Davis, L. C., Feldkamp, L. A.: Phys. Rev. *B23,* 6239 (1981)

116. Naegele, J. R., Manes, L.: Actinides 1981, LBL-report 12441, p. 69 (1981)
117. Courteix, D. et al.: Solid State Commun. *39,* 209 (1981)
118. Lohr, L. L., Hotokka, M., Pyykkö, P.: Quantum Chemistry Program Exchange, QCPE, *12,* 387 (1980)
119. Manes, L., Buijs, K., Schippa, B.: in: Transplutonium 1975, North-Holland, Amsterdam, 257 (1976)
120. Veal, B. W. et al.: Proc. 2nd Int. Conf. Electr. Struct. Actinides, Wroclaw 1977
121. Veal, B. W. et al.: in: Plutonium 1975 and Other Actinides (Blank, H., Lindner, R. eds.), North-Holland, Amsterdam, 383 (1976)
122. Schneider, W. D., Laubschat, C.: Phys. Rev. Letters *46,* 1023 (1981)
123. Cox, L. E.: J. Electron Spectrosc. *26,* 167 (1982)
124. Gubanov, V. A., Rosén, A., Ellis, D. E.: J. Phys. Chem. Solids *40,* 17 (1979)
125. Pireaux, J. J. et al.: Chem. Phys. *22,* 113 (1977)
126. Larsson, S.: Physica Scripta *16,* 378 (1977)
127. Bancroft, G. M. et al.: Phys. Chem. Letters *51,* 105 (1977)

Chapter F

The Theory of 5 f Bonding in Actinide Solids

M. S. S. Brooks

Commission of the European Communities, Joint Research Centre, Karlsruhe Establishment, European Institute for Transuranium Elements, Postfach 2266, D-7500 Karlsruhe, F.R.G.

The properties of the actinides in the first half of the series are deduced from the characteristics of 5 f wavefunctions, in many ways similar to the d-wavefunctions in transition metals.

Relativistic effects and the calculation of ground state properties are stressed.

I. Special Properties of Transition Metals

1. Introduction

Figure 1 shows the potential and the 3 d and 4 d wave functions of an isolated zirconium atom. Also shown is the wave function for a state with positive energy and angular momentum equal to 2. The 3 d state is well bound, has no nodes, and is spatially restricted. The 4 d state is loosely bound, has one node and is more extended. The state with positive energy (a scattering state) is, of course, not bound and extends to infinity. If we avoid the problem of normalizing a scattering state by using a constant, A, its wave function has the asymtotic form

$$\phi(r) \underset{r\to\infty}{\sim} A \sin(kr - l\pi_\alpha + \delta_l) \tag{1}$$

at large distances. Here r is the wave vector. δ_l is the l-wave phase shift, representing *completely* the effect of the potential at large distances. It is now possible to imagine a situation in which the zirconium atom is inserted into a simple metal matrix, for example aluminium. In Fig. 2 is shown how the potential and wave functions of Fig. 1 are modified in the solid. The energy E_0 is the 'muffing tin zero' of the aluminium matrix onto which the potential of the zirconium atom is smoothly joined. It is also approximately the minimum of the conduction band of Al since it is the energy at which free electrons begin to propagate. The wave functions of the zirconium atom are found by integrating the wave equation outwards to S (the radius of the atom) and joining them onto the solutions for Al. The nature of these solutions in the interstitial region depends upon whether or not $E \gtrless E_0$. If $E > E_0$ the solutions, for a given angular momentum l, are linear combinations of spherical Bessel and Neuman functions

$$\phi_l(kr) = \cos\delta_l j_l(kr) + \sin\delta_l \eta_l(kr) \tag{2}$$

where δ_l is the l-wave phase shift.

However, for $E < E_0$ they are exponentially decaying if there is no atomic Al bound state in the neighbourhood of E.

The 3 d wave function is unchanged although its energy may be changed slightly due to a constant shift in potential. In order to understand the change in the nature of the 4 d wave function it is necessary to observe the l-dependent effective potential (Fig. 1) which differs from the potential felt by s-states by a term $l(l+1)/r^2$ Ryd. – the centrifugal term entering the wave equation[1)]. In the free atom both 4 s and 4 d wave functions are bound but the centrifugal potential is responsible for a decaying 4 d wave function at both small and large distances. The 4 d charge density is therefore concentrated in a narrow region of space inside the atom. In the Al matrix the 4 d state is not bound because its energy is greater than E_0 but its wave function decays in the region where $E_{4d} < E_0 + l(l+1)/r^2$ Ryd. Consequently it appears very much like the atomic 4 d wave function but has a small tail at large distance since it can tunnel through the centrifugal potential barrier. Although the probability of finding a 4 d electron inside Zr has been reduced it is nevertheless very much greater than the probability of finding a 4 s electron inside Zr.

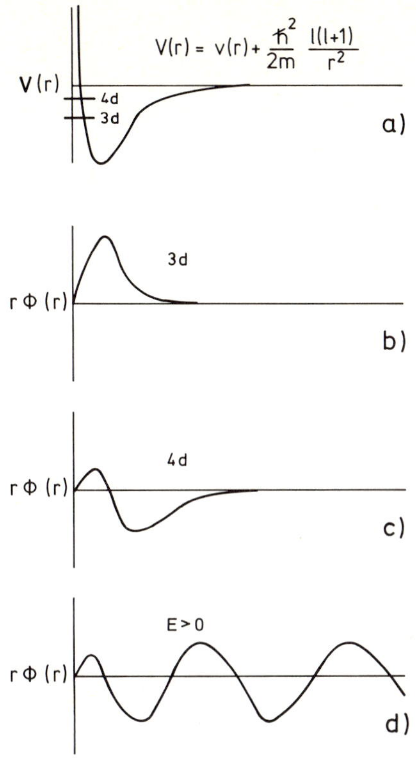

Fig. 1

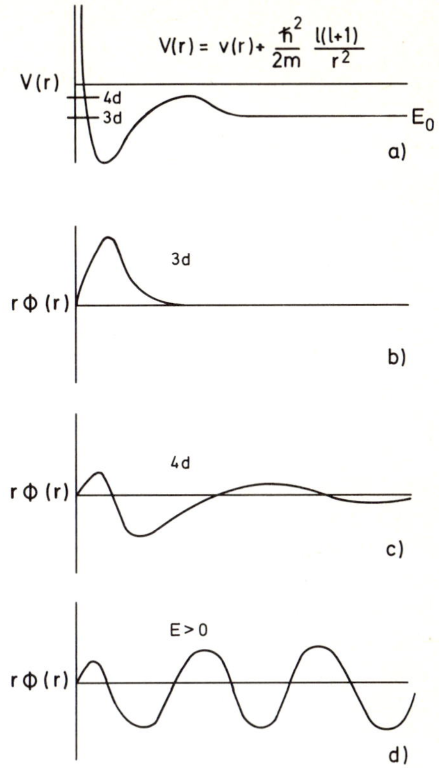

Fig. 2

Fig. 1. (a) The total potential of a free zirconium atom, which is the sum of potential and centrifugal parts. **(b)** The $l = 2$ radial wavefunction (multiplied by r^2) for an energy equal to the 3 d energy. **(c)** The $l = 2$ radial wavefunction (multiplied by r^2) for an energy equal to the 4 d energy. **(d)** The $l = 2$ radial wavefunction (multiplied by r^2) for an energy greater than zero

Fig. 2. (a) The total potential of a zirconium atom in an Al host. **(b)** The $l = 2$ radial wavefunction (multiplied by r^2) for an energy equal to the 3 d energy. **(c)** The $l = 2$ radial wavefunction (multiplied by r^2) for an energy equal to the 4 d energy. **(d)** The $l = 2$ radial wavefunction (multiplied by r^2) for an energy greater than zero

Such a state is called a resonant d-state. The phase shift of the 4 d electron with energy E_{4d} is given by

$$\tan \delta_d = \Gamma_d/2\,(E - E_{4d}) \tag{3}$$

where Γ_d is the resonance width. Thus δ_d changes by π as the energy passes through the resonance energy and the magnitude of the oscillatory tail is related to Γ. All transition metals have broad sp-conduction or nearly free electron bands with d-resonances which broaden into narrow bands pinned to the Fermi level. It was suggested some time ago by Friedel[2)] that the *light* actinide metals form another transition metal series, but with resonant f-electron states. Recent advances in the study of actinide metals have demonstrated that this hypothesis is correct[3–5)].

The trends in several ground state properties of transition metals have been shown in Figs. 2, 3 and 15 of Chap. A and Fig. 7 of Chap. C. The parabolic trend in the atomic volume for the 3–6 periods of the periodic table plus the actinides is shown in Fig. 3 of Chap. A. We note that the trend for the actinides is regular only as far as plutonium and that it is also broken by several 3 d metals, all of which are magnetic. Similar anomalies for the actinides would probably be found in Fig. 15 of Chap. A – the bulk modulus – and Fig. 7 of Chap. C – the cohesive energy if more measurements had been made for the heavy actinides.

The three basic ground state properties of the heavy actinides are more likely to follow those of the rare earths (Fig. 2 of Chap. A). The atomic volumes of the rare earth metals decrease monotonically with atomic number. This suggests, as will be explained more fully below, that the 4 f electrons make little or no contribution to cohesion. They are said to be on the low density side of a Mott transition – with the notable exception of one of the phases of cerium[6]. This is believed to be also the case for the second half of the actinide series[5,7].

It was noted earlier that the charge density of a narrow resonance band lies within the atoms rather than in the interstitial regions of the crystal in contrast to the main conduction electron density. In this sense it is sometimes said to be localized. However, the charge density from each state in the band is divided among many atoms and it is only when all states up to the Fermi level have contributed that the correct average number of electrons per atom is produced. In a rare earth such as terbium the 8 4 f electrons are essentially in atomic 4 f states and the number of 4 f electrons per atom is fixed without reference to the Fermi level. In this case the f-states are also said to be localized but in a very different sense. Unfortunately the two senses are often confused in literature on the actinides and, in order not to do so here, we shall refer to 'resonant' states and 'Mott-localized' states specifically.

2. *The Actinides as Transition Metals*

In 1956 Friedel[2] suggested that uranium was a transition metal with an f-resonance at the Fermi level. However, until a much later date, a chemical viewpoint based upon erroneous arguments concerning metallic radii[8] was pre-eminent. In this scheme the 5 f-electrons did not contribute to chemical bonding but were thought to be analogous to the 4 f-electrons in rare earths. More recently a rapid theoretical development[4,5] has produced a convincing theory of the light actinides (Ac-Pu) in which the f-states form a narrow band.

It is possible to characterize f-electron states in the actinides in quite a simple manner and to compare them with the states of other transition metal series. To this, we employ some simple concepts from energy band theory. Firstly, it is possible to express the real bandwidth in a simple close-packed metal as the product of two parts[9]. One factor depends only upon the angular momentum character of the band and the structure of the solid but not upon its scale. Therefore, since we shall use the fcc structure throughout, the scaling factor X is known once and for all.

For d-electrons $X_d = 23.4$ For f-electrons $X_f = 30.5$.

The second factor does depend upon the particular atomic potential and must be computed. This factor is the band mass μ_l and it is proportional to the reciprocal of the radius squared at the Wigner-Seitz sphere

$$\mu_l = 2/S^3\phi^2 \qquad (4)$$

where ϕ is the radial wave function evaluated at the Wigner-Seitz radius, S. The width of a band with a constant (rectangular) density of states is then

$$W_l = (X_l/\mu_l S^2) \qquad (5)$$

in Rydbergs where the band masses and Wigner-Seitz radii are tabulated for the transition metals, rare earths and actinides in Table 1. The masses for the transition metals were first tabulated by Andersen[9)] whilst the masses for rare earths and actinides were computed by the Author[10)] using overlapped crystal potentials and the $X_\alpha = 2/3$ approximation to exchange[11)].

Also shown in the Table are the relative positions, C_l, in energy of the different band centres (when hybridization is neglected) and the Wigner-Seitz radii in an assumed fcc structure. From Eq. (5) and Table 1 one estimates for iron, nickel, uranium, cerium, gadolinium and americium

Table 1a. Potential parameters for the d-transition series taken from Andersen and Jepsen (Ref. 9). C_l is the band centre and μ_l is the band mass. *S* is the Wigner-Seitz radius in atomic units

$3s^23p^6$	Sc	Ti	V	Cr	Mn	Fe	Co	Ni
S(A.U.)	3.427	3.052	2.818	2.684	2.699	2.662	2.621	2.602
$(C_p-C_s)S^2$	7.42	7.36	7.31	7.20	7.14	7.01	6.96	6.91
$(C_d-C_s)S^2$	3.18	2.28	1.67	1.17	1.29	1.11	0.92	0.75
μ_s	0.74	0.71	0.70	0.70	0.73	0.75	0.77	0.80
μ_p	0.85	0.83	0.82	0.83	0.84	0.88	0.90	0.92
μ_d	5.60	5.96	6.28	7.14	8.27	9.87	11.3	13.2
$4s^23d^{10}4p^6$	Y	Zr	Nb	Mo	Tc	Ru	Rh	Pd
S(A.U.)	3.761	3.347	3.071	2.922	2.840	2.791	2.809	2.873
$(C_p-C_s)S^2$	8.52	8.61	8.63	8.53	8.37	8.23	8.04	7.81
$(C_d-C_s)S^2$	4.07	2.79	1.84	1.21	0.74	0.33	−0.04	−0.09
μ_s	0.73	0.69	0.67	0.66	0.67	0.69	0.72	0.76
μ_p	0.81	0.77	0.76	0.76	0.77	0.79	0.81	0.85
μ_d	3.62	3.60	3.68	3.96	4.44	5.13	6.21	7.57
$5s^24d^{10}5p^6$	La	Hf	Ta	W	Re	Os	Ir	Pt
S(A.U.)	3.624	3.301	3.069	2.945	2.872	2.825	2.835	2.897
$(C_p-C_s)S^2$	9.27	9.47	9.61	9.63	9.57	9.37	9.26	9.09
$(C_d-C_s)S^2$	5.39	4.36	3.55	2.95	2.42	1.97	1.51	1.36
μ_s	0.78	0.73	0.71	0.70	0.71	0.73	0.76	0.81
μ_p	0.80	0.77	0.76	0.75	0.75	0.79	0.79	0.82
μ_d	3.29	3.24	3.22	3.40	3.71	4.10	4.82	5.74

Table 1b. Potential parameters for selected rare earths (hcp structure assumed) and actinides (fcc structure assumed) at fixed lattice parameters equivalent to an fcc lattice parameters of 5.0 Å and 4.8 Å, respectively

$4d^{10}5s^25p^6$	^{58}Ce	^{59}Pr	^{60}Nd	^{62}Sm	^{64}Gd	^{65}Tb	^{66}Dy	^{68}Er
S(A.U.)	3.69	3.69		3.69	3.69	3.69		
$(C_d–C_s)S^2$	2.93	3.25		4.05	4.60	5.51		
$(C_f–C_s)S^2$	−1.91	1.34		0.12	0.35	−0.69		
μ_s	0.62	0.63		0.67	0.69	0.72		
μ_d	2.30	2.40		2.60	2.72	2.82		
μ_f	43.0	56.40		108.0	152.6	205.6		

$5d^{10}6s^26p^6$	^{90}Th $s^2d^2f^0$	^{91}Pa $s^2d^2f^1$	^{92}U $s^2d^2f^2$ (s^2df^3)	^{93}Np $f^3s^2d^2$	^{94}Pu f^5s^2d	^{95}Am $f^6s^2d^1$	^{96}Cm f^7s^2d	^{97}Bk f^8s^2d
S(A.U.)	3.55	3.55	3.55	3.55	3.55	3.55	3.55	3.55
$(C_d–C_s)S^2$	3.89	4.15	4.39 (4.67)	4.93	5.18	5.42	5.70	5.94
$(C_f–C_s)S^2$	4.24	3.77	2.56 (4.62)	4.01	3.43	2.79	2.23	1.61
μ_s	0.60	0.62	0.63 (0.61)	0.63	0.64	0.66	0.67	0.69
μ_d	1.76	1.85	1.94 (1.82)	1.89	1.96	2.02	2.08	2.12
μ_f	12.00	15.98	21.00 (15.04)	19.02	23.78	29.80	36.76	45.42

iron W_d = 4.8 eV; nickel W_d = 3.7 eV;
uranium W_f = 2.3 eV; americium W_f = 1.1 eV;
gadolinium W_f = 0.2 eV; cerium W_f = 0.7 eV.

Immediately notable is that nickel and uranium have similar bandwidths as do americium and cerium whereas gadolinium would form a much narrower band. Some of the trends that are summarized in Table 1 are:

1. In the transition series the bands narrow and lower along any given series,
2. For a given angular momentum (d or f) the bands broaden as the principle quantum number increases. Thus 5d bands are much broader than the 3d bands. The reason for this is that states with principle quantum number n + 1 must have an additional node (the orthogonality node) to the states with principle quantum number n. This pushes the wave function, relatively, further from the nucleus. Hence the 5f wave functions are more extended than 4f wave functions which leads to their tendency to form bands.
3. In the d-transition series the conduction bands at the Fermi level are predominantly narrow d-bands superimposed upon a broad s–p band structure. In the actinides narrow f-bands at the Fermi level are superimposed upon an spd band structure, the d-bands now being broad conduction band states. The energy bands of the rare earths and heavy actinides are so narrow that it is energetically favourable for Mott-localized states to form. The reason is that polar states are present in bands and this costs

Coulomb correlation energy, U. The energy gain due to band formation is of the order of the bandwidth, W. Provided U can be calculated Table 1 can be used to examine the criterion U/W = *1* for Mott-localization[12)].

3. *Relativistic Effects*

Until now we have described electrons states as obeying the wave equation without specifying which wave equation. In general it is necessary to solve the Dirac equation[13)] even for a local potential. If the local potential is spherically symmetric the Dirac equation reduces to[13)]

$$g_{\varkappa'} = \frac{\varkappa + 1}{r} g_\varkappa + cf_\varkappa \left(1 + \frac{E - V}{c^2}\right) \tag{5.a}$$

$$cf_{\varkappa'} = (V - E) g_\varkappa + \frac{\varkappa - 1}{r} cf_\varkappa \tag{5.b}$$

where $g_\varkappa$, $f_\varkappa$ are the major and minor components of

$$\phi_{\varkappa\mu} = \begin{pmatrix} g_\varkappa \, \chi_{\varkappa\mu} \\ -if_\varkappa \, \sigma_r \, \chi_{\varkappa\mu} \end{pmatrix} \tag{6}$$

itself the solution of the Dirac equation. $\varkappa$ is the Dirac quantum number, $\chi_{\varkappa\mu}$ is a two component spinor and σ_r the radial component of the Pauli spin matrices. There are excellent reviews of the properties and solutions of the Dirac equation in a spherical symmetry available[14)], and we shall restrict discussion to a few relevant remarks[15)] for hydrogenic solutions in a pure Coulomb field.

(a) the relativistic wave functions are contracted towards the origin compared with non-relativistic wave functions. This effect is greatest for low angular momentum and falls off as the angular momentum increases.

(b) Zeros in the charge density do not occur since the nodes of the major and minor components overlap and the charge density is given by $\varrho(r) = g(r)^2 + f(r)^2$.

(c) the ratio of minor to major component is $\sim \alpha Z|2n$ where α is the fine structure constant, Z the nuclear charge and n the principal quantum number. Thus for actinides, Z = 90, $f/g \sim 0.3/n$ which is not negligible, especially for deep lying core states.

The pure Coulomb solutions only give a guide to the general behaviour for large atoms, however, since in all other cases than hydrogen the potential includes the field due to all other electrons[16)]

$$v(r) = -\frac{Ze^2}{r} + \int \varrho(r) v(r) dr + \mu_{xc}(r) \tag{7}$$

where $\mu_{xc}(r)$ is the exchange and correlation contribution which will be discussed later. Thus (5) must be solved self-consistently together with (6). Relativistic effects are perhaps best seen in this case by examining self-consistent atomic solutions for a particular heavy atom with, in one case, the velocity of light set equal to 10^{14} cm/s (non-relativistic limit). The outer energy levels for plutonium are shown in Fig. 3. The effects of relativity are

(a) to shift the 5f level up more than 0.8 Ry compared to the 7s (or vice versa) and 0.2 Ry compared to the 6d.

(b) to shift the 5f up 0.6 Ry compared to the 6d.

The origin of these very large changes may be traced to (a)–(c) above. Since penetrating orbitals are contracted most severely, the entire core shrinks; this increases the screening of the nuclear change for the f-electrons and their charge densities actually expand and energies increase. This is the indirect relativistic effect[17)], arising from the self-consistent field, which is actually the largest effect upon the valence electrons. The result is that, for actinides, the changes in energy levels are so large that non-relativistic calculations cannot yield even qualitatively correct results. This may be seen by studying Fig. 3 where it is clear that the low lying f-level would fill before the d and s levels (even

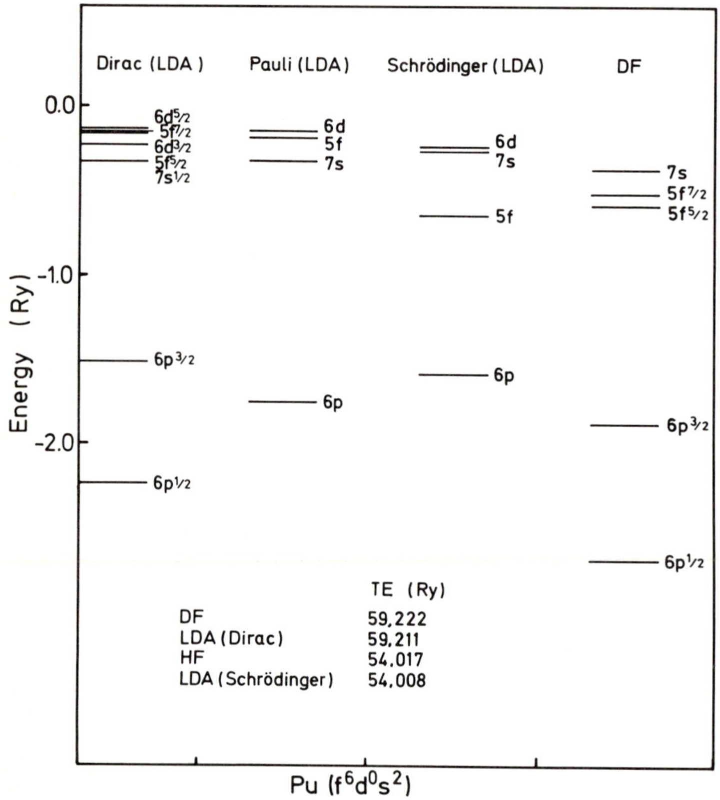

Fig. 3. A comparison of the eigenvalues of the outermost valence electrons for Pu using relativistic, semi-relativistic and non-relativistic kinematics and the local density approximation (*LSD*). Dirac-Fock eigenvalues after Desclaux[14)] are also shown. The total energies of the atoms (minus sign omitted), calculated with relativistic and non-relativistic kinematics are also shown. *HF* means Hartree Fock

allowing for energy changes due to greater f-occupation) and the actinides would have too many f-electrons.

In most atomic programs (5) is actually solved self-consistently either in a local potential[11)] or by the relativistic Hartree-Fock method[14)]. There is, however, an important time-saving device that is often used in energy band calculations for actinides where the same radial Eq. (5) must be solved[18)]. If (5.a) is substituted into (5.b) a single second order differential equation for the major component is obtained

$$-\frac{1}{2M}\left\{g''_\varkappa + \frac{2}{r}g'_\varkappa - \frac{1\,(1M)}{r^2}g_\varkappa\right\} - v'g'_\varkappa/4\,Mc^2 + vg_\varkappa - \frac{\varkappa+1}{r}v'g_\varkappa/4\,Mc^2 = Eg_\varkappa\,. \quad (8)$$

The only term depending upon $\varkappa$ is the last on the left hand side and this is just the spin-orbit coupling. If this term is dropped $\varkappa$-independent solutions, corresponding to zero spin-orbit coupling, are obtained. However, the large relativistic shifts, the mass velocity and Darwin terms[19)] are retained. In Fig. 3, for example, the relativistic levels remain in place but each of the spin-orbit split pairs is replaced by the average energy level

$$E_l = \{lE_{\varkappa=l} + (l+1)E_{\varkappa=l-1}\}/(2l+1) \quad (9)$$

In energy band calculations this procedure has two practical advantages:

1. The spin, in this case, is a good quantum number and spin polarized calculations are easy to perform,
2. The number of states to be dealt with is halved, leading to considerable time-saving in matrix manipulation.

The complete solutions to (5) will be referred to as 'Dirac' solutions whereas the solutions to (8) with spin-orbit coupling omitted will be referred to as 'modified Pauli' solutions.

4. *Schematic Models of Electronic Structure*

Armed with the band masses and relative positions of the band centres one may easily sketch the likely valence electron energies in actinide metals and compounds. This is useful because it is easy to understand, involves no computation and gives a rough idea of what to expect from large machine computations of electronic structure.

From Table 1 the scheme for the actinide metals shown in Fig. 4 is arrived at. The valence band structure is evidently more complicated in detail than that of the d-transition metals because there are now four different angular momentum states to deal with. However, the d bands are now broad conduction bands. This is not surprising since the broadening of d-bands is a systematic trend from the 3rd to the 5th transition metal series and has now passed a stage further. The reason for this is that the wave functions of each new d-series must be orthogonal to those of the earlier series. The necessary additional orthogonality mode extends the wave functions spatially and broadens the bands. Precisely the same phenomenon occurs between the 4f and 5f series. Thus d-electrons play the role of the major conduction electrons in the actinides and the relative population of the sp conduction bands is reduced. The narrow f-bands are pinned to the Fermi level

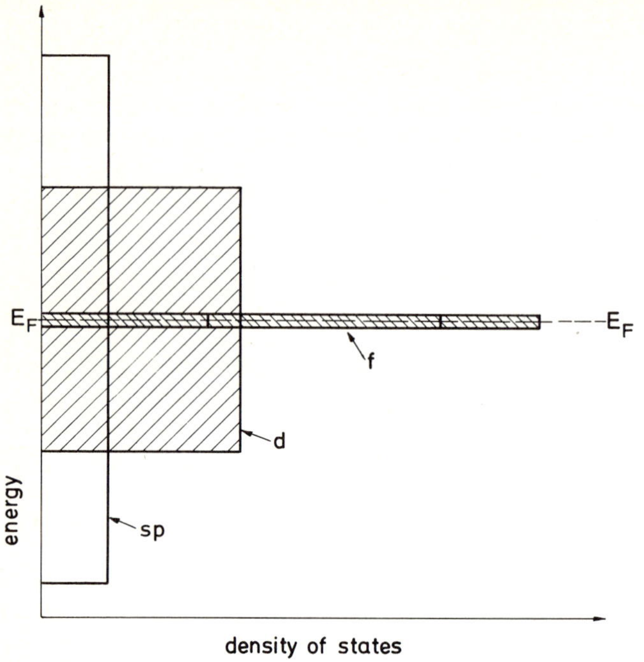

Fig. 4. Schematic partial density of states scheme for light actinide metals (s and p states are counted together) with f electrons delocalized

and are responsible for the transition-metal features of actinide chemical bonding. As the series is traversed from Ac to Am the f-bands narrow (and lower) taking the Fermi level with them. The number of electrons with f character increases almost monotonically in the process. Evidently an important change occurs at Am since its atomic volume is 40% larger than to be expected from the trend in the early actinides. The second half of the actinide series then follows from Am in a rare-earth-like fashion. In this case we expect the electronic structure to be represented better by Fig. 5 with the f-bands replaced by localized levels which are not pinned to the Fermi energy. The cohesive properties are then determined by the spd band structure as in the rare earths.

We have so far omitted complicating features such as hybridization between the bands but these will be dealt with later. The same modelling may be used for compounds as long as their crystal structure is simple enough. The simplest actinide compounds are the NaCl structured chalcogenides and pnictides which are also metallic. They are also particularly important since the U-Pu pnictides and chalcogenides are often magnetic whereas the metals are not. These compounds are the simplest light actinides systems in which magnetism occurs, therefore they have been the subject of intensive experimental and theoretical efforts. If UN is taken as an example we expect a valence band, derived from nitrogen 2p levels, to form below the uranium spdf band structure. The exact positions in energy depend upon the amount of charge transfer. The simple model in which three electrons are transferred to the nitrogen p states is incorrect as we shall see in detail later since UN will be our prototype for the NaCl compounds. A new feature in the compounds is the hybridization between cation and anion states. Without this hybridiza-

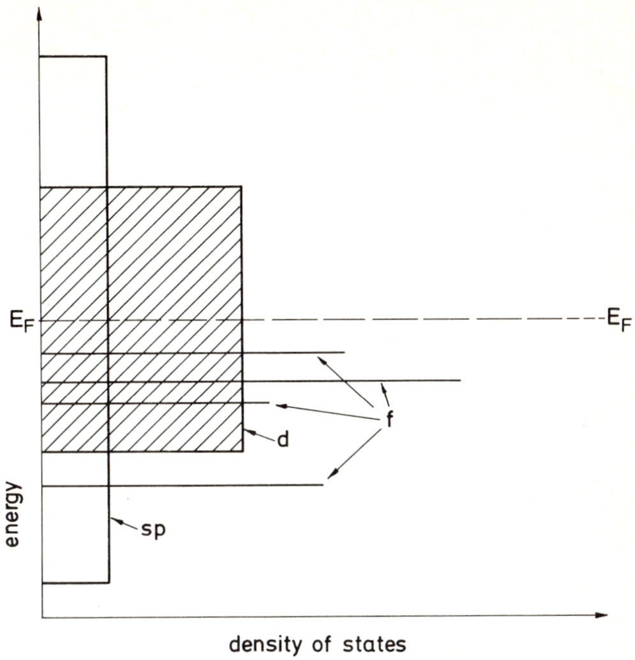

Fig. 5. Schematic partial density of states scheme for heavy actinide metals (s and p states are counted together) with f electrons delocalized

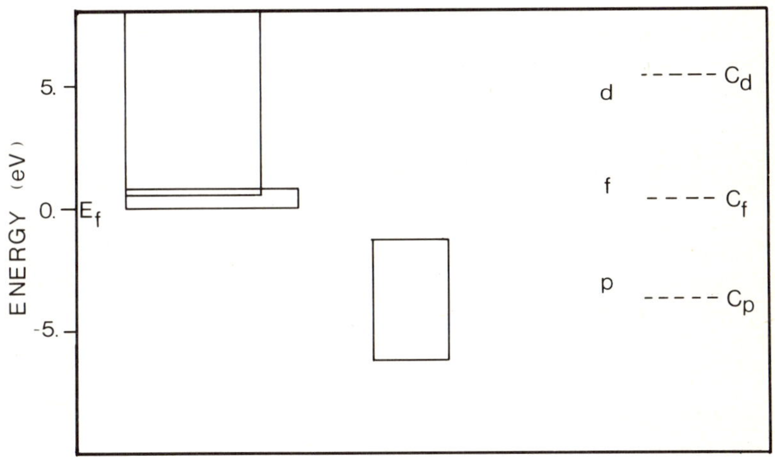

Fig. 6. Schematic partial density of states scheme for an NaCl-type (binary) compound (with UN as an example) with f electrons delocalized and unhybridized. Uranium is on the *left* and nitrogen on the *right*. In ascending order: nitrogen valence band; f-band tied to the Fermi level; the d conduction band. The Fermi level is at zero on the energy scale. The unhybridized band centres, C_{tl}, are shown on the right. This unhybridized model corresponds to the fully ionic model

tion the compound would be fully ionic but we shall see that this hybridization is strong and the f-p hybridization, in particular, has important effects. Once again, in a band model, the f-electrons are pinned to the Fermi level. If Table 1 is used together with similar information for the nitrogen p-states a first approximation to the electronic structure energy diagrams should appear as shown in Fig. 6.

II. Band Calculations in Metals and Compounds

1. *Density Functional Theory*

The simplest form of scattering theory is for a single particle moving in a local external potential. If we ignore relativistic effects and return to the Schrödinger equation

$$\{-\nabla^2/2 + V(r) - E\}\, \psi(r) = 0 \tag{10}$$

where V(r) is the potential. The Green function is defined by

$$\left\{-\frac{\nabla^2}{2} + V(r) - E\right\} G(r, r'; E) = \delta(r - r') \tag{11}$$

In general, however, in the reduction of the many electron to single-electron equations there appears an additional, non-local potential. Then (10) is replaced by

$$-\frac{\nabla^2}{2}\psi(r) + V(r)\,\psi(r) + \int U(r, r')\,\psi(r')dr' = E\psi(r) \tag{12}$$

where $U(r, r')$ is the non-local potential. In this case (11) is replaced by the Dyson equation

$$\left(-\frac{\nabla^2}{2} - E\right) G(r, r'; E) + \int dr'' \sum(r, r''; E)\, G(r, r''; E) = \delta(r - r') \tag{13}$$

where

$$\sum(r, r''; E) = V(r)\,\delta(r - r') + U(r, r'') \tag{14}$$

is the mass operator. Due to the non-local nature of the exchange potential the Hartree-Fock equations take the form (12), which is considerably more difficult to solve than (10). Furthermore, although the Hartree-Fock equations contain the exchange correctly, the Coulomb field due to the other electrons is treated in an average fashion after Hartree[20] and Coulomb correlations are dropped. A theorem due to Hohenberg and Kohn[21] and developed by Kohn and Sham[22] has placed the older Thomas Fermi[23] theory on a more rigorous basis. It was shown that the total energy of an interacting inhomogeneous electron gas in an external potential V(r) may be written in the form

$$E = \int V(r)n(r)dr + \int \frac{n(r)n(r')}{|r - r'|} drdr' + G\{n\} \tag{15}$$

where the second term is just the classical Coulomb repulsion energy and G{n} is a universal functional of the density. This expression for the energy is also a minimum for the correct density function n(r). The density G was then divided into

$$G\{n\} = T_s\{n\} + E_{xc}\{n\} \tag{16}$$

where T_s is the kinetic energy of a system of non-interacting electrons with the same density and E_{xc} is the exchange and correlation energy for an interacting system with density n(r). Now if n(r) is sufficiently slowly varying

$$E_{xc}\{n\} = \int n(r)\, \varepsilon_{xc}(n(r))dr \tag{17}$$

where E_{xc} is the exchange correlation energy per electron of a uniform electron gas of density n(r). Variation of energy with respect to the density then produces the local total potential

$$V(r) = v(r) + \int \frac{n(r')}{|r - r'|} dr' + \mu_{xc}(n(r)) \tag{18}$$

where

$$\mu_{xc} = \frac{d(n\varepsilon_{xc})}{dn} \tag{19}$$

is the exchange and correlation contribution to the chemical potential of a uniform electron gas of density n. In this way, known as the local density approximation (17), the many body problem reduces to a set of single particle equations. If the density functionals ε, μ are known the set of single particle equations may be solved self-consistently since the density is given by

$$n(r) = \sum_{i=1}^{N} |\psi_i(r)|^2 \tag{20}$$

where the ψ_i are solutions to (10) in the potential (18). In practice the density functionals are given in approximate, parametrized forms, derived from the many body theory of the free electron gas[24–28]. From (15) the total energy of both atoms and solids and its derivatives may be calculated *ab initio* from self-consistent calculations. In magnetic systems the energy is a functional, not of the ground state density, but of the spin density matrix[25]. This additional degree of variational freedom allows the spin magnetic moment to be calculated in spin polarized calculations. Spin polarized calculations must include twice as many states but, if spin-orbit coupling is neglected, the spin-up states are decoupled from the spin-down states and matrix sizes are not doubled. Spin-orbit coupling, however, couples spin-up and spin-down states leading to very large matrix problems in an actinide system.

The spin density functional formalism has very successfully described a wide range of solid state problems even in cases where it appears to be pushed beyond its validity. A review of the reasons for this and the cancellation of errors involved has been given by Gunnarson[29]. Formally, however, the rigorously computable quantities are all ground state properties and it is necessary always to remember that (10), in this context, is one of a set of 'effective' single particle equations. By this it is meant that the eigenvalues have no simple physical meaning, in particular they are not the excitation energies observed in experiments.

In order to find the single-particle-like excitation energies it is necessary to investigate the structure of the Green function, given by the Dyson Eq. (14). Sham and Kohn[30] here argued that

(a) $\sum$ is a short range Kernal, $|r - r'| \sim \lambda_F$

(b) if $\sum = \phi\,\delta(r - r') + M(r, r'; E\text{-}\phi)$

where ϕ is the electrostatic potential at $r_0 = (r_1 + r_2)/_2$, then the functional form of Σ depends only upon the density in the vicinity of r and r'. This immediately leads to a picture of excitations propagating with effective energy $(E - \phi(r))$ where ϕ is local and electrostatic, but acted upon by non local forces depending upon the density in the vicinity of r.

2. *Ground State Properties From Band Calculations*

The Hamiltonian

$$H = - \sum_{i} \nabla_i^2 + \sum_{i=j} e^2/r_{ij} + \sum_{i} V_{ext}(r_i) \tag{21}$$

contains kinetic energy, electron-electron interaction and external potential, respectively, and is usually written, as in (21), in Rydberg units. Through the local density approximation this is reduced to the sum of single electron Hamiltonians

$$H_i = - \nabla_i^2 + \mu_{xc}(r) + \int \frac{n(r)dr}{|r - r'|} + V_{ext}(r) \tag{22}$$

for which the eigenvalues E_i and wave functions $\psi_i(r)$ are to be found. The charge density is reconstructed from the wave functions and this is sufficient to recompute H_i. The process is iterated until the input and output charge densities agree to some specified accuracy. The total energy may also be written

$$E = \sum_{i=1}^{N} \varepsilon_i - \frac{1}{2} \iint \frac{n(r)n(r')}{|r - r'|} dr dr' + \int n(r) \{\varepsilon_{exc}(n(r) - \mu_{xc}(n(r))\} dr \tag{23}$$

For actinide atoms E is of the order of 60.000 Ryds. If the cohesive energy of a solid is required it is necessary first to compute the total energy of the atom, spin polarizing the valence electrons to include spin polarization energy and then to compute the total energy of the solid in a band structure calculation. The difference is the cohesive energy if the configuration of the atom and the solid are the same. If the ground state of the atom differs from that of the solid, the energy of atomization is calculated. Although it is not strictly correct, it is common practice in solid state studies to refer to either as the cohesive energy. Cohesive energies of metals are usually less than 160 kcal/mol (7 eV/atom) which is very small compared with the total energy of an actinide atom. Clearly this is because only a small number of valence electrons are involved in chemical bonding. There is a useful approximation, due to Gunnarsson et al.[31], which identifies that part of the total energy which changes – E_v – if the core charge densities are assumed to be frozen at their free atom values

$$E_v = \sum_n \int n\,\varepsilon_n - \int dr\, n_v(r) \left\{ \frac{1}{2}\phi_v(r) + \mu_{xc}(r) - \varepsilon_{xc}(r) \right\} + \int dr\, n_c(r)\,\{\varepsilon_{xc}(n(r)) - \varepsilon_{xc}(n_c(r))\} \tag{24}$$

E_v in an actinide atom varies typically from 3 Ryds for Ac to about 40 Ryds for Am. Even in this approximation very high standards of accuracy are required to obtain good values for the cohesive energy.

The atomic volumes or, equivalently, lattice parameters may be found accurately by calculating directly the zero temperature equation of state without intermediate computation of the total energy. The technique is based upon the virial theorem[32] which changes the volume integral for the energy into a surface integral for the pressure[33]. It is in general not easy to apply this technique to non-spherical systems but in the atomic sphere approximation (ASA) to the linear muffin tin orbital (LMTO) technique it may be shown that[34]

$$3PV = {}^{\delta}/\delta\, \mathrm{Lns} \int E n(r)\, d\varepsilon = 3V \sum_l P_L$$

$$3P_lV = \int S\phi_l^2(\varepsilon) n_l(\varepsilon) d\varepsilon\, \{(D_l + L + 1)(D_l - 1) + (E - \varepsilon_{xc})s^2\} \tag{25}$$

where $\phi_l(E)$ is the radial wave function, evaluated at the sphere boundary and energy E, $N_l(E)$ the partial density of states with character l; D the logarithmic derivative of ϕ at energy E, and S is the Wigner-Seitz radius. There are several major advantages to this formula. Firstly it may be seen that (25) depends only upon the *change* of single particle energies and does not include electron-electron interaction terms that enter the total energy. Secondly all of the quantities entering the formula are normally computed in a band structure calculation. No extra effort or time are required – in fact the convergence of the pressure is an excellent test of the convergence of the entire band structure calculation. The bulk modulus is computed by calculating the pressure at several lattice parameters (equation of state) and taking the numerical derivative at zero pressure.

In order to calculate the magnetic moment it is necessary to perform spin polarized band calculations. The extension of the local density approximation to cover the spin polarized case has been made by von Barth and Hedin[24] and Gunnarsson and Lundqvist[26].

If the modified Pauli equation is used, i.e. spin-orbit coupling is dropped, the spin-up and spin-down calculations are uncoupled. One may make separate band structure calculations for spin-up and spin-down bands which are shifted with respect to one another. The system in filled to the Fermi level and the net magnetic moment is the difference between spin-up and spin-down occupation numbers.

3. Band Calculations for Metals and Compounds

a. Non-Self-Consistent Energy Band Calculations for Metals

Early band structure calculations for the actinide metals were made both with[35)] and without[36)] relativistic effects. As explained above, at least the mass velocity and Darwin shifts should be included to produce a relativistic band structure. For this reason we shall discuss only the relativistic calculations[35)]. There were some difficulties with the f-band structure in these studies caused by the f-asymptote problem[37)], which have since been elegantly solved by linear methods[38)]. Nevertheless the non-self-consistent RAPW calculations for Th through Bk indicated some interesting trends that have also been found in more recent self-consistent calculations:

(a) superimposed upon the normal free electron (sp) band structure are broad d-bands which start about the bottom of the sp-bands. The f-states form a narrow set of bands at the Fermi level,
(b) the f-bands narrow along the actinide series going from Th to Bk,
(c) the f-bands also lower, being above the Fermi level for Th,
(d) an examination of the band structure for different values of the Slater X_α parameter for exchange showed that the f-bands were particularly sensitive to changes in X_α – which implies a sensitivity to the exchange-correlation approximation.

In addition Freeman and Koelling[35)] stressed the importance of the relationship between bandwidth and the Coulomb correlation integral.

b. Non-Self-Consistent Energy Band Calculations for Compounds

Early attempts to compute the band structures of actinide NaCl-type compounds by Kmetko and Waber[39)] and Adachi and Imoto[40)] suffered from being non-relativistic and from tight-binding approximations that were too restrictive. More recent and extensive studies by Davis[41)] (KKR) and Allen and Brooks[42)] (LMTO) have thrown light on the kind of band structures to be expected from self-consistent studies. Davis[41)] compared in detail the differences between the band structures of rare earth and actinide NaCl-type chalcogenides and pnictides. The valence p-bands of the monochalcogenide or pnictide were common to the energy bands of both series as was the overall spd conduction band structure arising from the cation. But the actinides are also expected to have a narrow set of f-bands at the Fermi level in contrast to the Mott-localized set of f-levels in the rare earths. Davis also pointed to the relatively large electronic specific-heat coefficients, γ, for AnX compounds in comparison with those observed for rare earth compounds. The enhancement of γ, it was suggested, was due to the increase in the density of states at the

Fermi level due to the appearance of narrow f-bands which are not present in rare earth compounds.

There was a remarkable disagreement regarding the f-bandwidth between the tight-binding calculations of Kmetko and Waber[39)] (·3 eV), Adachi and Imoto[40)] (·1 eV) and Davis[41)] (2–3 eV). Allen and Brooks[42)], in an effort to understand these different results varied the coefficient X_α in the exchange approximation and found that the f-bands narrowed as X_α was increased. For values of $\cdot 66 < X_\alpha < 1.0$ they obtained bandwidths of about 1–2 eV in agreement with Davis. The reason for the discrepancy appeared to be that the earlier tight binding calculations neglected hybridization between f and other band states which are responsible for much of the f-bandwidth. Of particular importance was the hybridization between cation-f and anion-p electrons. As in the calculations for the metals the f-bands were found to lie above the Fermi energy for thorium compounds but below for the later actinides.

At a different level of approximation, some of the trends for the entire uranium pnictide and chalcogenide series have been studied through single-site scattering phase shifts[43)]. The number of states is given by

$$n(E) = n_0(E) + \frac{1}{2\pi} \sum_j (2j+1)\, \delta_{\varkappa(E)} \tag{26}$$

for the scattering of partial waves off a given site. Here $\delta_\varkappa$ are the $\varkappa$-phase shifts and $n_0(E)$ is the free electron contribution. The phase-shifts change very rapidly going through a resonance and this produces large peaks in the density of states in positions where bands are expected to form. The trends that were noted are:

(a) Cation (f, d) and anion resonances (p) were close together: this should lead to extensive covalent bonding for both chalcogenides and pnictides.
(b) Cation resonances became less sharp as the atomic number of the cation increased. Therefore the covalent bonding should be greater for the heavier cations.
(c) The pnictides had broader p-resonances than the chalcogenides, therefore should be more covalent.
(d) The uranium $f^{5/2}$-$f^{7/2}$ spin-orbit splitting remained more or less constant in all the calculations.

Non-self-consistent cluster calculations[44)], based upon assumed ionic configurations, have been performed for oxides for both NaCl-type (AnO) and fluorite (AnO_2) structures[45)]. The manner in which the binding energy of the f-levels increases with atomic number of the actinide, becoming degenerate with the O-2p band at Am is a noticeable trend[44)] that is in agreement with experiment[46)]. The same trend was also reproduced by non-self-consistent LMTO calculations for the actinide oxides[47)]. The band calculations also gave reasonable estimates of the valence bandwidths of the oxides – a property that cluster calculations are not able to reproduce as they do not include long range O(p)–O(p) interactions[47, 48)]. Spin polarized versions[49)] of the molecule cluster calculations demonstrated that the exchange splitting of the 5f-levels is proportional to the unpaired spin. However, spin-orbit coupling should change the nature of the magnetism and, in order to include the effects of spin-orbit coupling and spin splitting simultaneously, "moment polarized" calculations[50)] for the Dirac equations have been developed.

c. *Self-Consistent Energy Band Calculations and the Equation of State of Actinide Metals*

Self-consistent energy band calculations for the actinide metals have been made by Skriver et al.[5], for the metals Ac-Am. The modified Pauli equation was used for this series of calculations but the corrections arising from use of the Dirac equation have recently been incorporated[51]. An fcc structure was assumed for all the metals in both series of calculations.

In the calculations of Skriver et al., the metals Ac-Pu were found to be paramagnetic (did not spontaneously spin polarize). The number of electrons in the f-bands was found to increase monotonically, being approximately zero for Ac and 5.2 for Pu. The measured trend in lattice parameter (Fig. 7) was reproduced very well from the calculated equations of state. As the f-band fills, the electrons contribute an increasingly negative pressure which is responsible for the rapid decrease of lattice parameter throughout the series. The d pressures are also negative, although not so large, and the sp-pressures are positive. Thus the equilibrium lattice parameters arise from a balance between the attractive d and f pressures and the repulsive sp pressures. This fact, that the f and d electrons would alone prefer a smaller atomic volume whereas the s and p electrons would prefer a larger volume, is basic to an understanding of chemical bonding in the actinide metals[52].

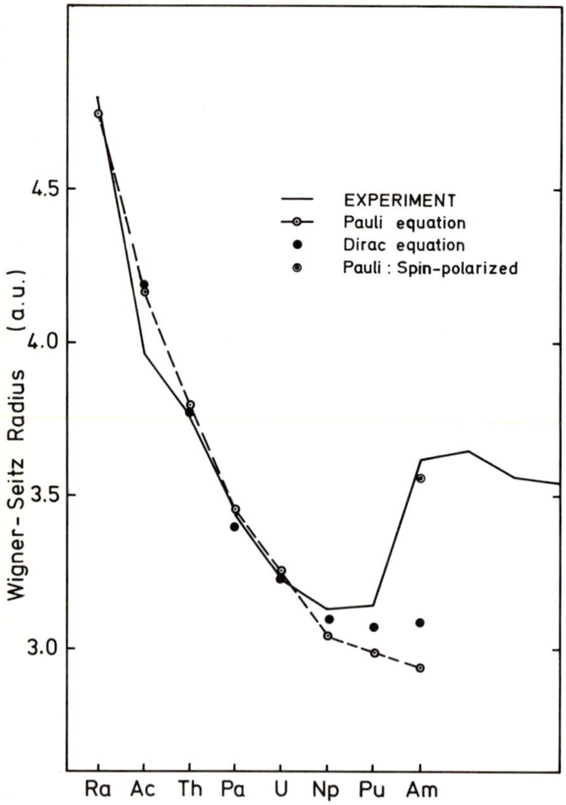

Fig. 7. Calculated and experimental Wigner-Seitz radii for the actinide metals. The Wigner-Seitz radii of the actinide metals from RLMTO calculations, compared with the radii from LMTO calculations and experiment (after Skriver et al.[5]). Also shown are the Wigner-Seitz radii of the actinide metals from RLMTO calculations (*full circles*)

The f-band width was found to be about 5 eV in Ac, about 3 eV for Th-Np and around 2 eV for Pu. In Am it is down to 1 eV. The Stoner parameter, I[53], was calculated to be about 0.5 eV and almost constant throughout the series. At Am, however, the product I $N(E_f)$ of the Stoner parameter and the f-density of states at the Fermi level exceeds one and spontaneous spin polarization occurs in the band calculation. Since Am has about 6.2 f-electrons and the moment saturates, this leads to an almost filled spin-up band and an empty spin-down band. The result is that the f-pressure all but vanishes leading to a large "jump" in atomic volume – in agreement with experiment. This has been interpreted[5] as Mott-localization of the f-electrons at Am and the f-electrons of all actinides heavier than Am are Mott-localized. The trend in their atomic volumes is then similar to those of the rare earths.

The trend in the f-pressure is almost parabolic with band filling and this is typical for a transition metal (with d replaced by f). The physical basis was given by the Friedel[54] who assumed that a rectangular density of states was being filled monotonically and thus was able to reproduce the parabolic trend in transition metal cohesive energies analytically. Pettifor[55] has shown that the pressure formula can similarly be integrated analytically.

If spin-orbit coupling is introduced the $l = 3$ band is replaced by the $j = 7/2$ and $j = 5/2$ bands. If spin-orbit coupling is large compared to the bandwidth these bands split, the $j = 5/2$ band filling first. The parabolic trend in atomic volume is then replaced by a double parabola and minima are expected (for f electrons) at $n_f = 3$ and $n_f = 10$ with maxima at $n_f = 6$ and 14. In fact spin-orbit coupling is not large enough to completely

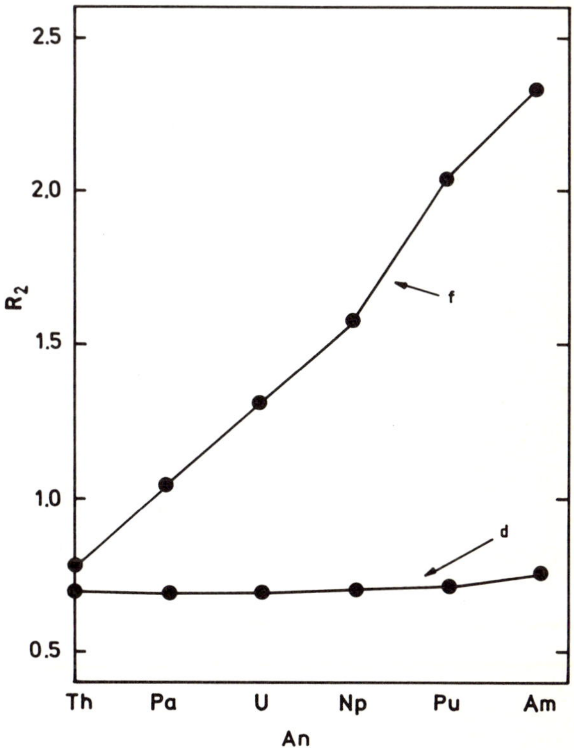

Fig. 8. The ratio of the number of electrons in the $j = l - 1/2$ and $j = l + 1/2$ bands ($= 2l/(2l + 2)$ for zero spin-orbit splitting) for both d and f electrons in the actinide metals

split the bands even at Am and it is small compared to the bandwidth for Ac-U. One way to measure the relative size of spin-orbit coupling is to plot the ratio (R) of f and d electron occupation of $j = 5/2$ and $j = 7/2$ bands and this is shown in Fig. 8[51]. When spin-orbit coupling is zero the ratio R is 2/3 and 3/4 (degeneracy weighting) for d and f electrons, respectively. R does not vary much from the ideal ratio for d-electrons due to their relatively large bandwidth (~ 1 Ryd.) but it does change for f-electrons along the actinide metal series due to the narrowing of the f-band and the increasing of spin-orbit coupling parameter. The effect upon the Wigner-Seitz radii is shown in Fig. 7. For Np, Pu and Am there are noticeable changes in volume but for Ac-U the effects are negligible[51].

d. Self-Consistent Calculations for Compounds

Self-consistent calculations for actinide clusters have been made using cellular multiple scattering techniques[56] and through linear combinations of atomic orbitals[44]. Band calculations have been made using the self-consistent RKKR[57], LMTO[38, 58] and RLMTO[59, 60] techniques.

In the cellular multiple scattering model[61], finite clusters of atoms are subjected to condensed matter boundary conditions in such a manner that a continuous spectrum is allowed. They are therefore not molecular calculations. An X_α type of exchange was used to create a local potential and different potentials for up and down spin-states could be constructed. For uranium pnictides and chalcogenides compounds the clusters were of 8 atoms (4 metal, 4 non-metal). The local density of states was calculated directly from the imaginary part of the Green function. The major features of the results are[61, 62]:

(a) the non-metal bands were mostly of s-p character with little d-f character,
(b) conversely, the metals bands were mostly of d-f character;
(c) an f-resonance around the Fermi level was split into spin-up and spin-down resonances separated by about 1 eV in UTe. The self-consistent moment in UTe was $m = 2.16\ \mu_B$ for UTe.
(d) The f-resonance produced a dip in the d-density of states at the Fermi level which led to a net d-spin anti-parallel to the f-spin.

In the finite cluster molecular orbital calculations, local exchange potentials were used[44] and embedding techniques have also been developed to simulate the crystalline environment[45]. Furthermore attempts have been made to take account of spin-orbit coupling effects upon the magnetism by "moment polarizing" rather than spin polarizing the calculations. This involves the splitting of Kramers doublets (or removing time-reversal symmetry in the Dirac solutions) rather than the pure spin-up and spin-down states. The thrust of these calculations has been the actinide (U-Am) monoxides and dioxides[44, 45] where they are particularly useful for studying defect clusters. AnO_2 was represented by the complex UO_8^{12-} and AnO by UO_6^{10-}. The dioxides were found to be not particularly ionic with ionicity from a Mulliken analysis, of about 1.3. Moment polarized calculations yielded a moment of $2.2\ \mu_B$ compared with an experimental value of $1.74\ \mu_B$. The valence bandwidth varied from 7–11 eV. Similar calculations for U-Pu carbides[44] produced a very broad valence band of up to 16 eV and a large number of cluster states near to E_f, indicating metallic or semi-metallic character.

A thorough series of calculations utilizing the fully relativistic KKR method have thrown considerable light upon chemical bonding in the uranium chalcogenides and pnictides, alloys and mixed crystals[57]. The frozen core approximation for the inner cores was adopted. The earlier calculations used an X_α-type of local density approximation but more recently the Hedin Lundqvist[25] parametization of exchange was used for US: Spin-orbit splitting was found to be appreciable compared with the small bandwidths (f- and d-like) lying near the Fermi level. A proper treatment of spin-orbit coupling therefore seems to be essential, especially if magnetic properties are to be described correctly. Charge transfer between anion and cation was also found to be significant[63], an effect which tended to cast doubt upon earlier non-self-consistent calculations. The f-bands were found to be particularly sensitive to exchange and correlation approximation, as is typical for very narrow bands.

The results of the LMTO and RLMTO calculations[58–60] are very similar to those of the RKKR calculations just described. The principle difference appears to come from the different exchange approximations used to create the potentials, rather than the band calculations themselves. The recent calculations by Podloucky and Weinberger[64] for US were made with a similar exchange and correlation approximation to the LMTO and RLMTO calculations. In Fig. 9 a and b we show the ΓX bands of the actinide nitrides NpN-PuN from self-consistent LMTO and RLMTO calculations, which may be compared with the RKKR calculations of Podloucky and Weinberger[64]. The agreement between the two sets of relativistic bands is very good.

III. Equation of State of Uranium Mononitride

Self-consistent energy band calculations have now been made through the LMTO method for all of the NaCl-type actinide pnictides and chalcogenides[58–60]. The equation of state is derived quite naturally from these calculations through the pressure formula[34] extended to the case of compounds[60]. The theoretical lattice parameter is then given by the condition of zero pressure.

Before trying to understand the trends in the lattice parameter across a series of compounds, we will study one compound, UN, in order to illustrate the method. We begin by studying the unhybridized band structure, especially the f, p, and d bands. We further replace the unhybridized densities of states by rectangular densities of states of the same area. In KKR or LMTO theory the structure constants contain all the structural information or equivalently, the shapes of the energy bands. These bands are then scaled and distorted by parameters depending solely upon the potential. The width of an unhybridized band for an atom of type t and angular momentum character l is given; approximately, by the simple expression

$$W_{tl} = (12\, N_{tl}^{-1} M_{tl}^{tl})^{1/2}\, \Delta_{tl} \qquad (27)$$

where $N_{tl} = (2l+1)N_t$ and N_t is the number of type t atoms in the unit cell. M_{tl}^{tl} is just the second moment of the structure constants. We have computed this, and other averages over the structure constants required in this section, and tabulated them (Table 2). These quantities are independent of the scale of the lattice, valid for any NaCl compound, and the same p-d, d-d moments have already been used[69] for the 3 d-oxides. The quantity Δ_{tl}

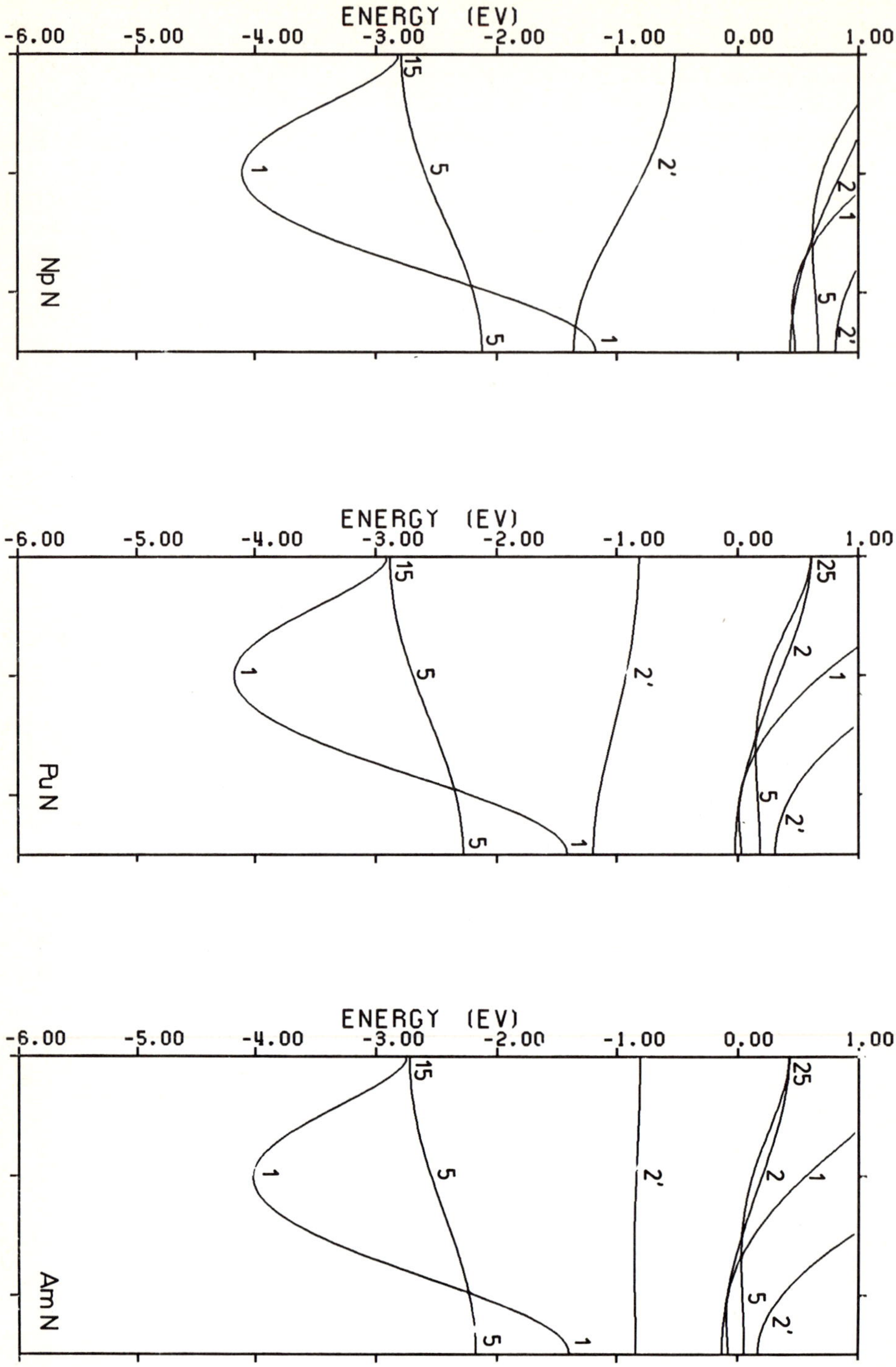

Fig. 9. a Energy bands, along the ΓX direction for NpN-AmN from modified Pauli calculations

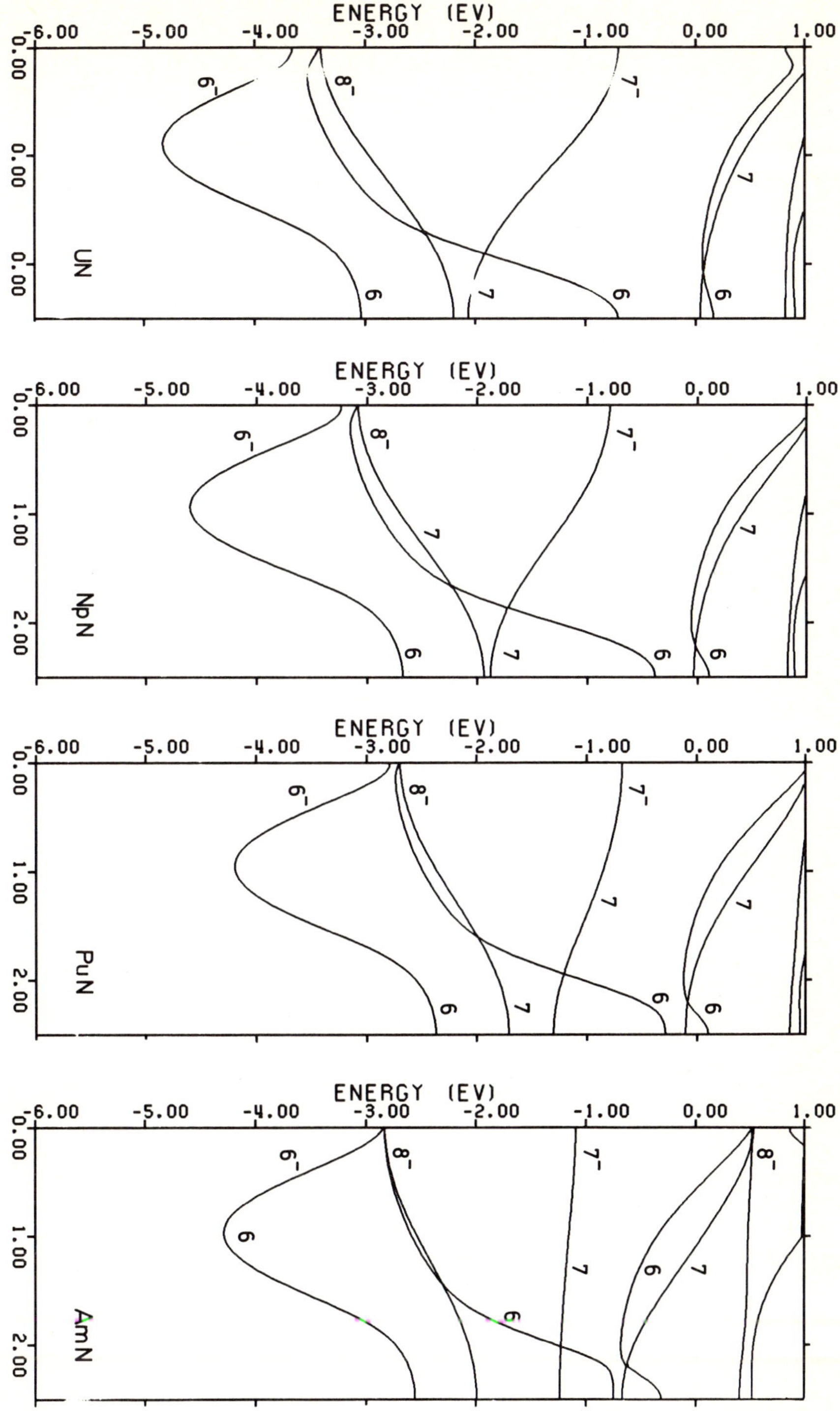

Fig. 9. b Energy bands, along the ΓX direction, of UN-AmN from RLMTO calculations

Table 2. NaCl canonical moments and averages over the densities of states in section 2: M = second moments; L = fourth moments

$M_{ff} = 542$	$M_{df} = 270$	$L_{pf} = 10^6$
$M_{dd} = 229$	$M_{pf} = 1209$	$L_{pd} = 5 \times 10^5$
$M_{pp} = 87$	$M_{pd} = 770$	

is the width parameter of the unhybridized tl band, depends upon the potential, and must be computed in a self-consistent band calculation for the compound under study. The width parameters of the uranium d and f, and anion p-states (in mRyds) are tabulated for all of the NaCl pnictides and chalcogenides in Table 3. The unhybridized bandwidths in UN are then trivially calculated from the tables and:

$$W_t = 70 \text{ mRyd}; \quad W_d = 680 \text{ mRyd}; \quad W_p = 370 \text{ mRyd}.$$

The centres, C_1, or the positions of the two bands are also potential dependent and must be computed. They are also given in Table 3. Figure 6 shows the rectangular f, p, and d bands and is drawn to scale so that the rectangles appear in approximately the correct places. Thus, we have a model unhybridized band structure (omitting, for the moment, s and p states) of a filled nitrogen 2p valence band, an f-band containing about two electrons, and an almost empty d band starting around the Fermi level.

The expression for the repulsion between two bands, in the weak hybridization limit, is given in[65]. The shift in the position of the tl band, due to hybridization with the t′l′ band, is

$$\delta_{tl} = N_{tl}^{-1} M_{tl}^{t'l'} \Delta_{tl} \Delta_{t'l'} (C_{tl} - C_{t'l'})^{-1} \tag{28}$$

Charge is also transferred by hybridization. The number of electrons with t′l′ character to be found in what was originally a filled tl band is

$$N_{t'l'}^{tl} = 2\,\delta_{tl} N_{tl} (C_{tl} - C_{t'l'}) \tag{29}$$

Table 3. Potential and parameters for the uranium chalcogenides and pnictides. Δ is the band width parameters and C the band centre (in mRyd.)

	Δ_f	Δ_d	Δ_p mRyd.	C_f	C_p	C_d
UN	2.2	29.1	20.0	48	−270	403
UP	1.1	18.2	30.7	56	− 79	347
UAs	0.7	13.7	24.4	15	− 89	363
USb	0.4	9.2	26.4	31	−179	376
US	1.1	17.7	21.1	49	−387	355
USe	0.8	14.4	23.1	40	−259	311
UTe	0.4	9.4	21.9	42	−320	359

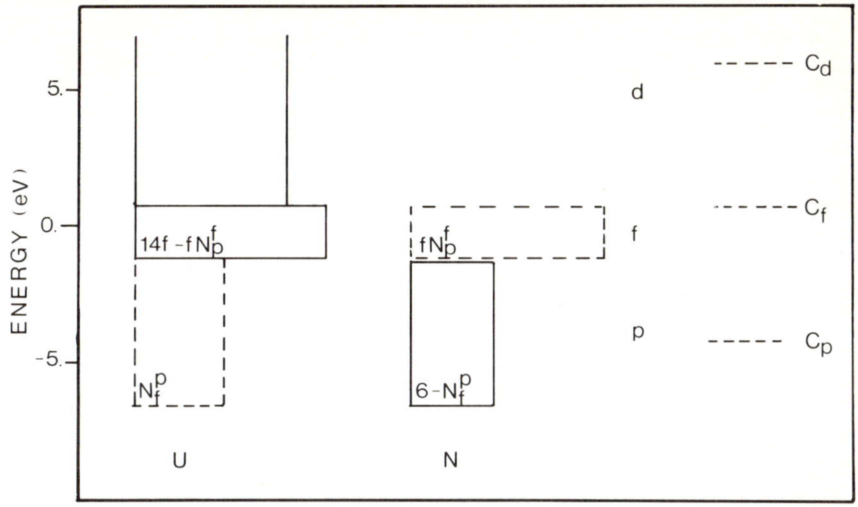

Fig. 10. Hybridized model densities of states for UN. The *full rectangles* are the original unhybridized densities of states (see Fig. 48). The *broken rectangles* are the additional projected densities of states due to hybridization. In a *vertical line* are the contributions to the local atomic and angular momentum projected densities of states. The electron transfer, in terms of the fractional occupancy (F) of the unhybridized f-band, is shown

Figure 10 is now replaced by Fig. 10. The valence band is predominantly of anion-p character but part of the charge density is to be found near the uranium atom with local d and f character. From (28), (29) and Tables 2 and 3 we estimate

$$N_f^p = 0.97\ ; \quad N_d^p = 1.99\ ; \quad fN_p^f = 0.14\ ; \quad fN_d^f = 0.26\ .$$

Here f is the fractional occupation of the unhybridized f band. At the top of the valence band, tied to the Fermi level, is an f-band containing a small d and p admixture. The total number of f-electrons has changed from 2 to approximately 2.5 and the number of nitrogen p electrons in the valence band is reduced from 6 to about 3.

The hybridization also broadens the bands to new widths W'_{tl}

$$W'_{tl} = W^2_{tl} + L^{t'l'}_{tl}\,\Delta^2_{tl}\Delta^2_{t'l'}\,(C_{tl} - C_{t'l'})^{1/2} \tag{30}$$

where $L^{t'l'}_{tl}$ is the fourth moment of the hybridization structure constants (Table 2). Most ineresting is the f-p interaction from which we find

$$W'_f = 180 \text{ mRyd}\ ; \quad W'_p = 400 \text{ mRyd}\ .$$

The f-bandwidth is increased by a factor of two by f-p hybridization, but it is changed little by the f-d interaction. Physically, itinerant f-electrons require the p-states on the anion to propagate throughout the crystal. The f-p contribution to bonding may also be estimated in this simple model. The change of total energy under compression – or electronic pressure – may be estimated from[32–34]

$$3\,PV = n_{tl}\,2(C_{tl} - \varepsilon_{xc})/m_{tl} + (\overline{E}_{tl} - C_{tl})(21 + 1), \tag{31}$$

where n_{tl} is the number of electrons with angular momentum l on type t atoms, ε_{xc} is the exchange and correlation potential at the WS sphere boundary (−600 mRyd) and m_{tl} is the band mass. For UN $m_f \approx 9$ and $m_p \approx 1.4$. Equation (31) divides the pressure into a contribution from the change of band centre energy with Wigner-Seitz radius, S_t, and a contribution depending upon the difference between the centre of gravity of the occupied states, $\overline{E}_{tl}$, and the unhybridized band centre. The band contribution vanishes for empty and filled bands and is negative for a less than half filled band. Equation (31) is easily evaluated if $\overline{E}_p$, $\overline{E}_f$ can be estimated. The p band is pushed downwards, δ_p, by the f-band and has $f(N_p - N_f^p)$ electrons in the anion spheres and N_f^p electrons in the cation spheres. The f-band is pushed upwards by δ_f and has $f(N_f - N_p^f)$ electrons in cation spheres and fN_p^f electrons in the anion spheres. The centres of gravities of the p-projected bands are then

$$\overline{E}_{p\ \mathrm{bond}} = C_p + \frac{N_f^p}{N_p}(C_p - C_f)$$

and

$$\overline{E}_{p\ \mathrm{antibond}} = C_f - (1-f)\,\frac{W_f}{2}\,.$$

Hence,

$$\underset{+}{3}\,PV_{p\ \mathrm{bond}} = 3\,N_f^p\left\{(C_f - C_p)\frac{fW_f}{2}\right\} \tag{32}$$

Similarly, the centres of gravity of the f-projected bonding and anti-bonding bands are

$$\overline{E}_{f\ \mathrm{bond}} = C_p\ ;$$

$$\overline{E}_{f\ \mathrm{antibond}} = C_f + \frac{N_p^f}{N_f}(C_f - C_p) - (1-f)\,\frac{W_f}{2}\,.$$

Hence,

$$3\,PV_{f\ \mathrm{bond}} = -7\left\{f\,\frac{W_f N_f}{2} + N_p^f(C_f - C_p)\right\}(1-f)\,. \tag{33}$$

The first term displays the usual parabolic behaviour with band filling. In UN it is about − 0.9 Ryd. The second part of (33) is the covalent f-contribution, is proportional to 1 − f, and in UN is also about −0.9 Ryd.

The charge transfer to s and p conduction band states cannot be obtained with the same accuracy in first-order perturbation theory. From the band calculation we know that the number of cation projected p-electrons is about 0.56. The repulsive pressure from these electrons may be estimated

$$3\,PV = -n_p\,5(\overline{V}_p - \varepsilon_{xc}) - 2(\overline{E}_p - \overline{V}_p)\,.$$

$\bar{E}_p$ is −0.07 Ryd and the square well pseudopotential, V_p, for the p-electrons is 0.6 Ryd, and ε_{xc} is 0.6 Ryd. Hence, 3PV = 2.8 Ryd. This is the largest single repulsive contribution to the equation of state and it balances most of the negative pressure coming from the d and f electrons.

This simple modelling has its limitations but it at least allows the f-contributions to bonding to be estimated throughout the uranium pnictide and chalcogenide series, given the data in Tables 2 and 3. However, other factors involving strong hybridization are neglected and the densities of states are in reality not rectangular. In particular the valence p-band is actually pushed up by hybridization (an effect of the cation core) so that the anion p-electrons contribute, at the equilibrium lattice parameter, a net repulsive pressure.

The band structure and electronic pressure for UN were computed in the lattice parameter range 4.75–4.90 Å. The resulting decomposed equation of state is shown in Fig. 11. The sp-pressures are drawn as one group, as are the d pressures from the two atoms. The attractive contributions are from uranium d and f electrons and the Madelung contribution. The theoretical equilibrium lattice parameter is at 4.86 Å compared with an experimental value of 4.89 Å. The accuracy of this result is probably fortuitous, as is seen by inspecting the trend of the lattice parameter across the uranium pnictide and chalcogenide series (Fig. 12). The trend in the lattice parameter is reproduced very well, but the theoretical lattice parameters fall progressively below experiment, the worst value being for USb which is 5.2% too small. The calculated bulk modulus for UN is 23% higher than the experimental value of 193 GPa and the high value may, at least in part, be due to our approximation of freezing the cores which include the nitrogen 2s and

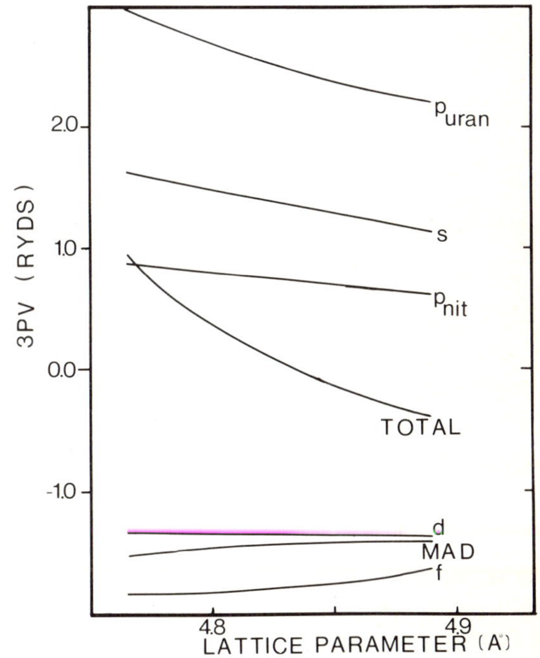

Fig. 11. The equation of state of UN. The curve labelled 'total' is the equation of state. The uranium d and f contributions are also shown. The curves labelled 'spd' are the s and p contributions plus the contribution of d electrons on the anion. The Madelung contribution is shown separately

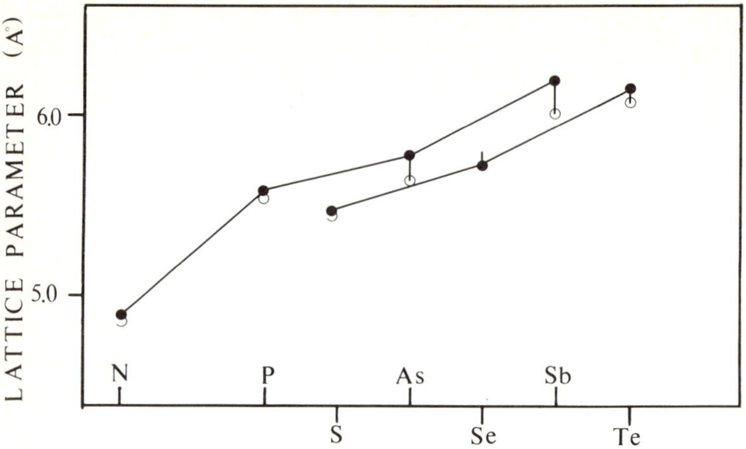

Fig. 12. The experimental *(full black circles)* and calculated lattice parameters of the uranium chalcogenides and pnictides

uranium 6p. These states would contribute small attractive pressures which increase in magnitude upon compression, thereby reducing the bulk modulus.

The equations of state across the series are similar to Fig. 11, the principal change being that the f-contribution is progressively reduced. The f-pressure for USb is a little more than half the value for UN and the d-pressure is relatively more important. The sp repulsive pressure is similarly reduced as the anion size is increased. The bulk modulus also decreases approximately inversely proportional to the lattice parameter.

Finally, we note that the lattice parameters appear to be computed relatively more accurately for the chalcogenides (ferromagnetic) than for the pnictides (antiferromagnet). This may well be related to the change of lattice parameter with magnetization[65]. The increase is greater for antiferromagnets than for ferromagnets as has been shown convincingly in by Andersen et al.[65].

Another way to approach the lattice parameter trend is to consider the actinide series of a given pnictide or chalcogenide. The f-p bond should then draw the minimum in lattice parameter towards the beginning of the series. A series of calculations[59, 60] for the actinide nitrides has been made which illustrate this effect quite well. The results of these calculations made (a) with the Pauli equation, (b) with the Dirac equation and (c) with the Pauli equation spin polarized (for UN-AmN) are shown in Fig. 13. The calculated lattice parameter for ThN always lies about 2% above the experimental value – possibly due to the frozen core approximation (the core of Th is particularly large). The computed lattice parameter for PaN lies about 4% below the experimental value[66]. The trend in Fig. 13 is then represented reasonably well by the theory until americium where the calculated lattice parameter is 4% too small.

Figure 13 also shows the results from the Dirac equation. Here the trend is reproduced very well as far as Pu with a minimum at UN. The reason that the theoretical lattice parameters have increased significantly for Pu and Am is that the fully relativistic f-bands consist of both $j = 5/2$ and $j = 7/2$ bands. Spin-orbit coupling is of the order of the band width and, with increasing atomic number, the $j = 5/2$ band fills preferentially. The

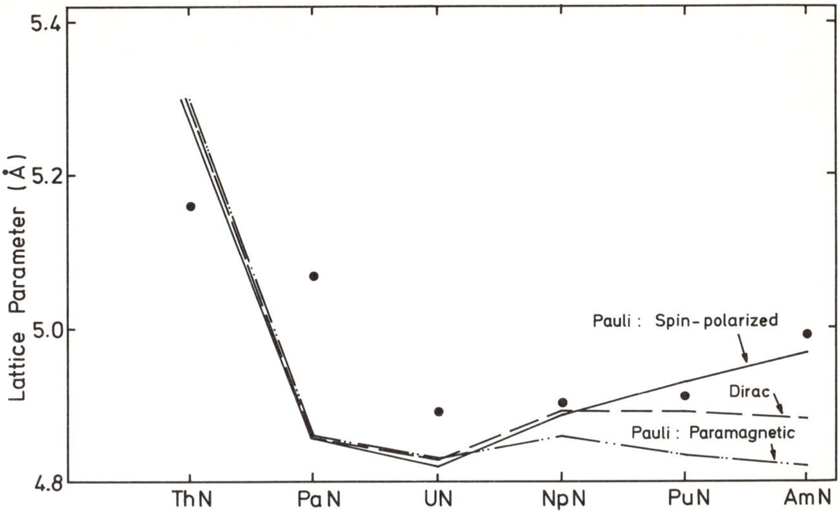

Fig. 13. Lattice parameters of the actinide nitrides from LMTO (labelled Pauli pramagnetic), RLMTO (labelled Dirac) and LMTO spin polarized (labelled Pauli: spin polarized) calculations. The *black filled circles* are the experimental lattice parameters

(negative) pressure from f-electrons is reduced when the $j = 5/2$ band is more than half-filled as occurs in PuN and AmN. However, a 2.5% difference between computed and experimental lattice parameters for AmN remains.

The product $IN_s(E_f)$ for the series is shown in Table 4. ThN and PaN should be non-magnetic and, as expected, self-consistent pure spin magnetic moments are obtained for UN-AmN. The effect of a moment upon the f band pressure is to change the first term in (33) to[65])

$$-49\ \{f(1-f) - m^2/4\}\ (30.5)\Delta_f$$

where m is the fractional spin moment (= moment/n_f). Clearly a non-vanishing moment produces a repulsive term in the f-band pressure which then vanishes for complete polarization and a half-filled band.

It becomes particularly, marked for AmN where the f-band is almost half-filled and polarizes completely. The result is that the lattice parameter of AmN from the spin polarized calculation is within 0.7% of the experimental value[5, 7)].

The trend in lattice parameter across the actinide nitride series is reproduced by an energy band theory in which it is assumed that the f-electrons are itinerant. The results with and without spin polarization do not differ greatly until AmN is reached but in this

Table 4. The product of the Stoner parameters and the density of states per spin at the Fermi level. $IN_S(E_F) \geq 1$ for ferromagnetism. UN is antiferromagnetic

	ThN	PaN	UN	NpN	PuN	AmN
$IN_S(E_F)$	0.09	0.18	1.8	3.6	1.9	3.8

case there is a sharp increase in lattice parameter due to the spin moment. This jump in lattice parameter may be interpreted in two ways. In the first it arises simply from the magnetic moment of an itinerant electron system. In the second, one notes that the spin-up band is almost filled and the spin-down band almost empty since AmN has just over six electrons with f-character. Now the many electron ground state from a filled band is equivalent to a localized state, as was noted some time ago by Mott and Stevens[67]. Therefore it is possible that the spin polarized calculation for a half or nearly half filled shell has produced the filled spin-up band required for a Mott-localized state. This is the same as the interpretation placed upon the sharp increase in lattice parameter for Am metal by Skriver et al.[5, 7], and, if this is correct, could also apply to AmN. High pressure studies of AmN would therefore be particularly interesting since a transition from the high volume low pressure phase to the low volume high pressure phase would be relatively easy to observe as no change in crystal structure is expected.

IV. References

1. Messiah, A. M.: Quantum Mechanics, North Holland 1958, Vol. I
2. Friedel, J.: J. Phys. Chem. Solids *1,* 175 (1956)
3. Freeman, A. J., Koelling, D. D.: in: The Actinides: Electronic Structure and Related Properties (eds. Freeman, A. J., Darby, J. B., Academic Press, New York and London, Vol. 1, p. 51 (1974)
4. Koelling, D. D., Freeman, A. J.: Phys. Rev. *B12,* 5622 (1975)
5. Skriver, H. L., Andersen, O. K., Johansson, B.: Phys. Rev. Letters *41,* 42 (1978)
6. Johansson, B.: Phil. Mag. *30,* 469 (1974)
7. Skriver, H. L., Andersen, O. K., Johansson, B.: Phys. Rev. Letters *44,* 1230 (1980)
8. Zachariasen, W. H.: Acta Crystallogr. *5,* 19, 660 and 664 (1952) and in "The Metal Plutonium" (1961)
9. Andersen, O. K., Jepsen O.: Physica *91 B,* 317 (1977)
10. Brooks, M. S. S.: J. de Physique *40,* C4–155 (1979)
11. Slater, J. C., Wood, J. H.: Int. J. Quantum Chem. *4,* 3 (1971)
12. Hubbard, J.: Proc. Roy. Soc. *A 276,* 238 (1963)
13. Rose, M. E.: Relativistic Electron Theory, Wiley, New York and London 1961
14. Desclaux, J. P.: Atomic and Nuclear Data Tables (ed. Way, K., Academic Press, New York and London, *12,* 311–406 (1973)
15. Burke, V. M., Grant, I. P.: Proc. Phys. Soc. *90,* 297 (1967)
16. Hartree, D. R.: The calculation of Atomic Structures Wiley, New York, 1957
17. Boyd, R. G., Larson, A. C., Waber, J. T.: Phys. Rev. *129,* 1629 (1963)
18. Andersen, O. K.: ibid. *B12,* 3060 (1975)
19. Koelling, D. D., Harmon, B. N.: J. Phys. *C10,* 3107 (1977)
20. Hartree, D. R.: Proc. Cambridge Phil. Soc. *24,* 426 (1928)
21. Hohenberg, P., Kohm, W.: Phys. Rev. *136,* 864 (1964)
22. Kohn, W., Sham, L. J.: ibid. *140A,* 1133 (1965)
23. Fermi, E.: Z. Physik *48,* 73 (1928)
24. von Barth, U., Hedin, L.: J. Phys. *C6,* 1629 (1972)
25. Hedin, L., Lundqvist, B. I.: ibid. *C4* (1971)
26. Gunnarsson, O., Lundqvist, B. I.: Phys. Rev. *B13,* 4274 (1976)
27. Gunnarsson, O.: J. Phys. *F6,* 587 (1976)
28. Gunnarsson, O.: Physica *91b,* 329 (1977)
29. Gunnarsson, O.: J. Appl. Phys. *49,* 1399 (1978)

30. Sham, L. J., Kohm, W.: Phys. Rev. *145,* 561 (1966)
31. Gunnarsson, O., Harris, J., Jones, R. O.: ibid. *B15,* 3027 (1977)
32. Slater, J. C.: Adv. Quantum Chem. *6,* 1 (1972)
33. Liberman, D. A.: Phys. Rev. *B3,* 2081 (1971)
34. Pettifor, D. G.: Commun. Phys. *1,* 141 (1976);
Mackintosh, A. R., Andersen, O. K.: in Electrons at the Fermi Surface (ed. Springford, M.) Cambridge Univ. Press, Cambridge 1979
35. Freeman, A. J., Koelling, D.: J. de Physique *33,* C3–37 (1972)
36. Kmetko, E., Hill, H. H.: Proc. 4th Int. Conf. on Plutonium and other Actinides (ed. Miner, W. N.) Nucl. Metallurgy *17(1),* 233 (1970)
37. Koelling, D. D.: J. de Physique *40,* C4–117 (1979)
38. Freeman, A. J.: Proceedings of Rare Earth and Actinides, Durham 1977, Inst. of Phys. Conf. Ser. *37,* 120 (1978)
39. Kmetko, E. A., Waber, J. T.: in Plutonium 1965 (ed. Kay, A. E.)
40. Adachi, H., Imoto, S.: J. Nucl. Sci. Technol. *6,* 371 (1969)
41. Davis, H. L.: in The Actinides: Electronic Structure and Related Properties (eds. Freeman, A. J., Darby, J. B.) Academic Press, New York 1974
42. Allen, R., Brooks, M. S. S.: J. de Phys. *C4,* 19 (1979)
43. Weinberger, P., Mallett, C. P.: Phys. Stat. Sol. (b) *94,* 257 (1979)
44. Gubanov, V. A., Rosen, A., Ellis, D. E.: Solid St. Commun *22,* 219 (1977)
45. Ellis, D. E.: in Actinides in Perspective (ed. Edelstein, N.) Pergammon, Oxford 1982, p. 123 (1982)
46. Veal, B. W., Lam, D. J.: Phys. Rev. *B10,* 1463 (1974)
47. Kelly, P. J., Brooks, M. S. S.: J. Phys. *C13,* L939 (1980)
48. Kelly, P. J.: Thesis Portsmouth Polytechnic 1980
49. Gubanov, V. A., Rosen, A., Ellis, D. E.: Solid St. Commun. *22,* 219 (1977)
50. Ellis, D. E.: J. Phys. *B10,* 1 (1977)
51. Brooks, M. S. S.: ibid. *F13,* 103 (1982)
52. Johansson, B., Skriver, H. L.: J. Mag. Magn. Mat. *29,* 217 (1982)
53. Stoner, E. C.: Proc. Roy. Soc. *165,* 372 (1938)
54. Friedel, J.: in: The Physics of Metals (ed. Ziman, J. M.) Cambridge Univ. Press, 341 (1969)
55. Pettifor, D. G.: J. Phys. *F7,* 613 (1977) and J. Phys. *F7,* 1009 (1977) and J. Phys. *F7,* 209 (1978)
56. Erbudak, M., Greuter, F., Meier, F., Reihl, B., Vogt, O., Keller, J.: J. Appl. Phys. *50,* 2099 (1979)
57. Weinberger, P., Podloucky, R., Neckel, A.: J. Mag. Magn. Mat. *29,* 247 (1982)
Weinberger, P., Podloucky, R., Mallett, C. P., Neckel, A.: J. Phys. *C12,* 801 (1979) and J. Phys. *C13,* 173 (1980)
58. Brooks, M. S. S., Glötzel, D.: J. Mag. Magn. Mat. *15–18,* 873 (1980); and Physica *102B,* 51 (1980)
59. Brooks, M. S. S.: J. Mag. Magn. Mat. *29,* 257 (1982)
60. Brooks, M. S. S.: J. Phys. *F14,* 639, 653, 857 (1984)
61. Erbudak, M., Keller, J.: Phys. Rev. Letters. *42,* 115 (1979)
62. Erbudak, M., Greuter, F., Meier, F., Reihl, B., Vogt, O., Keller, J.: J. Appl. Phys. *50,* 2099 (1979)
63. Mallett, C. P.: J. Phys. *C15,* 6361 (1982)
64. Podloucky, R., Weinberger, P.: Z. Phys. *B42,* 107 (1981)
65. Andersen, O. K., Skriver, H. L., Nohl, H., Johansson, B.: Pure Appl. Chem. *52,* 93 (1979)
66. Bohet, J., Müller, W.: J. Less Common Metals *57,* 185 (1978)
67. Mott, N. F., Stevens, K. W. H.: Phil. Mag. *2,* 1364 (1957)

General Conclusions and Trends in Actinide Research

It is worthwhile, as a conclusion to this book, to draw a balance of the to date understanding of actinide solids, emphasizing the areas of cross fertilisation with the kindred lanthanide and transition metals fields, and pointing out the new trends and ideas.

The actinide solid state properties are to a large extent based on the properties of the 5f wave-functions. Central to the actinide solid state research has been the co-existence of evidence and of concepts pointing clearly to the recognition of light actinides as being elements in which a metallic bond is enhanced by the overlapping of 5f wave-functions. The narrow band, itinerant character of the 5f's is similar to the one d-shells have: hence, the classification of these elements as 5f-transition metals.

However, much more than d-electrons in d-transition metals, in the actinides the 5f-shells lose their metallic bonding character when the half filling is reached (after Pu in elemental metals). This is seen in all ground state properties (see, e.g., the atomic radii, Fig. 3 of Chap. A) and in magnetism: actinide metals' paramagnetic susceptibilities can be discussed in terms of atomic magnetic moments from Am on. The 5f electron states become completely atomic (localized) states as for the 4f's in lanthanides. This is seen also by photoemission, in a sort of photographic way (Naegele et al.[1)], see Fig. 17 of Chap. E).

The concept of a Mott transition in the 5f shell arises almost naturally in the light of models first developed for d-transition metals (Stoner's and Mott-Hubbard's models, see Chap. A). This concept was first introduced by Johansson[2)] and later even more clarified, through the refined and successful calculations of ground state properties (see Chaps. C and F) of actinide metals and compounds performed, within the frame of modern band theories, by many authors (e.g. Skriver and Johansson[3)], Brooks and Glötzel[4)]). The intraatomic energy gain leading to localization is attributed to spin-polarisation, i.e. to the same phenomenon which leads to the atomic Hund's rules and is at the basis of the Stoner's and Hubbard's models. Spin-polarisation energies, either calculated within the spin-pairing theory (Jørgensen[5)]) or in an LSDA framework (Brooks and Johansson[6)]) explain well the transition and give fairly good answers for most of the important ground state properties.

Transitions from a localized to an itinerant state of an unfilled shell are not a special property of actinides: they can, for instance, be induced by pressure as they are in Ce and in other lanthanides or heavy actinides under pressure (see Chap. C). The uniqueness for the actinide metals series lies in the fact that the transition occurs "naturally" almost as a pure consequence of the increase of the magnetic moment due to unpaired spins, which is maximum at the half-filled shell. The concept has resulted in re-writing the Periodic Chart in such a way as to make the onset of an atomic magnetic moment the ordering rule (see Fig. 1 of Chap. E). Whether the spin-polarisation model is the only way to explain the transition remains an open question. In a very recent article by Harrison[7)], an Anderson type "impurity" model of resonant states is used to explain the transition between Pu and Am. In this model, an energy U, separating localised from polar states, is invoked and is, in principle, not only the spin-polarisation energy, but may include other forms of atomic correlation effects.

The great vivacity of the debate centered around the above ideas, has triggered a flourishing of experimental efforts in all fields, attracting on actinides the attention of non-actinide scientists. Uranium systems have been, of course, the most studied, since they do not present all the characteristics of hazards and the needs of special protections the other actinides do. Nevertheless, attention should be given and is given to Pu and Am, where the Mott-transition of the 5f metallic bond occurs.

A field in which the demand has constantly grown is the preparation of well characterised samples, especially of large-sized single crystals. Also, to be emphasized is the fact that the growing interest in actinide solid state physics has pushed experimentalists to the use of very refined techniques, some of which have proved essential to the understanding: as examples, we cite the high pressure X-ray diffraction electrical conductivity, magnetisation and many refined neutron scattering experiments as well as photoelectron spectroscopy.

It is in the metallic bond that the special character of the 5f wave-functions, leading to an easy formation of 5f bands, is better illustrated. However, since the early work of Friedel[8)], hybridisation of the 5f's with the 6d states already in metals was strongly emphasised. Practically because of this hybridisation light actinides do not present collective band magnetism (as d-transition metals do) before Cm (the magnetism of which is due to the ordering of atomic magnetic moments): hybridisation broadens the 5f band. The 5f orbitals in compounds, however, hybridise also with nonactinide electron states: e.g. with p anion states in the metallic NaCl-structure compounds (see Brooks and Glötzel[4)]).

In the much studied oxides, attempts have been made to modify the typical ionic models employed by assuming f-p covalent bonding[9)]. Hints of f presence as predicted e.g. in results of molecular orbitals cluster calculations[10)] in the ligand valence band of the oxides are reported by authors studying electron spectroscopic properties of these solids. However, all evidence (magnetism and spectroscopy) points to a very localised nonbonding 5f state (covalency is attributed by Brooks and Kelly[9)] to d-p mixing and not, or very little, to f-p mixing). The debate about covalency in oxides is not irrelevant: properties related to "local" bonding and strongly affected by directional properties of the bonding charges such as shear properties[9, 11)] or defect formation and clustering in nonstoichiometric oxides[12)] (all of them of great practical importance) may find clarification in the light of an accurate description of the oxide bond.

Future trends will be illustrated by two topics which are recent findings and that we consider are able to produce new developments in the future.

The two topics are quite different. The first is a purely theoretical result, which discovers a peculiar feature of actinides, and opens up new lines of research. The second is constituted by impressive experimental evidence, which is not yet fully understood by theory, and almost certainly will promote much theoretical effort.

i) As it was discussed in the concluding remarks of Chap. D, the analysis performed by Brooks and Kelly[13)] of the 5f band magnetism of UN in the frame of local spin density approximation (LSDA), but adding a spin-orbit coupling term to the Hamiltonian matrix of the band structure, reveals that in this solid a very large orbital moment exists, which is predominant over the spin moment.

The values of spin and orbital moments at two different lattice parameters, bracketing (and very close to) the experimental (normal pressure) value of 4.89 Å, are

a (Å)	$\mu_{orbital}$ (μ_B)	μ_{spin} (μ_B)	μ_{total} (μ_B)
5.03	−1.83	1.1	−0.73
4.78	−1.01	0.75	−0.26

The Stoner product of UN (see Chaps. A and D) is greater than one, in agreement with the antiferromagnetic behaviour of this solid. The antiferromagnetism was attributed to itinerant band magnetism (as in some d-metals and compounds but unlike light actinide metals). In fact, cohesive properties of this solid have been well explained in a pure spin-polarised picture[3], and Fournier et al.[14] have shown that the magnetic uranium sublattice moment and the Néel temperature have a similar pressure dependence. Discrepancies existed, however, between calculations and experiments:

- the calculated[13] spin only moment was about 30% greater than the experimental;
- the calculated[13] form factor of the spin densities was very different from the measured;
- the pressure dependence of the calculated reduced magnetic moment[13] was much smaller than the experimental one.

The discrepancies are resolved when the spin-orbit coupling term is introduced in the band calculation and a large orbital moment is evaluated. Since the spin-orbit interaction is known to be particularly high in all actinides (see Chap. A), many actinide compounds may show the same large orbital moment as in UN even in a band model. A large orbital moment was usually considered as a sign of electron localisation. The results of the calculation of[13] is in disagreement with this conclusion at least for actinides, showing that an itinerant state, as described in LSDA, may be consistent with a large orbital moment.

Furthermore, the orbital moment has a very marked directionality which is given by the angular properties of the wavefunctions. Its presence induces strong anisotropy in magnetic order and in the band structure of the magnetically ordered solid (in UN, the energy difference between the $\langle 110 \rangle$ and $\langle 111 \rangle$ ferromagnetic ground states is calculated to be 7×10^{10} ergs/cm^3!).

Since the application of pressure may modify strongly the charge density distribution in a solid, and therefore affects the orbital more than the spin moment, magnetic form factors and magnetic anisotropy may become much more pressure-dependent than usually assumed.

All this opens a vast new field of research for actinide compounds. Besides encouraging high pressure experiments, a search may be foreseen of appropriate compounds of actinides with other elements for which the peculiar properties deriving from these huge actinide orbital moments, and, in particular, their magnetic anisotropy, may result to be enhanced.

ii) Some actinide intermetallics show an *unexpectedly high* density of states, obviously of 5f origin, at the Fermi level, as evidenced, e.g., by measurements of the electronic γ (see Chap. A) in the low temperature specific heat. For one of them, UPt_3, a $\gamma = 420$ mg/

mol K^2 is reported[15] (compare with the already high γ's of elemental metals in Chap. A). "Heavy Fermions" is the name which is given to these conduction electrons of extremely high effective mass.

In UPt_3, a compound for which spin-fluctuations are known to exist, superconduction has recently been reported[16]. There is an apparent contradiction with the classical Bardeen-Cooper model for superconductivity, in which superconduction is hindered by the onset of ordered magnetic phenomena, and is usually not found in very narrow bands. UBe_{13}[17], which has a γ of ~1000 mg/mol K^2, exhibits similar behaviour.

This is an interesting „actinide" phenomenon which will call for more research, and which is already stimulating theoretical investigation for its understanding.

We have tried, in this section, and in the whole book, to present the various and most stimulating aspects of actinide solid state physics. New lines and problems open up when old problems seem to be reasonably understood: this, which is valid for all scientific research, is certainly true for the field which we have been exploring in this book.

References

1. Naegele, J., et al.: Phys. Rev. Lett. *52,* 1834 (1984)
2. Johansson, B.: Phil. Mag. *30,* 469 (1974)
3. Johansson, B., Skriver, H. L.: J. of Magn. and Magn. Mat. *29,* 217 (1982)
4. Brooks, M. S. S., Glötzel, D.: Physica *102 B*, 51 (1980)
5. Jørgensen, C. Kl.: Orbitals in Atoms and Molecules, Academic Press, New York (1962)
6. Brooks, M. S., Johansson, B.: J. Phys. F. Met. Phys. *13,* L 197 (1983)
7. Harrison, W. A.: Phys. Rev. *B 29*, 2917 (1984)
8. Friedel, J.: J. Phys. Chem. Solids *1,* 175 (1956)
9. Brooks, M. S. S., Kelly, P. J.: Solid State Communications *45,* 689 (1983)
10. Gubanov, V. A., Rosen, A., Ellis, D. E.: ibid. *22,* 219 (1977)
11. Blank, H.: in: Thermodynamics of Nuclear Materials 1974, *2,* IAEA, Vienna, 45 (1975)
12. Manes, L.: in: Nonstoichiometric Oxides (ed. Sørensen, O. T.) Academic Press, New York, 99 (1981)
13. Brooks, M. S. S., Kelly, P. J.: Phys. Rev. Letters *51,* 1708 (1983)
14. Fournier, J. M., et al.: Physica *102 B*, 282 (1980)
15. Frings, P. H., et al.: J. Magn. Magn. Mat. *31–34,* 240 (1983)
16. Stewart, G., et al.: Phys. Rev. Letters *52,* 679 (1984)
17. Ott, H. R., et al.: ibid. *50,* 1595 (1983)

Author-Index Volumes 1–60

Vahrenkamp, H.: Recent Results in the Chemistry of Transition Metal Clusters with Organic Ligands. Vol. 32, pp. 1–56.

Valach, F., Koreň, B., Sivý, P., Melník, M.: Crystal Structure Non-Rigidity of Central Atoms for Mn(II), Fe(II), Fe(III), Co(II), Co(III), Ni(II), Cu(II) and Zn(II) Complexes. Vol. 55, pp. 101–151.

Wallace, W. E., Sankar, S. G., Rao, V. U. S.: Field Effects in Rare-Earth Intermetallic Compounds. Vol. 33, pp. 1–55.

Warren, K. D.: Ligand Field Theory of Metal Sandwich Complexes. Vol. 27, pp. 45–159.

Warren, K. D.: Ligand Field Theory of f-Orbital Sandwich Complexes. Vol. 33, pp. 97–137.

Warren, K. D.: Calculations of the Jahn-Teller Coupling Costants for d^x Systems in Octahedral Symmetry via the Angular Overlap Model. Vol. 57, pp. 119–145.

Watson, R. E., Perlman, M. L.: X-Ray Photoelectron Spectroscopy. Application to Metals and Alloys. Vol. 24, pp. 83–132.

Weakley, T. J. R.: Some Aspects of the Heteropolymolybdates and Heteropolytungstates. Vol. 18, pp. 131–176.

Wendin, G.: Breakdown of the One-Electron Pictures in Photoelectron Spectra. Vol. 45, pp. 1–130.

Weissbluth, M.: The Physics of Hemoglobin. Vol. 2, pp. 1–125.

Weser, U.: Chemistry and Structure of some Borate Polyol Compounds. Vol. 2, pp. 160–180.

Weser, U.: Reaction of some Transition Metals with Nucleic Acids and their Constituents. Vol. 5, pp. 41–67.

Weser, U.: Structural Aspects and Biochemical Function of Erythrocuprein. Vol. 17, pp. 1–65.

Willemse, J., Cras, J. A., Steggerda, J. J., Keijzers, C. P.: Dithiocarbamates of Transition Group Elements in "Unusual" Oxidation State. Vol. 28, pp. 83–126.

Williams, R. J. P.: The Chemistry of Lanthanide Ions in Solution and in Biological Systems. Vol. 50, pp. 79–119.

Williams, R. J. P., Hale, J. D.: The Classification of Acceptors and Donors in Inorganic Reactions. Vol. 1, pp. 249–281.

Williams, R. J. P., Hale, J. D.: Professor Sir Ronald Nyholm. Vol. 15, pp. 1 and 2.

Wilson, J. A.: A Generalized Configuration-Dependent Band Model for Lanthanide Compounds and Conditions for Interconfiguration Fluctuations. Vol. 32, pp. 57–91.

Winkler, R.: Kinetics and Mechanism of Alkali Ion Complex Formation in Solution. Vol. 10, pp. 1–24.

Wood, J. M., Brown, D. G.: The Chemistry of Vitamin B_{12}-Enzymes. Vol. 11, pp. 47–105.

Woolley, R. G.: Natural Optical Activity and the Molecular Hypothesis. Vol. 52, pp. 1–35.

Wüthrich, K.: Structural Studies of Hemes and Hemoproteins by Nuclear Magnetic Resonance Spectroscopy. Vol. 8, pp. 53–121.

Xavier, A. V., Moura, J. J. G., Moura, I.: Novel Structures in Iron-Sulfur Proteins. Vol. 43, pp. 187–213.

Zumft, W. G.: The Molecular Basis of Biological Dinitrogen Fixation. Vol. 29, pp. 1–65.

Structure and Bonding

Editors: **M.J. Clarke, J.B. Goodenough, J.A. Ibers, C.K. Jørgensen, J.B. Neilands, D. Reinen, R. Weiss, R.J.P. Williams**

Volume 56

New Developments

With contributions by B. Dobiáš, I.R. Epstein, K. Kustin, S.K. Patil, V.V. Ramakrishna
1984. 44 figures, 37 tables. V, 155 pages
ISBN 3-540-13106-X

Volume 57

Complex Chemistry

With contributions by J. Emsley, R.D. Ernst, B. Hathaway, K.D. Warren
1984. 85 figures, 39 tables. V, 199 pages
ISBN 3-540-13411-5

Volume 50

Topics in Inorganic and Physical Chemistry

1982. 48 figures, 29 tables. V, 178 pages
ISBN 3-540-11454-8

Volume 51
A. Schweiger

Electron Nuclear Double Resonance of Transition Metal Complexes with Organic Ligands

1982. 47 figures, 19 tables. VIII, 128 pages
ISBN 3-540-11072-0

Volume 52

Structures versus Special Properties

1982. 90 figures, 19 tables. V, 202 pages
ISBN 3-540-11781-4

Springer-Verlag
Berlin
Heidelberg
New York
Tokyo

Structure and Bonding

Editors: **M. J. Clarke, J. B. Goodenough, J. A. Ibers, C. K. Jørgensen, J. B. Neilands, D. Reinen, R. Weiss, R. J. P. Williams**

Volume 55

Transition Metal Complexes – Structures and Spectra

With contributions by *M. Bacci, J. Fischer, M. H. Gubelmann, B. Koreň, F. Mathey, M. Melnik, J. H. Nelson, P. Sivý, F. Valach, A. F. Williams*
1984. 93 figures, 21 tables. V, 209 pages
ISBN 3-540-12833-6

Volume 58

Siderophores from Microorganisms and Plants

With contributions by *A. Chimiak, R. C. Hider, A. Liu, J. B. Neilands, K. Nomoto, Y. Sugiura*
1984. 35 figures, 16 tables. IX, 143 pages
ISBN 3-540-13649-5

Volume 53

Copper, Molybdenum, and Vanadium in Biological Systems

1983. 67 figures, 13 tables. V, 166 pages
ISBN 3-540-12042-4

Volume 54

Inorganic Elements in Biochemistry

With contributions by *P. H. Connett, H. Follmann, M. Lammers, S. Mann, J. D. Odom, K. E. Wetterhahn*
1983. 57 figures, 22 tables. V, 180 pages
ISBN 3-540-12542-6